The Intext Series in BASIC MATHEMATICS
under the consulting editorship of

RICHARD D. ANDERSON
Louisiana State University

ALEX ROSENBERG
Cornell University

Preparation for Calculus

Preparation for Calculus

WILLIAM L. HART

University of Minnesota

Intext Educational Publishers

SCRANTON   TORONTO   LONDON

ISBN   0-7002-2350-9

Copyright ©, 1971, International Textbook Company

Preface

This text provides content from which the teacher may select material for a one-semester college* course in various ways as preparation for a first course in calculus. It is assumed that the typical student has studied the equivalent of a standard second course in algebra and an introduction to trigonometry as given at the secondary level. However, it is inferred that his knowledge of precalculus mathematics is not as thorough as desirable in the fields of analytic geometry and analytic trigonometry. Also, it is expected that he needs a more mature acquaintance with all of the elementary functions of analysis, particularly in association with their analytic geometry. It is assumed that the student's command of basic algebra is sufficient so that, after its incidental use in the course, he will be properly prepared for successful performance of the manipulative parts of calculus. Hence, the text includes only brief remarks about the number system and a short initial review of the algebraic background, which may be considered merely as a convenient reference source, if the teacher does not see fit to spend class time on this content.

The text presents the following major content systematically.

- A foundation for plane analytic geometry, including a discussion of polar coordinates early in the text, and an introduction to solid analytic geometry in an advanced chapter.

- A limited treatment of the elements of trigonometry followed by detailed consideration of the standard trigonometric functions, or the circular functions of real numbers as used in calculus, omitting numerical trigonometry and emphasizing analytic aspects.

- The general exponential function, and then its inverse, the logarithm function, with only a brief review of logarithmic computation,† and with emphasis on the associated analytic geometry of the functions.

- Various topics useful in calculus, such as inequalities, mathematical induction, and certain facts about polynomial functions.

Each of the topics is given a rounded development, with brevity only where substantial previous acquaintance can be assumed. Hence, the text

---

*Or, with the text studied at a moderate tempo, the content could form the basis for a major part of a 12th grade course concluding with some months of elementary calculus.

†It is assumed that the student has studied logarithmic computation previously.

is self-contained, and avoids reliance on other sources. It is believed that, for many students, the book provides content suitable for a flexible pre-calculus course approximately of the variety referred to as *Mathematics Zero*, in recommendations of the Committee on the Undergraduate Program in Mathematics of the Mathematical Association of America.

On account of the naturally limited size and the major objectives of this book, evidently it cannot cover all of the topics expected in texts devoted solely to analytic geometry, trigonometry, and other fields from which content was selected. The relative brevity but completeness of the expositions of analytic geometry and trigonometry in the text are attained as a result of a realistic appraisal of the need for, or lack of importance of available content as preparation for the first course in calculus. Also, some of the classical subject matter of plane analytic geometry and trigonometry is omitted because the material is discussed more efficiently in calculus. For instance, in the plane analytic geometry: the book includes no discussion of tangent lines; omits transformation of coordinates by rotation of axes; gives only negligible optional attention to the minutiae of structural details concerning the conic sections; omits the more difficult aspects of graphing which become simple in calculus. In regard to triangles in the trigonometry: merely the laws of sines and of cosines are presented, and then are used numerically only to a very limited extent, without consideration of the general numerical solution of triangles.

The text has various flexible features which will aid the teacher in his selection of content for a one-semester or one-quarter course of some specified length for a class of any expected type. First there is an option as to the amount of review or semi-review, if any, from the first few chapters which should be considered in class periods. Beyond Chapter II, the relative independence of major topics, and the plainly indicated optional content,* permit the omission of many sections, and a few whole chapters, without destroying the continuity of the course. It is expected that the last two chapters will form part of any one-semester course only when the students do not require class discussion of the more elementary content, and when the course is a prerequisite for a substantial treatment of calculus. The arrangement of the content in the text would facilitate selection of material for a course of only moderate extent if the succeeding study of calculus is designed to have limited objectives. Also, the book would be suitable for use where only a few of its chapters would be needed at proper stages in a course on calculus.

The author expresses his sincere thanks for the many useful suggestions concerning the content and arrangement of the subject matter which he received from the consulting editors Professors Richard D. Anderson

---

*Marked with a star ★.

and Alex Rosenberg. Also, the courtesy and cooperation of the staff of the College Division, Intext Educational Publishers, during the publication of this text were deeply appreciated.

William L. Hart

LaJolla, California, November 1970

Contents

Chapter 5   Introduction to Polynomial Functions . . . 133

Chapter 6   Trigonometric Functions of Angles . . . 151

Chapter 7   The Standard Trigonometric Functions . . . 182

chapter 1

Review of Elementary Content

1. TERMINOLOGY ABOUT REAL NUMBERS

Real numbers are classed as positive, negative, or zero. A **rational number** is a real number which can be represented by a fraction p/q where p and q are integers and $q \neq 0$. A real number which is not rational is said to be **irrational**.

ILLUSTRATION 1.   5, -3, 2/5, and 0 are rational numbers. The endless nonrepeating decimals $\pi = 3.14159 \cdots$ and $\sqrt{2} = 1.414 \cdots$ are irrational numbers. Any nonrepeating endless decimal is an irrational number. All other decimals are rational numbers.

The **absolute value** of a real number b is defined as b itself if b is non-negative (that is, *zero* or *positive*), and as $-b$ if b is negative. We use " $|b|$ " to abbreviate "*the absolute value of b.*" Thus,

$$|b| = b \qquad if \qquad b \text{ is positive } or \text{ zero;} \tag{1}$$

$$|b| = -b \qquad if \qquad b \text{ is negative.} \tag{2}$$

Two numbers b and c are said to be *numerically equal* in case $|b| = |c|$. Then, $b = c$ or $b = -c$.

ILLUSTRATION 2. $|-9| = 9$; $|9| = 9$. Hence, 9 and -9 are numerically equal but "opposite in sign." Also, $|0| = 0$. From algebra, recall that

$$|bc| = |b| \cdot |c|; \qquad \left|\frac{b}{c}\right| = \frac{|b|}{|c|}; \qquad |b|^2 = b^2. \tag{3}$$

In this text, unless otherwise stated, any literal number symbol such as $a, x, b, \cdots$ will represent a *real number*. Later in the text, we shall discuss so-called *complex numbers*. If $\{a, b, c\}$ are any numbers of the system, R, of real numbers, the following laws hold for the operations of *addition* and *multiplication*.

   I. *Addition is **commutative**, or the sum of two numbers is the same in whatever order they are added. That is, $a + b = b + a$.*

   II. *Addition is **associative**. Or, in the addition of three (and then three or more) numbers, the numbers may be associated in any way in* *adding. Thus,*

$$a + (b + c) = (a + b) + c.$$

III. *Multiplication is* **commutative,** *or* $ab = ba$.

IV. *Multiplication is* **associative,** *or* $a(bc) = (ab)c$.

V. *Multiplication is* **distributive with respect to addition,** *or*

$$a(b + c) = ab + ac.$$

2. MULTIPLICATION AND DIVISION OF FRACTIONS

The following facts about fractions should be recalled. Any number occurring as a divisor is assumed to be *not zero,* because *division by zero is not defined in algebra.*

$$\frac{a}{b} \cdot \frac{c}{d} = \frac{ac}{bd}. \tag{1}$$

$$\frac{a}{b} \div \frac{c}{d} = \frac{a}{b} \cdot \frac{d}{c} = \frac{ad}{bc}. \tag{2}$$

ILLUSTRATION 1. $\dfrac{3}{5} \cdot \dfrac{2}{7} = \dfrac{6}{35}.$ $\dfrac{2}{3} \div \dfrac{5}{11} = \dfrac{2}{3} \cdot \dfrac{11}{5} = \dfrac{22}{15}.$

As a special case of (1) with $c = d = k$,

$$\frac{a}{b} = \frac{a}{b} \cdot \frac{k}{k} = \frac{ak}{bk}, \tag{3}$$

because $k/k = 1$. Thus, by (3), *if the numerator and the denominator of a fraction are multiplied by the same number, k, the value of the fraction is unaltered.* Since *division* by a number h means *multiplication* by $1/h$, it is seen that a fraction is unaltered in value if the numerator and the denominator are *divided* by the same number. We use the preceding fact in dividing out a common factor of the numerator and the denominator in a fraction.

ILLUSTRATION 2. $\dfrac{16}{14} = \dfrac{2 \cdot 8}{2 \cdot 7} = \dfrac{8}{7}$, where we divided by 2 in both the numerator and the denominator.

A fraction is said to be in *lowest terms* if the numerator and denominator have *no common factor** except ± 1. To change a fraction to lowest terms, divide out all common factors from the numerator and the denominator. In Illustration 2, we changed 16/14 to lowest terms.

---

*At present, *"factor"* will have a meaning which can be inferred from the simple nature of the fractions involved.

ILLUSTRATION 3.   The following fraction is changed to lowest terms by dividing both numerator and denominator by $2bc$.

$$\frac{6abc}{22bcx} = \frac{3a(2bc)}{11x(2bc)} = \frac{3a}{11x}.$$

Since any number $r = r/1$,

$$r \cdot \frac{a}{b} = \frac{r}{1} \cdot \frac{a}{b} = \frac{ra}{b}; \qquad (4)$$

$$\frac{a}{b} \div r = \frac{a}{b} \div \frac{r}{1} = \frac{a}{b} \cdot \frac{1}{r} = \frac{a}{br}. \qquad (5)$$

Hence, to *multiply* a fraction by r, multiply the numerator by r. To *divide* a fraction by r, multiply the denominator by r.

ILLUSTRATION 4. $6 \cdot \dfrac{a}{b} = \dfrac{6a}{b}.$ $\left(\dfrac{3}{5} \div 7\right) = \dfrac{3}{35}.$

The **reciprocal** of a number $N \neq 0$ is defined as $1/N$. In particular, the reciprocal of a fraction is the fraction inverted, because

$$1 \div \frac{a}{b} = 1 \cdot \frac{b}{a} = \frac{b}{a}.$$

ILLUSTRATION 5.   The reciprocal of 6 is $\frac{1}{6}$; of $\frac{3}{4}$ is $\frac{4}{3}$.

3. A NUMBER SCALE AND INEQUALITIES

The word *length,* or the unqualified word *distance* will refer to a non-negative number which is the measure of some distance or straight-line segment between two points, in a plane or in space of three dimensions. On a given line,* as in Figure 1, select any point O, to be called the **origin,** and let it represent the number 0. Choose a unit for length in measuring distances on the line OX. Then, if p is any positive number, let it be represented by that point on OX which is p units to the *right* of O, with OX thought of as horizontal in Figure 1. Let $-p$ be represented by that point on OX which is p units to the *left* of O. Thus, each real number is represented by a point on OX. Conversely, if K is any point on OX,

FIGURE 1

Line will mean *straight line.*

then K represents just one real number, which is positive if K is to the right of O, and is negative if K is to the left of O. We refer to OX in Figure 1 as a **number scale.**

DEFINITION I.   To *state that b is* **less than** *c, or that c is* **greater than** *b means that* $(c - b)$ *is* **positive.**

The inequality sign "$<$" is used for "*is less than,*" and "$>$" for "*is greater than.*" Hence, by Definition I,

$$b < c \quad means\ that \quad (c - b)\ \textbf{is positive.} \tag{1}$$

Also, $b < c$ and $c > b$ have the same meaning. The statement "$b \leqq c$" is read "*b is less than or equal to c,*" and "*c $\geqq$ b*" is read similarly. We refer to each of $b < c, c > b, b \leqq c$, and $c \geqq b$ as an *inequality.*

ILLUSTRATION 1. $3 < 7$ because $(7 - 3) = 4$, which is *positive.*

$-8 < -2$ because $-2 - (-8) = 6$, which is *positive.*

$-9 < 0$ because $0 - (-9) = 9$, which is *positive.*

To state that $p > 0$ means that $(p - 0)$, or p is *positive.* To say that $h < 0$ means that $(0 - h)$, or $-h$ is positive, and hence that h is *negative.*

$$p > 0 \quad means\ that\ p\ is\ \textbf{positive;} \tag{2}$$

$$h < 0 \quad means\ that\ h\ is\ \textbf{negative.} \tag{3}$$

Consider a number scale OX as in Figure 2. If x is any real number, and P is the point representing x on the scale, we call x the *coordinate* of P. We shall write "$P:(x)$," to be read simply "P, x" or, more elaborately if desired, "*P with coordinate x.*"

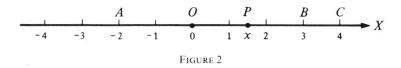

FIGURE 2

ILLUSTRATION 2.   In Figure 2, observe $A:(-2)$, $B:(3)$, $C:(4)$, and $P:(x)$, where points are labeled above the scale, and coordinates are shown below the scale.

Let $P_1:(x_1)$ and $P_2:(x_2)$ be any two distinct points on a number scale, as in Figure 3. Then $P_1 P_2$ will represent the *line segment* with P_1 as the initial point and P_2 as the terminal point, where $P_1 P_2$ is thought of as *directed from P_1 to P_2.* We shall refer to $P_1 P_2$ as a *directed line segment.* Segment $P_1 P_2$ is said to have *positive direction,* and will be assigned a *positive value,* if $P_1 P_2$ is directed to the *right* on OX. Segment $P_1 P_2$ is said to have *negative direction,* and will be assigned a *negative value,* if $P_1 P_2$ is directed to the *left* on OX. If P_1 and P_2 are the same point, we let $P_1 P_2$ represent a segment with *no direction,* consisting of the single point P_1.

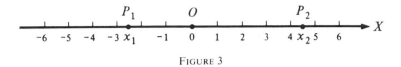

FIGURE 3

If P_1 and P_2 are any points on a number scale, let $\overline{P_1 P_2}$ represent the *directed distance from P_1 to P_2,* defined by

$$\overline{P_1 P_2} = x_2 - x_1. \tag{4}$$

We read "$\overline{P_1 P_2}$" as "*P one P two bar,*" and call $\overline{P_1 P_2}$ the *value* of $P_1 P_2$. From (4), $\overline{P_1 P_2}$ is either plus or minus the number of units of distance between P_1 and P_2 on OX. Then

$$(length\ of\ P_1 P_2) = |\overline{P_1 P_2}| = |x_2 - x_1|. \tag{5}$$

On the basis of the preceding agreements, we refer to a number scale as a *directed line* For any point $P:(x)$ on the scale, by use of (4) we verify that $\overline{OP} = x - 0$, or $\overline{OP} = x$. That is, the coordinate of P is the directed distance $\overline{OP}$, *positive* when P is to the *right* of O and *negative* when P is to the *left* of O.

The number scale is a convenient background where any real number r may be referred to as a *point,* meaning the point representing r. Thus, to remark that *b is close to c* means that the scale distance $|b - c|$ is small.

ILLUSTRATION 3.   With $A:(-2)$, $B:(3)$, and $C:(4)$ as in Figure 2,

$$\overline{AB} = 3 - (-2) = 5; \quad (length\ of\ AB) = |\overline{AB}| = 5. \tag{6}$$
$$\overline{BA} = -2 - 3 = -5; \quad (length\ of\ BA) = |\overline{BA}| = 5. \tag{7}$$
$$\overline{OC} = 4 - 0 = 4; \quad (length\ of\ OC) = |\overline{OC}| = 4.$$
$$\overline{CA} = -2 - 4 = -6; \quad (length\ of\ CA) = |\overline{CA}| = 6.$$

In (6) and (7), notice that

$$\overline{AB} = -\overline{BA}. \tag{8}$$

In general, for any points A and B on a number scale, (8) is true because reversal of the direction of a segment multiplies its value by -1.

From (1), recall that "$b < c$" is equivalent to stating that $(c - b)$ is *positive.* Then, if $P:(b)$ and $Q:(c)$ are the corresponding points on a number scale, we have $\overline{PQ} = c - b$, which is *positive,* and PQ is directed to the *right.* Thus, geometrically,

$b < c$ *means that b is to the* **left** *of c on the number scale.* $\qquad$ (9)

ILLUSTRATION 4.   Since $H:(-7)$ is to the left of $K:(-2)$ on a number scale, it follows that $-7 < -2$.   Also, we verify that $-2 - (-7) = 5$, which is *positive*.

4. SETS, VARIABLES, AND CONSTANTS

At many places in mathematics, the *undefined* concept of a *set* of objects is useful. The students in a class studying this text form a *set of people*. Each object in a set is called an *element* or a *member* of it. For any set *S*, it is assumed that, if any object *e* is described to us, we shall be able to decide whether or not *e* belongs to *S*. This means that any set *S* to which we refer will be *well defined*.   It is convenient to introduce the **empty set,** or **null set,** represented by ϕ, which has no members. If a set *S* has exactly *n* elements, where *n* is a nonnegative integer, we call *S* a *finite set*. If *S* is *not* a finite set, we call *S* an *infinite set*. Then, corresponding to any positive integer *n,* there exist more than *n* members of *S*.

ILLUSTRATION 1.   The positive integers form an infinite set of numbers.

A **subset** *S* of a set *T* is a set consisting of some (*possibly all*) of the members of *T*. If *S* is a subset of *T,* we say that *S* is *included* in *T,* and write "$S \subset T$," which we read "*S is included in T*."   We have $T \subset T$. Also, we agree to say that the empty set is included in every set, or $\phi \subset T$. If all members of *S* are in *T,* and if all members of *T* are in *S,* then *S* and *T* have the same members and we say $S = T$. If $S \neq T$ and $S \subset T$, then *S* is called a **proper subset** of *T*.   In such a case, there is at least one element of *T* which is not in *S*.

ILLUSTRATION 2.   The set *P* of positive integers is a proper subset of the set *T* of all integers, or $P \subset T,$ and $P \neq T$.

ILLUSTRATION 3.   To describe a set, we sometimes write symbols for the elements within braces. Such a device is referred to as the *roster method* for describing a set. Thus, if *H* is a set of numbers indicated by $H = \{1, 2, 3, 8, 9, 11, 12\}$, and $S = \{3, 11, 12\}$, then *S* is a proper subset of *H*, or $S \subset H$ and $S \neq H$.

A **variable** is a symbol, such as *x,* which is free to represent any element of a nonempty set *S,* which is called the **domain** of the variable. Each element of *S* is referred to as a *value* of the variable. The domain *S* may not consist of numbers. However, in this text, the domain of any variable will be a set of numbers, unless otherwise specified.

ILLUSTRATION 4.   We may use *y* to represent any resident of New York City. Then, the population of the city forms the domain of *y*.

ILLUSTRATION 5.   Let *T* be the set of numbers $x > 4$. We could also define *T* as the set of numbers $v > 4$. The letter *x* or *v* used for the arbitrary member of *T* is immaterial.

In a given exposition, a **constant** is a symbol for a *fixed number*. A constant *c* also can be referred to as a variable whose domain is a single element. Whenever a variable is mentioned, we shall infer that it is *not* a constant, unless otherwise specified.

ILLUSTRATION 6.   In the formula $V = \frac{4}{3}\pi r^3$ for the volume of a sphere of radius *r*, if all spheres are under consideration, then *V* and *r* are variables and π is a constant, $\pi = 3.14159\ldots$.

In this text, unless otherwise indicated, if a variable *x* is introduced in some symbol for a number, it will be understood that the domain of *x* consists of all real values of *x* for which the symbol has meaning.

ILLUSTRATION 7.   If *x* occurs in $(2x + 5)/(x - 4)$, then *x* is a variable whose domain consists of all real numbers $x \neq 4$, where 4 is eliminated because division by zero is not defined in algebra.

5. INTEGRAL EXPONENTS

If *x* and *h* are numbers, we refer to x^h as the *h*th *power* of *x*, and to *h* as the *exponent* of the power. In algebra, the student has employed powers where the exponents were positive or negative rational numbers, or zero. Until otherwise specified, any exponent which we use will be an *integer*. If *n* is a positive integer, recall that, by definition,

$$x^n = x \cdot x \cdots x. \qquad (n\ factors\ x) \qquad (1)$$

With $x \neq 0$,
$$x^0 = 1; \qquad x^{-n} = \frac{1}{x^n}. \qquad (2)$$

For a reason which is discussed in calculus, x^0 is *not defined if* $x = 0$. However, $0^n = 0$ for every positive integer *n*. The familiar theorems concerning exponents are as follows, where $x \neq 0$ if necessary, and the exponents are integers, which may be positive, negative, or zero.

$$\left. \begin{array}{c} x^n x^m = x^{m+n}; \quad (x^n)^k = x^{nk}; \quad (xy)^n = x^n y^n; \\ \dfrac{x^m}{x^n} = x^{m-n} = \dfrac{1}{x^{n-m}}. \end{array} \right\} \qquad (3)$$

Note 1. Unless otherwise specified, it may be assumed that any literal number symbol such as $a, b, x, y, \cdots$ which we employ is a variable, representing any corresponding number for which the given expression has meaning as a number symbol.

ILLUSTRATION 1. $x^3 x^4 = x^7;$ $x^{-3} = \dfrac{1}{x^3};$ $6^{-3} = \dfrac{1}{6^3} = \dfrac{1}{216}.$

With x^m / x^n in (3), we may choose to use the result x^{m-n} when $m - n > 0$, and $1/x^{n-m}$ when $n - m > 0$. Thus,

$$\frac{u^2}{u^6} = \frac{1}{u^{6-2}} = \frac{1}{u^4}; \qquad \frac{u^6}{u^2} = u^{6-2} = u^4.$$

Also, we might prefer

$$\frac{u^2}{u^6} = u^{2-6} = u^{-4}.$$

ILLUSTRATION 2. $\dfrac{(2x^2 y^2)^3}{(3xy^4)^2} = \dfrac{2^3 x^6 y^6}{3^2 x^2 y^8} = \dfrac{8x^4}{9y^2}.$ (4)

In (4), the fraction was changed to lowest terms by use of (3).

ILLUSTRATION 3. $\dfrac{x^{-2} y^3}{x^4} = \dfrac{y^3 \cdot \dfrac{1}{x^2}}{x^4} = \dfrac{\dfrac{y^3}{x^2}}{\dfrac{x^4}{1}} = \dfrac{y^3}{x^2} \cdot \dfrac{1}{x^4} = \dfrac{y^3}{x^6}.$

EXERCISE 1

Find the value of the symbol, and read it.
1. $|5|$.   **2.** $|-7|$.   **3.** $|-2|$.   **4.** $|0|$.   **5.** $|-2|^3$.

6. Mark $R{:}(3)$, $S{:}(-2)$, $T{:}(-5)$, and $U{:}(7)$ on a number scale and calculate $\overline{RS}$; $\overline{ST}$; $\overline{TU}$; $\overline{UR}$; $\overline{US}$; $|\overline{UT}|$; $|\overline{SR}|$; $|\overline{RS}|$.

Insert the proper sign, $<$ or $>$, between the numbers.
7. 3 and 9.   **8.** -7 and 3.   **9.** -11 and -4.   **10.** 5 and -2.
11. 0 and 6.   **12.** 0 and -4.   **13.** 0 and -1.   **14.** 5 and -6.

Visualize the number x on a number scale. State the property of x by use of an inequality.
15. x is to the right of 5.   **16.** x is to the right of 0.
17. x is to the left of -3.   **18.** x is to the left of 4.
19. x is to the right of -9.   **20.** x is to the left of 0.

21. If T represents the set of integers $\{1, 2, 3, 4, 5, 6\}$, specify six proper subsets of T consisting of five members each.

22. If T is the set of all integers, describe in words two infinite proper subsets of T; three finite subsets of T.

23. If x is a variable in the expression $(3x + 5)/(2x + 7)$, what value of x is excluded?

24. Obtain the reciprocal of 4; -3; $\frac{4}{5}$; $1\frac{2}{3}$; $5\frac{1}{2}$.

25. For what values of x is it true that that $0 < x^2$; $x^2 \leqq 0$; $x^3 < 0$; $0 \leqq x^3$?

Specify the value of the symbol without use of an exponent.

26. 3^{-1}.　　**27.** 5^{-2}　　**28.** 10^{-3}.　　**29.** $\left(\frac{2}{3}\right)^{-1}$.

30. 19^0.　　**31.** $\left(\frac{1}{5}\right)^{-2}$.　　**32.** $\left(\frac{1}{10}\right)^{-1}$.　　**33.** $\left(\frac{3}{4}\right)^{-2}$.

Perform the indicated operation. Eliminate any negative exponents. If the result is a fraction, change it to lowest terms.

34. $\dfrac{-14}{21}$.　　**35.** $\dfrac{3}{-7}$.　　**36.** $\dfrac{-15}{35}$.　　**37.** $\dfrac{-21}{-35}$.

38. $\dfrac{2}{3} \cdot \dfrac{5}{7}$.　　**39.** $\dfrac{5}{4} \cdot \dfrac{8}{3}$.　　**40.** $4\left(\dfrac{5}{7}\right)$.　　**41.** $\dfrac{2}{7} \div 3$.

42. $8\left(\dfrac{3}{5}\right)$.　　**43.** $\dfrac{5}{7} \div \dfrac{3}{5}$.　　**44.** $\dfrac{3}{4a} \div \dfrac{5c}{3}$.　　**45.** $\dfrac{a}{2c} \cdot \dfrac{4d}{x}$.

46. $\dfrac{25abc}{10bd}$.　　**47.** $\dfrac{3x^2y}{15yz}$.　　**48.** $6 \div \dfrac{2a}{b}$.　　**49.** $\dfrac{9x}{y} \div 2$.

50. a^2aa^3.　　**51.** $(5x)^3$.　　**52.** $(2a^2b^3)^2$.　　**53.** $(-x^2y^3)^3$.

54. $\left(\dfrac{3}{2}\right)^4$.　　**55.** $\left(\dfrac{2x}{3y}\right)^3$.　　**56.** $\left(\dfrac{-2}{5x}\right)^3$.　　**57.** $\left(\dfrac{ax^2}{by^3}\right)^4$.

58. $\dfrac{a^8}{a^2}$.　　**59.** $\dfrac{b^3}{b^7}$.　　**60.** $\dfrac{6a^3b}{9a^2b^4}$.　　**61.** $\dfrac{5x^{-2}}{x^3}$.

62. $\dfrac{(2x^2y)^3}{(3xy^2)^2}$.　　**63.** $\dfrac{(5a^2b)^3}{(3ab^2)^4}$.　　**64.** $\left(-\dfrac{3}{5}\right)^{-2}$.　　**65.** $\left(\dfrac{3ab^2}{5a^3b}\right)^3$.

66. $x^{-2}y$.　　**67.** $\dfrac{y^{-1}z^{-2}}{xy}$.　　**68.** $\left(\dfrac{x^{-1}y^2}{x^2y^{-1}}\right)^3$.　　**69.** $\left(\dfrac{3a^{-1}}{5b^{-1}}\right)^4$.

6. MONOMIALS AND POLYNOMIALS

　　A **monomial** in certain variables $x, y, z, \cdots$ is defined as a nonzero constant, called the *coefficient,* multiplied by powers of the variables with exponents which are nonnegative integers. If each of these exponents is zero, the monomial is a constant, not zero. A sum of monomials, or terms, is called a **polynomial** in the variables. A polynomial is named a *binomial* or a *trinomial* according as the polynomial has two or three terms, respectively. The **degree** *of a monomial* in certain variables is defined as the sum of the exponents of their powers which are factors of the monomial. The *degree of a polynomial* in the variables is the degree of the monomial of *highest degree* in the polynomial. A polynomial in any variables is said to be a *linear,* or *quadratic,* or *cubic,* or *quartic* polynomial according as its degree in the variables is 1, 2, 3, or 4, respectively.

ILLUSTRATION 1.   If b is a constant, then $9bx^2y^3$ is a monomial of degree 5 in the variables x and y with coefficient $9b$.

A polynomial of degree n in a single variable x is of the form*

$$a_0 + a_1x + a_2x^2 + \cdots + a_nx^n,$$

where $n \geqq 0$ and $(a_0, a_1, \cdots, a_n)$ are constants with $a_n \neq 0$.

ILLUSTRATION 2.   Any linear polynomial in x is of the form $(a + bx)$ where a and b are constants, with $b \neq 0$. Any quadratic polynomial in x is of the form $(ax^2 + bx + c)$, where $\{a, b, c\}$ are constants with $a \neq 0$. Any linear polynomial in two variables x and y is of the form $(ax + by + c)$, where a and b are not both zero.

ILLUSTRATION 3.   If $x \neq 0$, any constant $b \neq 0$ can be thought of as bx^0, which is of degree zero in x. Even if the domain of x contains the number zero, we agree to say that any constant $b \neq 0$ is of degree zero in x.

7. SQUARE ROOTS

If R and A are real numbers such that $R^2 = A$, then R is called a *square root* of A. If R is real then $R^2 \geqq 0$. Hence, if $A < 0$ then no real number R exists so that $R^2 = A$, or a negative number has no real square root. Thus, at present, with only real numbers available, we can refer to a square root of A only when $A \geqq 0$. Consideration of square roots for negative numbers in a later section will lead to a discussion of imaginary numbers.

Any positive number P has two square roots, one positive and one negative, where the roots have the same absolute value. The positive square root of P is represented by $\sqrt{P}$, and the negative square root of P is $-\sqrt{P}$. Any reference to THE square root of P will mean its *positive* square root. Hence, we read "$\sqrt{P}$" as THE square root of P, and "$-\sqrt{P}$" as "*minus the square root of P.*" The symbol $\sqrt{P}$ is called a **radical,** whose **radicand** is P. By the definition of $\sqrt{P}$,

$$(\sqrt{P})^2 = P. \tag{1}$$

*The three dots $\cdots$ may be read "and so forth."

The number 0 has just one square root, 0, because $0^2 = 0$ and $R^2 \neq 0$ if $R \neq 0$, so that no number $R \neq 0$ can be a square root of 0. We let $\sqrt{0} = 0$.

ILLUSTRATION 1.   The square roots of 25 are $\pm\sqrt{25}$, or $+5$ and -5, because $5^2 = (-5)^2 = 25$.

If $b \neq 0$, then b^2 has the square roots b and also $-b$, because $(-b) \cdot (-b) = +b^2$. Thus, if $b > 0$ the positive square root of b^2 is b. Hence,

$$\text{if } b \geqq 0 \quad \text{then} \quad \sqrt{b^2} = b, \tag{2}$$

where the special case $b = 0$ in (2) is a consequence of the fact that $\sqrt{0} = 0$. If $b < 0$, the positive square root of b^2 is $-b$, or

$$\text{if } b < 0 \quad \text{then} \quad \sqrt{b^2} = -b. \tag{3}$$

If b is *any* real number, from (2) and (3) we obtain $\sqrt{b^2} = |b|$. In order to avoid recalling (3) in routine problems, we agree that, unless otherwise stipulated, in any radical $\sqrt{N}$ all literal number symbols not in exponents will represent positive numbers, and be such that $N \geqq 0$. Then, (2) will apply in all cases except where mentioned otherwise.

If $H \geqq 0$ and $K \geqq 0$, then

$$\sqrt{HK} = \sqrt{H}\sqrt{K}, \tag{4}$$

because $(\sqrt{H}\sqrt{K})^2 = (\sqrt{H})^2(\sqrt{K})^2 = HK$. If $H \geqq 0$ and $K > 0$, then

$$\sqrt{\frac{H}{K}} = \frac{\sqrt{H}}{\sqrt{K}}, \tag{5}$$

because

$$\left(\frac{\sqrt{H}}{\sqrt{K}}\right)^2 = \frac{(\sqrt{H})^2}{(\sqrt{K})^2} = \frac{H}{K}.$$

ILLUSTRATION 2. $\sqrt{20} = \sqrt{4(5)} = \sqrt{4}\sqrt{5} = 2\sqrt{5}.$

$$\sqrt{\frac{25}{36}} = \frac{\sqrt{25}}{\sqrt{36}} = \frac{5}{6}.$$

A nonnegative integer is said to be a **perfect square** if the integer is the square of an integer. A rational number is said to be a perfect square if the number is the square of a rational number.

ILLUSTRATION 3. $0, 81,$ and $\dfrac{4}{25}$ are perfect squares because

$$0 = 0^2; \quad 81 = 9^2; \quad \frac{4}{25} = \left(\frac{2}{5}\right)^2.$$

Suppose that n is an even integer, so that $n = 2h$ where h is an integer. Then, if $p > 0$,

$$\sqrt{p^n} = \sqrt{p^{2h}} = p^h = p^{n/2}, \quad \text{because} \quad (p^h)^2 = p^{2h}.$$

Thus we have the following result:

If n is an even integer and $p > 0$, $\sqrt{p^n} = p^{n/2}$.   (6)

ILLUSTRATION 4.   By use of (4), (5), and (6),

$$\sqrt{\frac{81a^2b^6}{4c^4}} = \frac{\sqrt{81a^2b^6}}{\sqrt{4c^4}} = \frac{9ab^3}{2c^2}.$$

In this chapter, the coefficients in any monomials or polynomials in the variables will be rational numbers. Then, to state that any monomial or polynomial of this type is a *perfect square* will mean that the monomial, or polynomial, is the square of an expression of the same type. In a monomial which is a perfect square, each exponent is an *even integer* because, in squaring a monomial, each exponent is multiplied by 2.

To obtain the square root of a monomial which is a perfect square, we use (4) and (6).

ILLUSTRATION 5. $\sqrt{25x^2z^4w^6} = \sqrt{25}\,\sqrt{x^2}\sqrt{z^4}\,\sqrt{w^6} = 5xz^2w^3.$

If N and D are monomials which are perfect squares, we obtain $\sqrt{N/D}$ by use of (4), (5), and (6), as in Illustration 4.

To simplify a radical, we may use (4) to remove from the radicand any factor which is a perfect square. This simplification may be of aid in calculating the square root of a number which is not a perfect square.

ILLUSTRATION 6. $\sqrt{28} = \sqrt{4(7)} = \sqrt{4}\,\sqrt{7} = 2\sqrt{7}$. Then, by use of Table I for $\sqrt{7}$, we obtain $\sqrt{28} = 2(2.646) = 5.292.$

ILLUSTRATION 7.   To simplify the following radical, (4) is used.

$$\sqrt{98x^3y^5} = \sqrt{49x^2y^4(xy)} = \sqrt{49x^2y^4}\,\sqrt{xy} = 7xy^2\sqrt{xy}.$$

To rationalize the denominator in a radical where the radicand is a fraction, or in a fraction involving radicals, means to obtain a new form where the denominator will *not* involve a radical. To rationalize the denominator in a radical $\sqrt{N/D}$, multiply both numerator and denominator of N/D, if necessary, by a factor which causes the denominator to become a perfect square.

ILLUSTRATION 8.   To rationalize the denominator in $\sqrt{\frac{2}{3}}$, we multiply both numerator and denominator of $\frac{2}{3}$ by 3:

$$\sqrt{\frac{2}{3}} = \sqrt{\frac{2}{3} \cdot \frac{3}{3}} = \frac{\sqrt{6}}{\sqrt{3^2}} = \frac{\sqrt{6}}{3}.$$

EXERCISE 2

Obtain the square roots of the number.

1. 49.   **2.** 144.   **3.** .64.   **4.** .81.   **5.** $\dfrac{4}{25}$.

Find the square root.

6. $\sqrt{121}$.   **7.** $\sqrt{.25}$.   **8.** $\sqrt{\dfrac{1}{81}}$.   **9.** $\sqrt{225}$.

10. $\sqrt{\dfrac{4}{169}}$.   **11.** $\sqrt{\dfrac{49}{225}}$.   **12.** $\sqrt{.64}$.   **13.** $\sqrt{\dfrac{100}{9}}$.

14. $\sqrt{x^6}$.   **15.** $\sqrt{4u^4}$.   **16.** $\sqrt{49a^4}$.   **17.** $\sqrt{9x^{12}}$.

18. $\sqrt{\dfrac{u^8}{4x^2}}$.   **19.** $\sqrt{\dfrac{9x^2}{b^4}}$.   **20.** $\sqrt{\dfrac{u^2v^2}{81x^4}}$.

21. $\sqrt{\dfrac{49a^4}{b^2c^6}}$.   **22.** $\sqrt{\dfrac{9x^2y^4}{u^6v^8}}$.   **23.** $\sqrt{\dfrac{4a^4b^2}{9c^6}}$.

24. $\sqrt{\dfrac{16a^4b^2}{9x^2u^6}}$.   **25.** $\sqrt{\dfrac{225a^4}{196c^6}}$.   **26.** $\sqrt{\dfrac{16x^2z^4}{81u^4b^8}}$.

Calculate.

27. $(\sqrt{133})^2$.   **28.** $(\sqrt{5xz^{17}})^2$.   **29.** $(\sqrt{156ab^3})^2$.

Simplify by rationalizing any denominator, and by removing any perfect square which is a factor of a radicand. Compute by use of Table I if possible.

30. $\sqrt{18}$.   **31.** $\sqrt{48}$.   **32.** $\sqrt{72}$.   **33.** $\sqrt{500}$.

34. $\sqrt{\dfrac{2}{3}}$.   **35.** $\sqrt{\dfrac{1}{5}}$.   **36.** $\sqrt{\dfrac{7}{2}}$.   **37.** $\sqrt{\dfrac{5}{12}}$.

38. $\sqrt{8a^3}$.   **39.** $\sqrt{50x^2y^3}$.   **40.** $\sqrt{4x^5y^6}$.

41. $\sqrt{\dfrac{3x}{2y}}$.   **42.** $\sqrt{\dfrac{2a}{5z}}$.   **43.** $\sqrt{\dfrac{18x^3}{5y^3}}$.

44. $\dfrac{3}{2\sqrt{5}}$.   **45.** $\dfrac{5}{3\sqrt{2}}$.   **46.** $\dfrac{7}{5\sqrt{3}}$.

8. SPECIAL PRODUCTS AND POWERS, AND FACTORING

It is desirable to cultivate the ability to obtain the right-hand side of each of the following equations quickly.

$$(x + y)(x - y) = x^2 - y^2. \tag{1}$$
$$(x + y)^2 = x^2 + 2xy + y^2. \tag{2}$$

$$(x - y)^2 = x^2 - 2xy + y^2. \tag{3}$$
$$(x + a)(x + b) = x^2 + (ax + bx) + ab. \tag{4}$$
$$(ax + b)(cx + d) = acx^2 + (ad + bc)x + bd. \tag{5}$$
$$(x + y)^3 = x^3 + 3x^2y + 3xy^2 + y^3. \tag{6}$$
$$(x - y)^3 = x^3 - 3x^2y + 3xy^2 - y^3. \tag{7}$$
$$x^3 + y^3 = (x + y)(x^2 - xy + y^2). \tag{8}$$
$$x^3 - y^3 = (x - y)(x^2 + xy + y^2). \tag{9}$$

On the right in (2), and in (3), we observe a trinomial which is a perfect square. When (1)–(7) are read from right to left, they give formulas for factoring corresponding polynomials.

ILLUSTRATION 1.   By use of certain equations in (1)–(9), as read from left to right, we obtain the following results.

$$(x^2 - 2y^3)(x^2 + 2y^3) = x^4 - 4y^6.$$
$$(2c + 3d)^2 = (2c)^2 + 2(2c)(3d) + (3d)^2 = 4c^2 + 12cd + 9d^2.$$
$$(2a + b)^3 = (2a)^3 + 3(2a)^2(b) + 3(2a)b^2 + b^3$$
$$= 8a^3 + 12a^2b + 6ab^2 + b^3.$$
$$8a^3 - 27b^3 = (2a)^3 - (3b)^3 = (2a - 3b)(4a^2 + 6ab + 9b^2).$$

In any use of factoring, we agree that we shall be dealing with the factors of a polynomial where all of its coefficients are integers.   Any factor will be a polynomial of the nature just described.

A positive integer N is said to be **prime number** in case $N > 1$, and N has no positive integer as a factor except N and 1.   In factoring a positive integer, we shall employ only positive integers as factors.   To factor a negative integer $-N$, we shall use -1 as one factor, and take all other factors as positive integers.   In factoring an integer, it is desirable usually to express it as a product of powers of prime factors, and -1 if necessary. We do not refer to any negative integer as being a prime number.

A polynomial, P, will be called a **prime polynomial** if $P \neq \pm 1$ and if P has no factor except itself or ± 1.

ILLUSTRATION 2.   The first few prime numbers, in increasing order, are $\{2, 3, 5, 7, 11\}$. $500 = 2 \cdot 2 \cdot 5 \cdot 5 \cdot 5 = 2^2 5^3$.

To factor a polynomial P will mean to express P as a product of prime factors.

ILLUSTRATION 3.   By use of (1), (2), (7), and (8), the following results are obtained.

$$9x^2 - 25y^2 = (3x - 5y)(3x + 5y).$$
$$4x^2 + 12xz + 9z^2 = (2x + 3z)^2.$$

$$(2 - 3a)^3 = 2^3 - 3(2^2)(3a) + 3(2)(3a)^2 - (3a)^3$$
$$= 8 - 36a + 54a^2 - 27a^3.$$
$$8a^3 + b^3 = (2a)^3 + b^3 = (2a + b)(4a^2 - 2ab + b^2).$$
$$x^4 - y^4 = (x^2 - y^2)(x^2 + y^2) = (x - y)(x + y)(x^2 + y^2). \tag{10}$$

In (10), $(x^2 + y^2)$ is a prime polynomial.

There is no convenient or elementary method for recognizing when a polynomial is prime. Thus, $(x^4 + y^4)$ is prime. The somewhat similar polynomial $(u^6 + v^6)$ is *not* prime, as shown below.

ILLUSTRATION 4. $u^6 + v^6 = (u^2)^3 + (v^2)^3$

$$= (u^2 + v^2)(u^4 - u^2v^2 + v^4), \text{ by use of (8).}$$

We refer to $(u^6 + v^6)$ as a sum of two *perfect cubes,* which then was factored by use of (8). Similarly, $(u^6 - v^6)$ can be factored by use of (9) in the first stage.

As in Illustration 4, a sum or a difference of any two perfect cubes can be factored by use of (8) or (9). A difference of two perfect squares can be factored by use of (1).

With a fraction N/D where $D = a\sqrt{b} + c\sqrt{d}$, we rationalize the denominator by introducing the factor $(a\sqrt{b} - c\sqrt{d})$, so that the new denominator is the difference of two squares, because

$$(a\sqrt{b} + c\sqrt{d})(a\sqrt{b} - c\sqrt{d}) = (a\sqrt{b})^2 - (c\sqrt{d})^2 = a^2b - c^2d$$

ILLUSTRATION 5.   To rationalize the denominator below on the left, we multiply both numerator and denominator by $(2\sqrt{3} + \sqrt{5})$:

$$\frac{\sqrt{3} - 2\sqrt{5}}{2\sqrt{3} - \sqrt{5}} = \frac{\sqrt{3} - 2\sqrt{5}}{2\sqrt{3} - \sqrt{5}} \cdot \frac{2\sqrt{3} + \sqrt{5}}{2\sqrt{3} + \sqrt{5}}$$

$$= \frac{2(\sqrt{3})^2 - 4\sqrt{3}\sqrt{5} + \sqrt{3}\sqrt{5} - 2(\sqrt{5})^2}{(2\sqrt{3})^2 - (\sqrt{5})^2} = \frac{-4 - 3\sqrt{15}}{12 - 5}$$

$$= -\frac{4 + 3\sqrt{15}}{7}.$$

EXERCISE 3

Multiply, or expand the power, and collect similar terms.

1. $(u - 3)(u + 5)$.
2. $(2x + 5)(3x - 2)$.
3. $(a - 2b)(a + 2b)$.
4. $(3x - y)(3x + y)$.
5. $(2x - 5y)(3x + 2y)$.
6. $(cd - 2)(cd + 2)$.
7. $(uv - 3)(3uv + 4)$.
8. $(5 - 2xy)(3 + 4xy)$.

9. $(u + 2v)^2$.   **10.** $(2a - b)^2$.   **11.** $(c + 2d)^3$.   **12.** $(2x - 3y)^3$.
13. $(u + 2v)(u^2 - 2uv + 4v^2)$.   **14.** $(2c - d)(4c^2 + 2cd + d^2)$.

Factor.

15. $3u + 5uw$.

16. $2ac + 3a^2c^2$.

17. $-3bx + 4by$.

18. $5x - 4x^2$.

19. $6ax^2 - 3x^3$.

20. $xy^2 + 3xy$.

21. $u^2 - v^2$.

22. $z^2 - w^2$.

23. $4x^2 - 9y^2$.

24. $81w^2 - u^2v^2$.

25. $9c^2 - d^2$.

26. $25 - 4u^2w^2$.

27. $u^2 + 18u + 81$.

28. $4x^2 + 20xy + 25y^2$.

29. $9 - 12w + 4w^2$.

30. $49a^2 - 28ab + 4b^2$.

31. $6xy + x^2 + 9y^2$.

32. $-10u + 25 + u^2$.

33. $x^2 - 3x - 28$.

34. $a^2 + 6a - 72$.

35. $x^2 - 6x - 27$.

36. $5 - 12x + 7x^2$.

37. $3u^2 - 8uv + 5v^2$.

38. $2a^2 - 7ab + 3b^2$.

39. $x^3 - w^3$.

40. $8 - u^3$.

41. $a^3 + 27b^3$.

42. $81a^2 - y^4$.

43. $u^4 - 16v^4$.

44. $81h^4 - 16k^4$.

45. $x^6 - w^6$.

46. $625 - 81x^4$.

47. $8u^3 + 27v^3$.

Factor by first expressing the polynomial as a difference of two squares, or as a sum or a difference of two cubes.

48. $x^8 - y^8$.   **49.** $16x^4 - 625u^4$.   **50.** $x^6 - y^6$.
51. $u^6 + 64w^6$.   **52.** $x^9 - y^9$.   **53.** $81a^4 - 256$.

54. By dividing $(x^5 + y^5)$ by $(x + y)$, prove that $(x^5 + y^5)$ has $(x + y)$ as a factor, and thus find the factor F so that $(x + y)F = x^5 + y^5$.

Comment. If n is an *odd* positive integer, then $(x + y)$ is a factor of $(x^n + y^n)$, and $(x^n + y^n) = (x + y)F$ where

$$F = x^{n-1} - x^{n-2}y + \cdots + y^{n-1}. \tag{1}$$

In (1) on the right, the coefficients are alternately $+1$ and -1; also, the exponents of x decrease by 1 and the exponents of y increase by 1 from term to term as read from left to right. A special case of this result is seen in (8) on page 14. The preceding facts will be proved in a later chapter.

55. Make a conjecture about two factors for $(x^n - y^n)$ where n is a positive integer. Verify the conjecture for $n = 3$ by use of (9) on page 14, and for $n = 4$ by multiplication.

Rationalize the denominator and simplify.

56. $\dfrac{2 - \sqrt{3}}{3 + 2\sqrt{3}}$.

57. $\dfrac{\sqrt{5} + 2}{3 - \sqrt{5}}$.

58. $\dfrac{6 - 5\sqrt{2}}{3\sqrt{2} - 4}$.

59. $\dfrac{\sqrt{2} - 3\sqrt{3}}{\sqrt{2} + \sqrt{3}}$.

60. $\dfrac{\sqrt{5} + 2\sqrt{3}}{3\sqrt{3} - \sqrt{5}}$.

61. $\dfrac{\sqrt{6} + 3\sqrt{2}}{2\sqrt{6} - \sqrt{2}}$.

9. SIMPLIFICATION OF FRACTIONS

Sometimes a bar, $\overline{}$, called a *vinculum*, is placed over a sum of terms to indicate that the sum should be treated as if it were a single term. Thus, a vinculum has the same effect as a pair of parentheses.

Illustration 1. $-\{2a - (4 - \overline{3a - 2})\} =$

$$-\{2a - (4 - 3a + 2)\} = -\{2a - 6 + 3a\} = -5a + 6.$$

A sum of fractions with a common denominator can be written as a single fraction. The new numerator is formed by adding the given numerators, with each placed within parentheses preceded by the sign before the corresponding fraction. The denominator of the result is the common denominator of the given fractions. A fraction bar has the force of a vinculum with respect to terms in the numerator.

Illustration 2. $\dfrac{3}{5} - \dfrac{2 - 4x}{5} + \dfrac{3 + 2x}{5}$

$$= \frac{3 - (2 - 4x) + (3 + 2x)}{5} = \frac{4 + 6x}{5}.$$

In the remainder of this section, we assume that the numerator and denominator are polynomials with integers as coefficients in any fraction being considered.

The *lowest common multiple* (LCM) of a set of polynomials is the polynomial with smallest coefficients which has each given polynomial as a factor. The *lowest common denominator* (LCD) of a set of fractions is the LCM of their denominators. To express a sum of fractions as a single fraction, first find the LCD of the given fractions, and express each of them as a new fraction having the LCD. Then form the sum of the new fractions.

Example 1.   Find the sum as a single fraction:

$$\frac{2x}{5a^3b} - \frac{7}{3ab^2}. \tag{1}$$

Solution.   LCD $= 15a^3b^2$. Then

$$\frac{15a^3b^2}{5a^3b} = 3b; \qquad \frac{15a^3b^2}{3ab^2} = 5a^2.$$

Hence, multiply the numerator and denominator by $3b$ in the fraction at the left in (1), and by $5a^2$ in the fraction at the right:

$$\frac{2x}{5a^3b} - \frac{7}{3ab^2} = \frac{3b(2x)}{3b(5a^3b)} - \frac{7(5a^2)}{3ab^2(5a^2)} = \frac{6bx - 35a^2}{15a^3b^2}.$$

ILLUSTRATION 3. $2a + 6 - \dfrac{3a - 5}{2a + 3} = \dfrac{2a + 6}{1} - \dfrac{3a - 5}{2a + 3}$

$$= \dfrac{(2a + 6)(2a + 3) - (3a - 5)}{2a + 3} = \dfrac{4a^2 + 15a + 23}{2a + 3}.$$

If no fraction appears in the numerator or denominator of a given fraction, it is called a *simple fraction,* and otherwise is called a *complex fraction.* To simplify a complex fraction, first express its numerator and denominator as simple fractions, and then perform the division to obtain a simple fraction as the final result.

ILLUSTRATION 4. $\dfrac{\dfrac{3}{2} - \dfrac{5a}{3}}{2a + \dfrac{1}{3}} = \dfrac{\dfrac{9 - 10a}{6}}{\dfrac{6a + 1}{3}}$

$$= \dfrac{9 - 10a}{6} \cdot \dfrac{3}{6a + 1} = \dfrac{9 - 10a}{12a + 2}.$$

ILLUSTRATION 5.   To change the following fraction to lowest terms, first we factor the numerator and denominator:

$$\dfrac{4a^2 - 4ab + b^2}{2a^2 + 9ab - 5b^2} = \dfrac{(2a - b)^2}{(2a - b)(a + 5b)} = \dfrac{2a - b}{a + 5b},$$

where we divided out the factor $(2a - b)$.

EXAMPLE 2.   Express S below as a single fraction:

$$S = \dfrac{3x}{4x^2 - 25} - \dfrac{5 - 3x}{6x^2 + 11x - 10}. \tag{2}$$

Solution.   1. Factor the denominators:

$$4x^2 - 25 = (2x - 5)(2x + 5).$$
$$6x^2 + 11x - 10 = (2x + 5)(3x - 2).$$
$$\text{LCD} = (2x - 5)(2x + 5)(3x - 2).$$

2. in (2),

for the 1st fraction, $[\text{LCD} \div (4x^2 - 25)] = (3x - 2);$

for the 2d fraction, $\{\text{LCD} \div [(2x + 5)(3x - 2)]\} = (2x - 5).$

Hence, to change to the LCD, multiply both numerator and denominator by $(3x - 2)$ in the fraction at the left in (2), and by $(2x - 5)$ in the fraction at the right in (2). Then add the fractions.

$$S = \dfrac{3x(3x - 2)}{\text{LCD}} - \dfrac{(5 - 3x)(2x - 5)}{\text{LCD}} = \dfrac{15x^2 - 31x + 25}{(3x - 2)(4x^2 - 25)}.$$

EXERCISE 4

Perform the indicated operation and express the result as a simple fraction in lowest terms.

1. $\dfrac{10 - \dfrac{2}{3}}{\dfrac{1}{4} + \dfrac{1}{5}}$.

2. $\dfrac{3x - \dfrac{2}{5}}{4x - \dfrac{3}{4}}$.

3. $\dfrac{\dfrac{2}{3x} + \dfrac{3}{y}}{\dfrac{3}{x} - \dfrac{2}{y}}$.

4. $\dfrac{\dfrac{4}{b} - b}{2 - \dfrac{1}{bc}}$.

5. $\dfrac{a^2 + 7a + 12}{a^2 - 9}$.

6. $\dfrac{6 + x - x^2}{4 - x^2}$.

7. $\dfrac{2y^2 - 5ay - 3a^2}{y^2 - 6ay + 9a^2}$.

8. $\dfrac{3}{2a - 1} + \dfrac{5}{3a + 3}$.

9. $1 - \dfrac{6}{x} + \dfrac{1}{x^2 - 4x}$.

10. $\dfrac{5x}{2x - 2y} + \dfrac{3}{x^2 - y^2}$.

11. $\dfrac{x - 2}{9 - x^2} - \dfrac{x + 2}{x^2 - 6x + 9}$.

12. $\dfrac{3a - \dfrac{2}{b}}{9a^2 - \dfrac{4}{b^2}}$.

13. $\dfrac{x^2 - \dfrac{4}{y^2}}{1 - \dfrac{2}{xy}}$.

14. $\dfrac{\dfrac{y^2}{x} - \dfrac{x^2}{y}}{\dfrac{1}{x} - \dfrac{1}{y}}$.

15. $\dfrac{3d - 2c}{\dfrac{4c}{3d} - \dfrac{3d}{c}}$.

16. $\dfrac{au - bu}{\dfrac{1}{a} - \dfrac{a}{b^2}}$.

17. $\dfrac{a^3 + b^3}{\dfrac{a}{b} - 1 + \dfrac{b}{a}}$.

18. $\left(1 + \dfrac{4}{u - 2}\right)\left(2 - \dfrac{8}{u + 2}\right)$.

19. $\left(\dfrac{4u}{v} - \dfrac{9v}{u}\right) \div \left(u - \dfrac{3v}{2}\right)$.

20. $\dfrac{1}{2u - 2} - \dfrac{u + 6}{u^2 + 3u - 4}$.

21. $\dfrac{u + 2v^2}{u^3 - v^3} + \dfrac{2}{u^2 + uv + v^2}$.

22. $\dfrac{3x + 2y}{x^2 - 4y^4} - \dfrac{x - y}{xy + 2y^3}$.

23. $\dfrac{y^2 - 9}{x - 2y} \div \dfrac{y^2 - 6y + 9}{8y^3 - x^3}$.

10. INTRODUCTION TO COMPLEX NUMBERS

If $A < 0$ and r is a square root of A, then $r^2 = A$. But, if r is a real number then $r^2 \geqq 0$. Hence, *no real number r satisfies $r^2 = A$ if $A < 0$.* Or, a negative number A has *no real square root*. Thus, if negative numbers are to have square roots, it is essential that numbers of a new type should be introduced to represent such roots. We proceed to describe the appropriate new numbers.

From preceding remarks, it is seen that -1 has no real square root. Then, first, we enlarge the system, R, of real numbers so that -1 will have a square root. Thus, we let $\sqrt{-1}$ be a symbol for a new number, to be called an *imaginary number,* which we shall represent also by the symbol i. Let H be the new system of numbers consisting of *all real numbers and i.* Since i is a new number, it is necessary to define the product $(i)(i)$, which we take as follows:

$$i \cdot i = -1, \quad or \quad i^2 = -1. \tag{1}$$

Hence, in the number system H, the number -1 has i as one square root. For every pair of real numbers a and b, we expand H by adjoining to it a new number bi to represent the *product* of b and i, and another new number $(a + bi)$ to represent the *sum* of a and bi. We agree that $(a + 0i)$ is an optional symbol for the real number a, and $(0 + bi)$ is an optional symbol for bi. The preceding expansions of the number system, commencing with the real number system R, has produced a system, C, of numbers as follows:

$$\left. \begin{array}{l} C \text{ consists of all numbers } (a + bi) \text{ where } a \text{ and } b \text{ are} \\ \text{real numbers.} \end{array} \right\} \tag{2}$$

For any a and b, the number $(a + bi)$ is called a **complex number** whose **real part** is a and **imaginary part** is b. If $b \neq 0$, then $(a + bi)$ is called an **imaginary number.** If $a = 0$ and $b \neq 0$, then $(a + bi)$ is called a **pure imaginary number.** We refer to i as the *imaginary unit* and to 1 as the *real unit.*

ILLUSTRATION 1. $(2 - 3i)$ is a complex number which is an imaginary number, with real part 2 and imaginary part -3. The real number 6 can be thought of as $(6 + 0i)$. The number $4i$, or $(0 + 4i)$, is a pure imaginary number. We have $0 = 0 + 0i$.

To state that two complex numbers $(a + bi)$ and $(c + di)$ are *equal* will mean that their real parts are the same, and the imaginary parts are the same:

$$a + bi = c + di \quad means \, that \quad a = c \text{ and } b = d. \tag{3}$$

ILLUSTRATION 2.   If $a + bi = 3 - 5i$, then $a = 3$ and $b = -5$.

In this text, unless otherwise specified, any literal number symbol except i will represent a real number, and i always will denote the imaginary unit. In the system, C, of complex numbers, addition and multiplication* can be defined as follows:

$$\left. \begin{array}{l} \textit{The sum and the product of two complex numbers} \\ (a + bi) \textit{ and } (c + di) \textit{ are the results obtained if} \\ (a + bi) \textit{ and } (c + di) \textit{ are treated as if they are poly-} \\ \textit{nomials in a real number i, for which } i^2 = -1. \end{array} \right\} \tag{4}$$

*We shall not consider division for complex numbers in this text.

From (4), we obtain

$$(a + bi) + (c + di) = (a + c) + (b + d)i;$$
$$(a + bi)(c + di) = ac + adi + bci + bdi^2$$
$$= ac + bd(-1) + i(ad + bc), \text{ or}$$
$$(a + bi)(c + di) = (ac - bd) + i(ad + bc).$$

ILLUSTRATION 3. $(2 - 5i) + (3 + 4i) = 5 - i.$

$(2 - 5i)(3 + 4i) = 6 + 8i - 15i - 20i^2 = 6 - 7i - 20(-1) = 26 - 7i.$

ILLUSTRATION 4.   By use of (4), $(-i)^2 = (-i)(-i) = i^2 = -1.$ Therefore -1 has the square root $-i$ as well as the square root i.

If P is positive, then

$$(i\sqrt{P})^2 = i^2(P) = -P; \qquad (-i\sqrt{P})^2 = i^2P = -P.$$

Hence, by the definition of a square root, $-P$ has the two square roots $i\sqrt{P}$ and $-i\sqrt{P}$, where we let

$$\sqrt{-P} = i\sqrt{P}. \tag{5}$$

We shall read "$\sqrt{-P}$" as "*the square root of* $-P$." In particular, from (5), $\sqrt{-1} = i$, which agrees with our original statements concerning i.

ILLUSTRATION 5.   The two square roots of -9 are $\pm\sqrt{-9}$ or $\pm i\sqrt{9}$, or $\pm 3i$.

It is necessary to recall that the formula

$$\sqrt{H}\sqrt{K} = \sqrt{HK} \tag{6}$$

was stated to apply only when both* H and K are not negative. If (6) is used when $H < 0$ and $K < 0$, a wrong result will be obtained.

ILLUSTRATION 6.   By use of (5),

$$\sqrt{-2}\sqrt{-50} = i\sqrt{2}(i\sqrt{50}) \tag{7}$$
$$= i^2\sqrt{2}\sqrt{50} = (-1)\sqrt{100} = -10.$$

If (6) is used immediately in (7), we obtain

$$\sqrt{-2}\sqrt{-50} = \sqrt{(-2)(-50)} = \sqrt{100} = 10,$$

which is *wrong*.

As a consequence of (4), addition and multiplication in the system of complex numbers obey the commutative, associative, and distributive laws mentioned for real numbers on pages 1 and 2. Hence, the laws of exponents apply with positive integral powers of i.

*Formula (6) is valid if just one of H and K is negative.

ILLUSTRATION 7. $i^3 = i^2(i) = (-1)i = -i.$

$$i^4 = i^2(i^2) = (-1)(-1) = 1.$$

$$i^{11} = i^4 i^4 i^3 = (1)(1)(-i) = -i.$$

Similarly, if n is any positive integer, i^n is equal to i, $-i$, 1, or -1.   Any power i^n with $n \geq 4$ can be obtained conveniently by recalling that $i^4 = 1$.

Our introduction of the system, C, of complex numbers was motivated by the desire to obtain numbers which would serve as square roots of negative numbers.   It is interesting to note that C not only applies for the preceding purpose but also is satisfactory as a basis for all of mathematical analysis, short of the most advanced stages.

EXERCISE 5

Introduce the imaginary unit i and simplify the final radical if possible. Any literal number symbol except i represents a positive number.

1. $\sqrt{-16}$.   **2.** $\sqrt{-81}$.   **3.** $\sqrt{-27}$.   **4.** $\sqrt{-36}$.

5. $\sqrt{-90}$.   **6.** $\sqrt{-13}$.   **7.** $\sqrt{-\dfrac{25}{9}}$.   **8.** $\sqrt{-\dfrac{1}{64}}$.

9. $\sqrt{-\dfrac{4}{49}}$.   **10.** $\sqrt{-\dfrac{49}{16}}$.   **11.** $\sqrt{-.09}$.   **12.** $\sqrt{-.36}$.

13. $\sqrt{-\dfrac{2}{5}}$.   **14.** $\sqrt{-\dfrac{1}{3}}$.   **15.** $\sqrt{-\dfrac{3}{8}}$.   **16.** $\sqrt{-4x^4}$.

17. $\sqrt{-36u^2}$.   **18.** $\sqrt{-128z^2}$.   **19.** $\sqrt{-50a^6}$.

20. $\sqrt{-\dfrac{a^2 b^2}{9x^4}}$.   **21.** $\sqrt{-\dfrac{3u^4}{25x^2}}$.   **22.** $\sqrt{-\dfrac{27x^3}{20u^2}}$.

Give the two square roots of the number.

23. -49.   **24.** -100.   **25.** -144.   **26.** -125.

Perform the indicated operation and simplify by use of $i^2 = -1$.

27. i^6.   **28.** i^7.   **29.** i^{13}.   **30.** i^8.   **31.** i^{15}.

32. $(2 - i)(2 + i)$.   **33.** $(2i - 3)(5i - 1)$.   **34.** $(2 - 3i)(1 + 4i)$.

35. $(3i + 2)^2$.   **36.** $(2 - 5i)^2$.   **37.** $(i - 3)^2$.

38. $(2 + i)^3$.   **39.** $(3 - 2i)^3$.   **40.** $(3i + 2)^3$.

41. $(2i^2 - i^3 + 6i + 5)(2 - 3i^2 + i)$.

42. $\sqrt{-3}\sqrt{-27}$.   **43.** $\sqrt{-5}\sqrt{-125}$.   **44.** $\sqrt{-2}\sqrt{-20}$.

45. $\sqrt{-4}(2 - \sqrt{-64})$.   **46.** $\sqrt{-3}(5 - \sqrt{-243})$.

47. By substitution, show that $x = -1 + 2i$ is a solution of the equation $x^2 + 2x + 5 = 0$.

11. RADICALS OF ANY ORDER

Suppose that n is a positive integer. Then, to state that a number R is an nth root of a number A means that

$$R^n = A. \tag{1}$$

An nth root of A is called a **square root** of A when $n = 2$ and a **cube root** when $n = 3$. Otherwise, no special name is given to an nth root.

ILLUSTRATION 1. -3 is a cube root of -27 because $(-3)^3 = -27$. Two fourth roots of 16 are $+2$ and -2 because $2^4 = 16$ and $(-2)^4 = 16$.

In this section, we agree that the number system is the set of all complex numbers $(a + bi)$. In college algebra, it is proved that *any complex number $(a + bi)$, not zero, has n distinct nth roots*. In this text we shall restrict our remarks about nth roots of A to the case where A is *real*. In any reference to an nth root, it will be implied that n is a positive integer. Then, we recall the following facts from algebra.

   I. *If A is real and $A \neq 0$, then A has n distinct nth roots, where some or all of them may be imaginary numbers.*

   II. *If n is an even integer and $A > 0$, then A has exactly two real nth roots, of opposite signs, and equal absolute values.*

   III. *If n is an odd integer and A is real, then A has just one real nth root, which is of the same sign as A.*

   IV. *If n is an even integer and $A < 0$, then A has no real nth root.*

If $A = 0$, *its only nth root, R, is $R = 0$ because $0^n = 0$ and $R^n \neq 0$ if $R \neq 0$.*

If $A > 0$, the positive nth root of A is called its **principal nth root.** If $A < 0$ and n is odd, the negative nth root of A is called its *principal nth root.* If $A = 0$, its only nth root, 0, is called the principal nth root of A. If $A < 0$ and n is even, so that all nth roots of A are imaginary, then it is said that A has no principal nth root.

ILLUSTRATION 2.   The real 4th roots of 81 are ± 3, where $+3$ is the principal 4th root of 81. The principal cube root of -8 is -2. It can be proved that $(1 \pm i\sqrt{3})$ also are cube roots of -8.

The **radical** $\sqrt[n]{A}$, which we read as "*the nth root of A,*" is used to represent the principal nth root of A when A has a real nth root. In $\sqrt[n]{A}$, we call A the **radicand** and n the **index** of the radical. If $n = 2$, the index is not written, and thus, as before, we use $\sqrt{A}$ instead of $\sqrt[2]{A}$ for the square root of A. If A has *no* real nth root, then $\sqrt[n]{A}$ can be used to represent any convenient nth root. In particular, if $P > 0$, we have agreed previously to let $\sqrt{-P} = i\sqrt{P}$. Our preceding agreements about principal roots and radicals can be summarized as follows:

$\sqrt[n]{A} > 0$ *if* $A > 0$.

$\sqrt[n]{A} < 0$ *if* $A < 0$ *and n is odd.*

$\sqrt[n]{A}$ *is imaginary if* $A < 0$ *and n is even.*

ILLUSTRATION 3. $\sqrt[4]{625} = 5$; $\sqrt[3]{-64} = -4$; the two real 4th roots of 625 are $\pm\sqrt[4]{625}$ or ±5; $\sqrt[4]{-8}$ is imaginary.

ILLUSTRATION 4.   All 4th roots of -16 are imaginary.

In a radical $\sqrt[n]{A}$, suppose that n is an even integer. Then, except as otherwise indicated, in this text any literal number symbol used in the radicand A and not in an exponent will represent a *positive* number which is such that $A > 0$.

For any positive integer n, if $A \geqq 0$ when n is even,* then

$$\sqrt[n]{A^n} = A, \tag{2}$$

This result is true because, if $A \neq 0$, A has the same sign as A^n, is an nth root of A^n, and thus is the *principal* nth root of A^n.

We emphasize the following properties of radicals.

$$(\sqrt[n]{A})^n = A. \tag{3}$$

$$\sqrt[n]{A^n} = A, \text{ if } A > 0 \text{ when } n \text{ is even.} \tag{4}$$

$$\sqrt[n]{AB} = \sqrt[n]{A}\sqrt[n]{B}. \tag{5}$$

$$\sqrt[n]{\frac{A}{B}} = \frac{\sqrt[n]{A}}{\sqrt[n]{B}}. \tag{6}$$

If m, n, and m/n are integers with n > 0 and A > 0, then

$$\sqrt[n]{A^m} = A^{m/n}. \tag{7}$$

The result in (3) is true because $\sqrt[n]{A}$ is an nth root of A. We obtained (4) in (2). To prove (5), raise the right-hand side of the nth power, and thus obtain

$$(\sqrt[n]{A}\sqrt[n]{B})^n = (\sqrt[n]{A})^n(\sqrt[n]{B})^n = AB.$$

Hence, the right-hand side of (5) is *an* nth root of AB. Also, by considering possibilities as to A and B being positive or negative, we find that the two sides of (5) are both *positive,* or both *negative,* or both *zero.* Hence, (5) is true. Similarly (6) is true. In (7), by the laws of exponents,

$$(A^{m/n})^n = A^{(m/n)\cdot n} = A^m,$$

and thus (7) is true.

*If $A < 0$ and n is even, then $A^n > 0$ and the positive nth root of A^n is $-A$ or $\sqrt[n]{A^n} = -A$. For *any* real number A, if n is even then $\sqrt[n]{A^n} = |A|$.

ILLUSTRATION 5.   By use of (5) and (7),

$$\sqrt[4]{16y^8} = \sqrt[4]{16}\sqrt[4]{y^8} = 2y^{8/4} = 2y^2; \tag{8}$$

$$\sqrt[3]{y^{-6}} = y^{-6/3} = y^{-2}.$$

By use of (6), then (5), and then (7),

$$\sqrt[3]{\frac{8y^9}{27x^6z^3}} = \frac{\sqrt[3]{8y^9}}{\sqrt[3]{27x^6z^3}} = \frac{2y^3}{3x^2z}.$$

Let us suppose that all numerical coefficients are rational numbers in any polynomial (including the case of a monomial) to which we may refer. Then, if n is a positive integer, a polynomial is said to be a **perfect nth power** in case the polynomial is the nth power of a polynomial. By use of (4), (5), and (7), we may obtain the nth root of any monomial which is a perfect nth power, as in (8).

ILLUSTRATION 6.   By use of (4) and (5), we may remove from the radicand in $\sqrt[n]{A}$ any factor which is a perfect nth power. Thus,

$$\sqrt[3]{54x^3y^3} = \sqrt[3]{27x^3y^3(2x^2)} = \sqrt[3]{27x^3y^3}\sqrt[3]{2x^2} = 3xy\sqrt[3]{2x^2}.$$

$$\sqrt[5]{64} = \sqrt[5]{2^6} = \sqrt[5]{2^5(2)} = 2\sqrt[5]{2}.$$

To rationalize the denominator in $\sqrt[n]{N/D}$, we multiply both numerator and denominator of N/D by a number which gives a *perfect nth power in the denominator*. This type of operation was considered on page 12 for the case where $n = 2$.

ILLUSTRATION 7. $\displaystyle\sqrt[3]{\frac{2x^2y}{3z^2}} = \sqrt[3]{\frac{2x^2y(9z)}{27z^3}} = \frac{\sqrt[3]{18x^2yz}}{3z}.$

In simplifying any radical $\sqrt[n]{A}$ where A involves fractions, we agree, first, to express A as a *simple fraction* in lowest terms, and then to rationalize the denominator. With any result $\sqrt[n]{H}$, we shall remove any perfect nth power which is a factor of H, and insert a corresponding factor before the radical. Also, it is advised to eliminate negative exponents.

ILLUSTRATION 8.   To simplify the following fraction, first we use (6), and then rationalize the denominator.

$$\frac{\sqrt[3]{4x^3y^2}}{\sqrt[3]{6x^5y^3}} = \sqrt[3]{\frac{4x^3y^2}{6x^5y^3}} = \sqrt[3]{\frac{2}{3x^2y} \cdot \frac{9xy^2}{9xy^2}} = \frac{\sqrt[3]{18xy^2}}{\sqrt[3]{27x^3y^3}} = \frac{\sqrt[3]{18xy^2}}{3xy}.$$

ILLUSTRATION 9. $\displaystyle\sqrt{3 + \frac{2}{x} + \frac{x^{-2}}{3}} = \sqrt{3 + \frac{2}{x} + \frac{1}{3x^2}}$

$$= \sqrt{\frac{9x^2 + 6x + 1}{3x^2} \cdot \frac{3}{3}} = \frac{\sqrt{3(3x + 1)^2}}{\sqrt{9x^2}} = \frac{(3x + 1)\sqrt{3}}{3x}.$$

EXERCISE 6

State the principal square root of the number.

1. 121.   **2.** 400.   **3.** 225.   **4.** $\dfrac{64}{9}$.   **5.** $\dfrac{121}{4}$.

State the principal cube root of the number.

6. -27.   **7.** 64.   **8.** -1.   **9.** $\dfrac{1}{8}$.   **10.** $\dfrac{125}{27}$.

State the principal 4th root of the number.

11. 16.   **12.** 256.   **13.** 81.   **14.** 625.   **15.** 10,000.

Obtain the indicated root or power.

16. $\sqrt[3]{125}$.   **17.** $\sqrt{49}$.   **18.** $\sqrt[4]{256}$.   **19.** $\sqrt[4]{.0016}$.

20. $\sqrt[3]{-216}$.   **21.** $\sqrt[3]{-125}$.   **22.** $\sqrt[3]{-27}$.   **23.** $\sqrt[5]{-32}$.

24. $\sqrt[3]{w^3 z^6}$.   **25.** $(\sqrt[5]{x})^5$.   **26.** $(\sqrt[3]{3z})^3$.   **27.** $\sqrt[4]{x^4 y^8}$.

28. $\sqrt[3]{8a^6 b^3}$.   **29.** $\sqrt[4]{16a^4 w^4}$.   **30.** $\sqrt[3]{-125x^3 y^6}$.

31. $\sqrt[3]{\dfrac{8x^3 y^6}{125z^3}}$.   **32.** $\sqrt[3]{\dfrac{-1}{125x^3}}$.   **33.** $\sqrt[4]{\dfrac{16z^{12}}{x^8 y^4}}$.

Simplify. Where possible, remove factors from any radicand.

34. $\sqrt{6}\sqrt{12}$.   **35.** $\sqrt{12a^3 y^5}$.   **36.** $\sqrt[3]{24x^5 y^3}$.

37. $\sqrt[3]{81a^5}$.   **38.** $\sqrt[3]{-128a^4}$.   **39.** $\sqrt[4]{32x^5 y^2}$.

40. $\sqrt[3]{-\dfrac{16}{a^3 b^6}}$.   **41.** $\sqrt[4]{\dfrac{81x^4 a^6}{y^8}}$.   **42.** $\sqrt{4 + 4y^2}$.

43. $\sqrt[4]{5}\sqrt[4]{3}$.   **44.** $\sqrt[3]{2}\sqrt[3]{20}$.   **45.** $\sqrt[3]{6}\sqrt[3]{12}$.

46. $\dfrac{\sqrt{10}}{\sqrt{2}}$.   **47.** $\dfrac{\sqrt[3]{20}}{\sqrt[3]{5}}$.   **48.** $\dfrac{\sqrt[3]{-16}}{\sqrt[3]{2}}$.   **49.** $\dfrac{\sqrt[3]{15x}}{\sqrt[3]{3x}}$.

50. $\dfrac{\sqrt[3]{54a^4 b}}{\sqrt[3]{2ab^4}}$.   **51.** $\dfrac{\sqrt[3]{5a^4}}{\sqrt[3]{40a}}$.   **52.** $\dfrac{\sqrt[3]{2x^8}}{\sqrt[3]{6x^5}}$.

53. $\sqrt[3]{2x}\sqrt[3]{8x^4 y^4}$.   **54.** $\sqrt[3]{4x^2}\sqrt[3]{6x^2 y^3}$.   **55.** $(3\sqrt[3]{2x})^3$.

56. $\sqrt[3]{\dfrac{a}{25}}$.   **57.** $\sqrt[4]{\dfrac{9}{4b}}$.   **58.** $\sqrt[3]{\dfrac{9}{5x}}$.   **59.** $\sqrt[4]{\dfrac{c}{2a^2}}$.

60. $\sqrt{\dfrac{d}{c^2} + \dfrac{2d}{ch} + \dfrac{d}{h^2}}$.   **61.** $\sqrt{3 + \dfrac{6u}{v^2} + \dfrac{3u^2}{v^4}}$.

62. $\sqrt{\dfrac{1}{3} + 2a^{-4}}$.   **63.** $\sqrt{u^{-3} + v^{-2}}$.

12. RATIONAL EXPONENTS

We proceed to define powers of numbers where the exponents are not necessarily integers, as assumed previously in this text.

ILLUSTRATION 1.   If a power such as $K^{2/3}$ is to obey the law of exponents for integral exponents as stated by $(x^m)^n = x^{mn}$, then it should be true that

$$(K^{2/3})^3 = K^{3 \cdot (2/3)} = K^2.$$

Hence, by the definition of a cube root, $K^{2/3}$ should be a *cube root of* K^2. The preceding remarks motivate the following definition of *a power with a rational exponent* as an optional form for a radical.

DEFINITION II.   *If m and n are integers, with* * $n > 0$, *then*

$$A^{m/n} = \sqrt[n]{A^m}; \tag{1}$$

[*when* $m = 1$ *in* (1)] $$A^{1/n} = \sqrt[n]{A}. \tag{2}$$

Notice that (1) is in agreement with (7) on page 24 in case m/n is an integer.

ILLUSTRATION 2. $6^{\frac{1}{4}} = \sqrt[4]{6}.$ $27^{\frac{1}{3}} = \sqrt[3]{27} = 3.$

$8^{\frac{2}{3}} = \sqrt[3]{8^2} = \sqrt[3]{64} = 4;$ $(-125)^{\frac{1}{3}} = \sqrt[3]{-125} = -5.$

For convenience, we agree *not* to use a symbol $A^{m/n}$ when A has no real nth root, that is, when $A < 0$ and n is even. If we arrange to have the exponent m/n as a fraction in lowest terms, we can avoid many instances of the form $A^{m/n}$ which we have excluded.

ILLUSTRATION 3.   We have excluded use of $(-32)^{\frac{6}{10}}$, but we may understand this to be given as $(-32)^{\frac{3}{5}}$. Then,

$$(-32)^{\frac{3}{5}} = (-2^5)^{\frac{3}{5}} = \sqrt[5]{(-2^5)^3} = \sqrt[5]{-2^{15}} = -2^{15/5} = -2^3 = -8.$$

THEOREM I.   *If m and n are integers, then*

$$A^{m/n} = (\sqrt[n]{A})^m. \tag{3}$$

Proof (*for the case* $A > 0$, *for convenience*).   1. By (1), $A^{m/n}$ is the principal nth root of A^m, and thus is *positive*.

2. By the laws for use of integral exponents,

$$[(\sqrt[n]{A})^m]^n = (\sqrt[n]{A})^{mn} = [(\sqrt[n]{A})^n]^m. \tag{4}$$

*In case m/n is negative, we have thus agreed to consider m as negative, with $n > 0$. It is understood that $A \neq 0$ if $m < 0$.

Since $(\sqrt[n]{A})^n = A$, from (4) we obtain $[(\sqrt[n]{A})^m]^n = A^m$. Also, $\sqrt[n]{A} > 0$ and thus the right-hand side of (3) is the *positive* nth root of A^m. Therefore the two sides of (3) are equal, because both represent the same root of A^m.

ILLUSTRATION 4.   By use of (3),

$$(-32)^{\frac{3}{5}} = (\sqrt[5]{-32})^3 = (-2)^3 = -8. \tag{5}$$

Notice the convenience of (5) as compared with the computation of $(-32)^{\frac{3}{5}}$ by use of (1) in Illustration 3.

In this chapter, we have defined A^k when k is any rational number. Without proof, from algebra we accept the fact that the laws for integral exponents remain true if the exponents involved are any rational numbers, when the base for any power is not zero. A proof of the preceding fact would make use of the properties of radicals.

ILLUSTRATION 5. $(x^8)^{\frac{3}{2}} = x^{8 \cdot \frac{3}{2}} = x^{12}.$
$$u^{\frac{1}{2}}u^{\frac{3}{4}} = u^{\frac{2}{4}}u^{\frac{3}{4}} = u^{\frac{5}{4}}. \tag{6}$$

In (6), we may express the result in radical form:

$$u^{\frac{5}{4}} = u^{\frac{4}{4}}u^{\frac{1}{4}} = u\sqrt[4]{u}.$$

In operations on radicals, frequently it is advisable to change to powers with rational exponents, and then make use of laws for exponents.

ILLUSTRATION 6.   By use of exponential forms,

$$\sqrt{u^5}\,\sqrt[3]{u^2} = \sqrt{u^4 u}\,\sqrt[3]{u^2} = u^2 u^{\frac{1}{2}} u^{\frac{2}{3}} = u^2 u^{\frac{3}{6}} u^{\frac{4}{6}}$$
$$= u^2 u^{\frac{7}{6}} = u^3 u^{\frac{1}{6}} = u^3 \sqrt[6]{u}. \tag{7}$$

ILLUSTRATION 7. $\sqrt[3]{\sqrt[4]{x}} = (x^{\frac{1}{4}})^{\frac{1}{3}} = x^{\frac{1}{12}} = \sqrt[12]{x}.$

Suppose that a product of powers of numbers involves rational exponents which are not integers, and that we desire a final radical form. Then, we may proceed as follows.

A. Express each exponent as an integer plus a positive fraction which is less than 1, *and then express all of these fractions with their lowest common denominator (suppose that this denominator is* n).

B. Write the result as a product of two factors where only integral powers are involved in one factor and nonintegral exponents in the remaining factor. Then change this factor to a radical with index n.

ILLUSTRATION 8. $\sqrt[3]{3}\,\sqrt[6]{2} = 3^{\frac{1}{3}}2^{\frac{1}{6}} = 3^{\frac{2}{6}}2^{\frac{1}{6}} = (3^2 2)^{\frac{1}{6}} = \sqrt[6]{18}.$

ILLUSTRATION 9.   We obtain a radical of lower order below:

$$\sqrt[6]{16} = \sqrt[6]{2^4} = 2^{\frac{2}{3}} = \sqrt[3]{2^2} = \sqrt[3]{4}.$$

ILLUSTRATION 10. $\sqrt[3]{4}\sqrt{2} = \sqrt[3]{2^2}\sqrt{2} = 2^{\frac{2}{3}}2^{\frac{1}{2}}$

$$= 2^{\frac{4}{6}}2^{\frac{3}{6}} = 2^{\frac{7}{6}} = 2^1(2^{\frac{1}{6}}) = 2\sqrt[6]{2}.$$

ILLUSTRATION 11. $\dfrac{\sqrt[3]{4x^2}}{\sqrt{3x}} = \dfrac{\sqrt[3]{2^2x^2}}{\sqrt{3x}} = \dfrac{2^{\frac{2}{3}}x^{\frac{2}{3}}}{3^{\frac{1}{2}}x^{\frac{1}{2}}} = \dfrac{2^{\frac{4}{6}}x^{\frac{4}{6}}}{3^{\frac{3}{6}}x^{\frac{3}{6}}}$

$$= \left(\frac{2^4x^4}{3^3x^3}\right)^{\frac{1}{6}} = \sqrt[6]{\frac{16x}{3^3}\cdot\frac{3^3}{3^3}} = \frac{\sqrt[6]{432x}}{3}.$$

EXERCISE 7

Find the value of the symbol by first expressing it in radical form, or by immediate recognition of the result.

1. $4^{\frac{1}{2}}$.   **2.** $25^{\frac{1}{2}}$.   **3.** $8^{\frac{1}{3}}$.   **4.** $27^{\frac{1}{3}}$.   **5.** $64^{\frac{1}{3}}$.

6. $144^{\frac{1}{2}}$.   **7.** $16^{-\frac{1}{2}}$.   **8.** $32^{\frac{1}{5}}$.   **9.** $81^{\frac{1}{4}}$.   **10.** $9^{\frac{1}{2}}$.

11. $4^{\frac{3}{2}}$.   **12.** $8^{\frac{2}{3}}$.   **13.** $81^{\frac{3}{4}}$.   **14.** $125^{\frac{2}{3}}$.   **15.** $16^{\frac{3}{4}}$.

16. $4^{-\frac{1}{2}}$.   **17.** $\left(\dfrac{2}{3}\right)^{-1}$.   **18.** $.09^{\frac{1}{2}}$.   **19.** $9^{-\frac{3}{2}}$.   **20.** $25^{-\frac{3}{2}}$.

Write without a denominator by use of a negative exponent.

21. $\dfrac{3}{x^2}$.   **22.** $\dfrac{5}{c^4}$.   **23.** $\dfrac{2x}{x^{\frac{1}{3}}}$.   **24.** $\dfrac{8}{3a^2}$.

Express each radical as a power, and each power with a fractional exponent as a radical, if needed. Use laws for exponents when possible.

25. $y^{\frac{3}{5}}$.   **26.** $c^{\frac{4}{3}}$.   **27.** $au^{\frac{3}{5}}$.   **28.** $5y^{\frac{2}{3}}$.

29. $3a^{\frac{1}{4}}$.   **30.** $ax^{\frac{3}{4}}$.   **31.** $\sqrt[4]{a^3}$.   **32.** $\sqrt[3]{cy}$.

33. $\sqrt[5]{x^7}$.   **34.** $(9x^2)^{\frac{3}{2}}$.   **35.** $(8u^6)^{\frac{2}{3}}$.   **36.** $(16a^8)^{\frac{3}{4}}$.

Simplify by use of the laws for exponents. Obtain the result in a form not involving any negative exponent.

37. $x^{\frac{1}{2}}x^{\frac{2}{3}}$.   **38.** $u^{\frac{3}{5}}u^{\frac{1}{2}}$.   **39.** $y^{\frac{1}{3}}y^{\frac{3}{4}}$.   **40.** $(x^{\frac{2}{3}})^6$.

41. $\dfrac{a^{-3}x^{\frac{2}{3}}}{a^{-4}x^{-2}}$.   **42.** $\dfrac{(u^{-1}v^2)^3}{u^{\frac{1}{2}}v^{\frac{1}{2}}}$.   **43.** $\dfrac{y^5}{y^{\frac{2}{3}}}$.   **44.** $\left(\dfrac{zx^2}{3y^{\frac{1}{2}}}\right)^3$.

First change to exponential form, and then express the result as a radical in simple form. Any explicit numeral should be expressed as a power.

45. $\sqrt[12]{y^4}$. **46.** $\sqrt[8]{y^2}$. **47.** $\sqrt[6]{x^3}$. **48.** $\sqrt[6]{8}$.

49. $\sqrt[4]{9x^2}$. **50.** $\sqrt[4]{25y^2}$. **51.** $\sqrt[6]{27a^3}$. **52.** $\sqrt[6]{125y^9}$.

53. $\sqrt{\sqrt[3]{3}}$. **54.** $\sqrt[3]{\sqrt{x}}$. **55.** $(\sqrt{3a})^4$. **56.** $(3\sqrt[3]{2})^4$.

57. $\sqrt{\sqrt[3]{x^2}}$. **58.** $\sqrt{y}\,\sqrt[4]{y}$. **59.** $\sqrt[3]{x}\sqrt[4]{x}$.

60. $\sqrt[3]{x^2}\sqrt[4]{x^3}$. **61.** $\sqrt[4]{\dfrac{9}{25}}$. **62.** $\sqrt[3]{\sqrt[4]{8x^3}}$.

63. $\dfrac{\sqrt[3]{9y^2x}}{\sqrt{2x}}$. **64.** $\dfrac{\sqrt{3ab}}{\sqrt[3]{4a^2b^2}}$. **65.** $\dfrac{\sqrt[3]{4uv^2}}{\sqrt[4]{25u^2v^2}}$.

Express by use of a radical in simple form.

66. $x^{\frac{2}{3}}x^{\frac{1}{4}}$. **67.** $4y^{\frac{1}{2}}y^{\frac{2}{3}}$. **68.** $9x^{\frac{1}{3}}x^{\frac{1}{4}}$.

13. THE BINOMIAL THEOREM

By definition, if n is a positive integer,

$$(x + y)^n = (x + y)(x + y) \cdots (x + y). \qquad [n \text{ factors } (x + y)]$$

The following results can be verified by multiplication of repeated factors, where each is $(x + y)$.

$$(x + y)^1 = x + y. \tag{1}$$
$$(x + y)^2 = x^2 + 2xy + y^2. \tag{2}$$
$$(x + y)^3 = x^3 + 3x^2y + 3xy^2 + y^3. \tag{3}$$
$$(x + y)^4 = x^4 + 4x^3y + 6x^2y^2 + 4xy^3 + y^4. \tag{4}$$

In each of the equations above, the first term on the right may be considered to have the factor y^0, which is 1, and the last term on the right may be written with x^0 as a factor. Then, each term is a product of a power of x and a power of y where *the sum of the exponents is n,* with the left-hand member written as $(x + y)^n$.

On the basis of the special cases for $n = 1, 2, 3,$ and 4 in (1)–(4), we accept the fact that the following results are true concerning the expansion of $(x + y)^n$.

I. *In any term, the sum of the exponents of x and y is n.*

II. *The first term is x^n; in each succeeding term the exponent of x is 1 less than in the preceding term.*

III. *The second term is $nx^{n-1}y$; in each succeeding term the exponent of y is 1 greater than in the preceding term.*

IV. *If we multiply the coefficient of any term by the exponent of x in that term, and divide by the number of the term (counting from the left), the result is the coefficient of the next term.*

V. *The coefficients of terms equidistant from the ends of the expansion are equal.*

EXAMPLE 1.   Expand $(c + d)^5$.

Solution. By use of I, II, and III, we obtain

$$(c + d)^5 = c^5 + 5c^4d + \quad c^3d^2 + \quad c^2d^3 + \quad cd^4 + d^5,$$

where spaces are left for coefficients. By use of IV, the coefficient of c^3d^2 is $(5 \cdot 4) \div 2$, or 10. Then, we obtain the remaining coefficients by use of V, and thus

$$(c + d)^5 = c^5 + 5c^4d + 10c^3d^2 + 10c^2d^3 + 5cd^4 + d^5.$$

Note 1. If n is a positive integer, *the product of all integers from 1 to n inclusive* is called "**n factorial**" and is represented by $n!$ as below:

$$n! = 1 \cdot 2 \cdot 3 \cdots n.$$

Separately, we define $0! = 1$. For instance,

$$3! = 1 \cdot 2 \cdot 3 = 6; \qquad 5! = 1 \cdot 2 \cdot 3 \cdot 4 \cdot 5 = 120.$$

By use of I–V, we obtain

$$\left. \begin{aligned} (x + y)^n &= x^n + nx^{n-1}y + \frac{n(n - 1)}{2!}x^{n-2}y^2 + \cdots \\ &+ \frac{n(n - 1) \cdots (n - r + 1)}{r!}x^{n-r}y^r + \cdots + y^n. \end{aligned} \right\} \tag{5}$$

The term involving $x^{n-r}y^r$ in (5) is called the *general term* of the expansion, and can be verified by r repetitions of IV. The result in (5) is referred to as the **binomial formula**. The fact that this formula is correct is called the **binomial theorem**.

The $(n + 1)$th row of the following array gives the coefficients in the expansion of $(x + y)^n$. To obtain any row after a particular row, place 1 at the left; the second number of the new row is the sum of the first and second numbers of the preceding row; the third number of the new row is the sum of the second and third numbers of the preceding row; etc. The array of coefficients is known as **Pascal's triangle.** Blaise Pascal (1623–1662) was a brilliant French mathematician and philosopher.

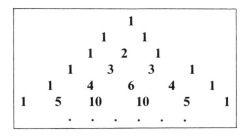

In the expansion (5), let

$$_nC_r = \frac{n(n-1)(n-2)\cdots(n-r+1)}{r!}. \tag{6}$$

Then, in (5), the general term involving y^r is

$$_nC_r x^{n-r} y^r. \tag{7}$$

We refer to $_nC_r$ as the *rth binomial coefficient* in the expansion of $(x + y)^n$. Also, in the study of finite mathematics, it is found that $_nC_r$ is equal to *the number of subsets of r elements each which can be formed from a set of n elements.*

EXAMPLE 2.   Obtain the term involving x^2 in the expansion of $(y + 2x^{\frac{1}{2}})^{12}$.

Solution. The desired term will involve the 4th power of $2x^{\frac{1}{2}}$. By use of (6) and (7) with $n = 12$ and $r = 4$, the term is

$$_{12}C_4 \, y^{12-4} \, (2x^{\frac{1}{2}})^4 = \frac{12 \cdot 11 \cdot 10 \cdot 9}{1 \cdot 2 \cdot 3 \cdot 4} y^8 (16x^2) = 7920x^2y^8.$$

EXERCISE 8

Expand. Pascal's triangle may be used.

1. $(u - v)^5$.   2. $(c - 2)^4$.   3. $(3 + x)^6$.

4. $(x + 2y)^3$.   5. $(x + 3a)^4$.   6. $(a^2 - c)^5$.

7. $(u - 2x)^7$.   8. $(x^{\frac{1}{2}} - y)^6$.   9. $(\sqrt{x} - y)^5$.

10. $(a^{-1} - y^{-1})^3$.   11. $(a^{\frac{1}{3}} + b^{\frac{1}{2}})^4$.   12. $(2x - y^{\frac{1}{2}})^6$.

By use of I–IV *of the preceding section, obtain only the first four terms of the expansion of the power in descending powers of the first term of the binomial.*

13. $(c + 3)^{10}$.   14. $(x - 2)^{20}$.   15. $(a^2 + d)^{25}$.

Calculate the power correct to three decimal places by use of only as many terms as necessary from the expansion of the power of the binomial in ascending powers of r, after the base of the power is written in the form $(1 + r)$.

16. $(1.01)^{15}$.   **17.** $(1.02)^{12}$.   **18.** $(1.03)^{10}$.   **19.** $(1.04)^{20}$.

Hint. $(1.01)^{15} = (1 + .01)^{15} = 1 + 15(.01) + \cdots$.

Obtain the specified term in the expansion of the power without finding the other terms.

20. The term involving x^6 in the expansion of $(u + x)^9$.

21. The term involving a^{10} in the expansion of $(x + a^2)^{16}$.

22. The term involving u^{-4} in the expansion of $(x - u^{-1})^9$.

chapter 2

Equations, Inequalities, and Sets

14. TERMINOLOGY FOR EQUATIONS

An *equation* is a statement that two numbers are equal. An *open equation* is an equation involving at least one variable. Hereafter, the unqualified word *equation* will refer to an *open equation*.

A **solution** of an equation consists of a value of each variable in the equation such that it becomes a true statement when each variable is assigned the specified value. *To solve* an equation means to find all of its solutions. A solution of an equation in a single variable, say x, also is called a **root** of the equation. For brevity, the set of all solutions of an equation may be called its **solution set.**

ILLUSTRATION 1.   The equation $2x^2 - x - 1 = 0$, or

$$(2x + 1)(x - 1) = 0,$$

has the solutions $x = 1$ and $x = -\frac{1}{2}$, for which $x - 1 = 0$ and $2x + 1 = 0$, respectively. The solution set is $\{-\frac{1}{2}, 1\}$.

ILLUSTRATION 2.   A solution of $x + y = 3$ is a pair of corresponding values of x and y which satisfy the equation. Thus, $(x = 2, y = 1)$ is a solution. This equation has infinitely many solutions.

To say that an equation is **inconsistent** means that it has no solution, and otherwise the equation is said to be **consistent.** A **conditional equation** is one which is *not* satisfied by *all* values of the variables. If an equation is satisfied by all values of any variables involved, it is called an **identity,** or **identical equation.** Two equations are said to be **equivalent** if they have *the same solutions.*

ILLUSTRATION 3.   The equations $x - 2 = 3$ and $x(x - 2) = 3x$ are *not* equivalent because the second equation has the solution $x = 0$, not possessed as a solution by the first equation.

Let n be a positive integer. Then, a **polynomial equation** of degree n in a variable x is of the type $P = Q$ where each of P and Q is a polynomial in x or is 0, and $(P - Q)$ is a polynomial of degree n in x. Thus, the equation is equivalent to one of the form

$$a_0 + a_1x + a_2x^2 + \cdots + a_nx^n = 0, \tag{1}$$

where the coefficients $(a_0, a_1, \cdots, a_n)$ are constants and $a_n \neq 0$. Also, (1) may be referred to simply as an equation of degree n in x. We call (1) a linear equation if $n = 1$; a quadratic equation if $n = 2$; a cubic equation if $n = 3$. In (1), suppose that the coefficients are any complex numbers. Then, in Chapter 5, on the basis of a famous theorem which will not be proved, it will be shown that the solution set of (1) is not empty, and consists of at most n numbers, real or imaginary.

ILLUSTRATION 4.   Any linear equation in x is equivalent to an equation $ax = b$ where $a \neq 0$. The single solution of $ax = b$ is $x = b/a$.

EXAMPLE 1.   Solve: $\dfrac{17}{30} = \dfrac{x + 3}{10} - \dfrac{3x - 2}{15}.$ (2)

Solution.   In (2), the LCD $= 30$. To clear of fractions, multiply both sides of (2) by 30, to obtain

$$30 \cdot \frac{17}{30} = 30 \frac{x + 3}{10} - 30 \frac{3x - 2}{15};$$

$$17 = 3x + 9 - 6x + 4, \quad or \quad 17 = -3x + 13.$$

Hence, $3x = 13 - 17 = -4$, or $x = -4/3$ is the solution of (2).

15. QUADRATIC EQUATIONS IN A SINGLE VARIABLE

Any quadratic equation in one variable x is equivalent to an equation

$$ax^2 + bx + c = 0,$$ (1)

where $\{a, b, c\}$ are constants and $a \neq 0$. To solve (1) if $b = 0$, we first solve for x^2 and extract square roots.

EXAMPLE 1.   Solve: $7x^2 - 20 = 4x^2.$ (2)

Solution. $7x^2 - 4x^2 = 20,$ or $3x^2 = 20;$ $x^2 = \frac{20}{3}.$

Hence, $x = \pm \sqrt{\dfrac{20}{3}} = \pm \sqrt{\dfrac{20}{3} \cdot \dfrac{3}{3}} = \pm \dfrac{1}{3} \sqrt{60}.$

Since $\sqrt{60} = \sqrt{4 \cdot 15} = 2\sqrt{15}$, we obtain $x = \pm \frac{2}{3} \sqrt{15}$. From Table I,

$$x = \pm \tfrac{2}{3}(3.873), \quad or \quad x = \pm 2.582.$$

ILLUSTRATION 1.   The solutions of $x^2 = -8$ are

$$x = \pm \sqrt{-8} = \pm i \sqrt{8}, \quad or \quad x = \pm 2i \sqrt{2}.$$

Recall the following important fact from elementary algebra.

The product of two or more real numbers is zero when and only when one or more of the factors is equal to zero.   (3)

The result in (3) is true also for a product of any *complex* numbers.* We shall use (3), and sometimes its extension to complex numbers, in solving equations by use of factoring, as in the next example.

EXAMPLE 2.   Solve: $2x^2 + 5x - 3 = 0.$ (4)

Solution.   1. Factor: $(2x - 1)(x + 3) = 0.$ (5)

2. By use of (3), equation (5) is satisfied if and only if

$$2x - 1 = 0 \quad or \quad x + 3 = 0.$$

Hence, the solutions are $x = \frac{1}{2}$ and $x = -3$.

EXAMPLE 3.   Solve: $4x^2 - 12x + 9 = 0.$ (6)

Solution.   1. Factor: $(2x - 3)^2 = 0$, *or* $(2x - 3)(2x - 3) = 0.$ (7)

2. If $2x - 3 = 0$ then $x = \frac{3}{2}$. Since each factor in (7) gives the same value for x, it is said that (6) has *two equal solutions*. However, the *solution set* of (6) consists of the *single number 3/2*.

Let $M = N$ represent an equation in the single variable x. If both sides of the equation are divided by H, where H involves x, and $H = 0$ for some value or values of x, it is possible that solutions of the original equation may be lost. Hence, in solving an equation in x, try to avoid dividing both sides by a number H where H involves x.

EXAMPLE 4.   Solve: $5x^2 = 8x.$ (8)

Correct solution. $5x^2 - 8x = 0.$

Factor: $x(5x - 8) = 0.$ (9)

Hence, $x = 0$, or $5x - 8 = 0$ and $x = \frac{8}{5}$. The solutions are $x = 0$ and $x = \frac{8}{5}$.

Incorrect solution.   Divide both sides by x in (8), to obtain $5x = 8$ and $x = \frac{8}{5}$ as the only solution. In this method, the division by x caused *the loss of the solution $x = 0$.*

EXERCISE 9

Solve the equation. First clear of fractions if any appear.

1. $\dfrac{2x}{3} + 2 - \dfrac{3x + 9}{4} = 0.$ **2.** $\dfrac{5}{24} - \dfrac{2y - 5}{8} = \dfrac{y + 4}{12}.$

3. $3x^2 = 12.$ **4.** $x^2 = -9.$ **5.** $9x^2 = -25.$

6. $2x^2 = 3.$ **7.** $9 - 7z^2 = 6.$ **8.** $\frac{1}{2}x^2 - 1 = \frac{1}{5}x^2.$

---

*A proof of this extension of (3) is accomplished most conveniently by use of a trigonometric form for complex numbers.

9. $x^2 - 3x = 28$.   **10.** $y^2 = 2y + 8$.   **11.** $z^2 + z = 6$.
12. $z^2 + 3z = 10$.   **13.** $5x = 10x^2$.   **14.** $4x^2 - 81 = 0$.
15. $5x^2 - 9x = 0$.   **16.** $2x^2 = 3x$.   **17.** $25x^2 - 1 = 0$.
18. $x^2 + x = 12$.   **19.** $2x^2 + x = 6$.   **20.** $6x^2 + x = 2$.
21. $6x^2 + 13x = 5$.   **22.** $7x^2 + 20x = 3$.   **23.** $2x^2 + 9x = -10$.
24. $9x^2 + 30x + 25 = 0$.   **25.** $9x^2 - 6x + 1 = 0$.
26. $x^2 - 6x + 9 = 0$.   **27.** $x^2 + 4x + 4 = 0$.
28. $4z^2 - 20z + 25 = 0$.   **29.** $8x^2 - 14x - 15 = 0$.

Solve by use of factoring.

30. $x^4 - 5x^2 + 4 = 0$.   **31.** $y^4 + 10y^2 + 9 = 0$.

Hint. In Problem 30 the equation is said to be in the *quadratic form* in x^2 because, if we let $y = x^2$, then $y^2 = x^4$ and the equation becomes $y^2 - 5y + 4 = 0$. Such substitution is *not* required in solving. We obtain $(x^2 - 1)(x^2 - 4) = 0$; hence $x^2 - 1 = 0$ or $x^2 - 4 = 0$, which give four solutions.

32. $x^4 - 16 = 0$.   **33.** $16y^4 - 625 = 0$.   **34.** $81z^4 = 16$.

Comment. The four solutions in Problem 32 are the four fourth roots of 16.

35. By solving an equation, find the four fourth roots of 625.

36. By solving an equation, find the four fourth roots of 1.

Solve by factoring.

37. $(x + 3)(2x + 5)(3x - 7) = 0$.   **38.** $6x^3 - x^2 - 15x = 0$.

16. THE QUADRATIC FORMULA

A binomial $(x^2 + px)$ becomes a perfect square trinomial, or we *complete a square*, if we add

$$\text{the square of one-half of the coefficient of } x, \tag{1}$$

that is, if we add $\left(\tfrac{1}{2}p\right)^2$, or $p^2/4$. This is true because, by (2) on page 13,

$$\left(x + \frac{p}{2}\right)^2 = x^2 + 2x\left(\frac{p}{2}\right) + \frac{p^2}{4} = x^2 + px + \frac{p^2}{4}. \tag{2}$$

EXAMPLE 1.   Solve by completing a square:

$$9x^2 - 12x - 1 = 0. \tag{3}$$

Solution.   1. $$9x^2 - 12x = 1. \tag{4}$$

Divide by 9: $$x^2 - \frac{4}{3}x = \frac{1}{9}. \tag{5}$$

2. To complete a square on the left in (5), add $\left[\tfrac{1}{2}\left(\tfrac{4}{3}\right)^2\right]$, or $\left(\tfrac{2}{3}\right)^2$ to

both sides:

$$x^2 - \frac{4}{3}x + \left(\frac{2}{3}\right)^2 = \frac{4}{9} + \frac{1}{9}, \quad or \quad \left(x - \frac{2}{3}\right)^2 = \frac{5}{9}. \tag{6}$$

Hence,

$$x - \frac{2}{3} = \pm\sqrt{\frac{5}{9}} = \frac{\pm\sqrt{5}}{3};$$

$$x = \frac{2 + \sqrt{5}}{3} \quad or \quad x = \frac{2 - \sqrt{5}}{3}. \tag{7}$$

From Table I, $\sqrt{5} = 2.236$. The solutions are $x = 1.412$ and $x = .079$.

EXAMPLE 2.   Solve by completing a square, where $a \neq 0$:

$$ax^2 + bx + c = 0. \tag{8}$$

Solution.   1. Subtract c, and divide by a, on both sides:

$$ax^2 + bx = -c; \quad x^2 + \frac{b}{a}x = -\frac{c}{a}. \tag{9}$$

To complete a square on the left, add $\left(\frac{1}{2} \cdot \frac{b}{a}\right)^2$ to both sides:

$$x^2 + \frac{b}{a}x + \left(\frac{b}{2a}\right)^2 = \frac{b^2}{4a^2} - \frac{c}{a}, \quad or \quad \left(x + \frac{b}{2a}\right)^2 = \frac{b^2 - 4ac}{4a^2}. \tag{10}$$

2. Extract square roots in (10):

$$x + \frac{b}{2a} = \pm\sqrt{\frac{b^2 - 4ac}{4a^2}} = \pm\frac{\sqrt{b^2 - 4ac}}{2a}. \tag{11}$$

Hence

$$x = -\frac{b}{2a} \pm \frac{\sqrt{b^2 - 4ac}}{2a}, or$$

$$x = \frac{-b \pm \sqrt{b^2 - 4ac}}{2a}. \tag{12}$$

We refer to (12) as the **quadratic formula** for the solution of the general quadratic equation (8).

EXAMPLE 3.   Solve: $2x^2 - 6x - 3 = 0$.

Solution.   In the standard form (8), we have $a = 2$, $b = -6$, and $c = -3$. With these values used in (12), we obtain

$$x = \frac{6 \pm \sqrt{36 + 4(2)(3)}}{4} = \frac{6 \pm \sqrt{60}}{4}.$$

With $\sqrt{60} = \sqrt{4(15)} = 2\sqrt{15}$, and $\sqrt{15} = 3.873$, from Table I,

$$x = \frac{3 \pm \sqrt{15}}{2} = \frac{3 \pm 3.873}{2}; \quad x - 3.436 \quad and \quad x = -.436.$$

Hereafter, any particular quadratic equation in one variable, x, should be solved by merely extracting square roots, if no linear term in x is involved; by factoring if factors can be found easily; otherwise by use of the quadratic formula. The method of completing a square should not be used unless it is explicitly requested. This method was introduced primarily as a means for deriving the quadratic formula.

EXAMPLE 4.   Solve: $\qquad x^2 + 4x + 5 = 0.$

Solution. For use in (12), we have $\{a = 1, b = 4, c = 5\}$:

$$x = \frac{-4 \pm \sqrt{16 - 20}}{2} = \frac{-4 \pm \sqrt{-4}}{2}.$$

Hence, $x = \frac{1}{2}(-4 \pm 2i)$, or $x = -2 + i$ and $x = -2 - i$.

In (12), observe that the solutions are *real and unequal* if $b^2 - 4ac > 0$; *real and equal* if $b^2 - 4ac = 0$; *imaginary and unequal* if $b^2 - 4ac < 0$; *rational* if $\{a, b, c\}$ are rational numbers and $(b^2 - 4ac)$ is a *perfect square.* Hence, $(b^2 - 4ac)$ is called the **discriminant** of $ax^2 + bx + c = 0$, because the nature of its roots in the preceding respects is determined by the value of $(b^2 - 4ac)$.

ILLUSTRATION 1.   The discriminant of $3x^2 - 5x + 6 = 0$ is $(5^2 - 4 \cdot 3 \cdot 6) = -47$. Hence, without solving for x, we can state that the solutions of the given equation are imaginary.

EXERCISE 10

Solve the equation by completing a square.

1. $x^2 - 2x - 8 = 0.$ 2. $2x^2 - 3x - 5 = 0.$
3. $9x^2 - 12x + 4 = 0.$ 4. $x^2 - 2x + 4 = 0.$

Solve for x or y by use of the quadratic formula. Any letter other than x or y is a constant.

5. $4x^2 + 4x + 1 = 0.$ 6. $4y^2 + 12y + 9 = 0.$
7. $2x^2 = 3 + 5x.$ 8. $8x - 3 = 2x^2.$
9. $3x^2 = 10x - 3.$ 10. $2x = 3 - x^2.$
11. $3y + 1 = 5y^2.$ •   12. $24y^2 - 2y = 15.$
13. $y^2 - 6y + 13 = 0.$ 14. $4x^2 + 8x = -9.$
15. $4x^2 - 8x + 5 = 0.$ 16. $5x^2 = 2x - 4.$
17. $4x^2 + 25 = 0.$ 18. $9x^2 - 64 = 0.$
19. $4x^2 - 12x + 13 = 0.$ 20. $16x^2 - 34x = 15.$

Compute the discriminant and determine the nature of the solutions of the equation without solving it. Then, find the solutions, to check the preceding results.

21. $y^2 - 7y + 10 = 0.$ 22. $5y^2 - 2y + 1 = 0.$
23. $9x^2 - 12x + 4 = 0.$ 24. $30 - 9y^2 = 25y.$

25. $x^2 - 2x - 3 = 0$.   **26.** $1 = 2x - 2x^2$.

By use of the discriminant, find the values of the constant k so that the equation in the variable x will have equal solutions.

27. $3kx^2 - 5x + 2 = 0$.   **28.** $5x^2 - 2kx - k = 0$.

29. $kx^2 - 3kx + 5 = 0$.   **30.** $x^2 + kx^2 - 5kx + 3k = 0$.

31. Solve by factoring, by use of (9) on page 14: $x^3 - 8 = 0$.

Comment. In Problem 31, $x^3 = 8$ and the solutions are the *three cube roots* of 8.

By solving an equation, find the specified roots.

32. The three cube roots of 1; of -1.

33. The three cube roots of 125; of -125.

17. IRRATIONAL EQUATIONS

If an equation involves at least one radical (or an equivalent power with a rational exponent) where the radicand involves the variable, or variables, the equation is said to be an *irrational equation.*

ILLUSTRATION 1.   The equation $\sqrt{x + 3} = 5$ is an irrational equation.

On squaring both sides of an equation $U = V$, we obtain $U^2 = V^2$. This equation is true if $U = V$ or if $U = -V$. Thus, the solution set of $U^2 = V^2$ consists of the *union* of the solution sets of $U = V$ and of $U = -V$. Suppose, now, that we wish to solve an equation $U = V$ in x and start by squaring both sides, which gives $U^2 = V^2$. If we continue with this equation, we cannot be certain that any solution $x = c$ obtained will satisfy $U = V$. Instead, it might happen that $x = c$ satisfies $U = -V$. Thus, after squaring both sides of an equation in x, we may find that the new equation is satisfied by values of x which are *not solutions of the original equation.*

ILLUSTRATION 2.   The only solution of $x - 2 = 1$ (1)

is $x = 3$. On squaring both sides of (1) we obtain

$$x^2 - 4x + 4 = 1, \quad or \quad x^2 - 4x + 3 = 0. \quad (2)$$

From (2), $(x - 3)(x - 1) = 0$. Hence, (2) has the solutions $x = 3$ and $x = 1$. Thus, by squaring both sides in (1), we obtained (2) which has the solution $x = 1$ besides the solution $x = 3$ of (1).

When an operation for solving an equation in x produces a value for x which is *not* a solution of the *given equation*, this value is referred to as an **extraneous solution,** or **extraneous root** of the given equation. In Illustration 1, it was seen that, if both sides of an equation are squared,* extraneous roots may be introduced. Hence, the following caution must be observed.

Caution. If both sides of an equation in x are squared in solving an equation, extraneous solutions may be introduced. Hence, each value obtained for the variable should be tested by substitution in the original equation in order to determine whether the value is a solution, or is an extraneous solution.

EXAMPLE 1.   Solve each of equations (*a*) and (*b*).

(*a*) $2x + 1 = \sqrt{2x^2 + 1}$.	(*b*) $2x + 1 = -\sqrt{2x^2 + 1}$.
Solution. 1. Square both sides: $4x^2 + 4x + 1 = 2x^2 + 1$, *or* $$2x^2 + 4x = 0;$$ $$2x(x + 2) = 0.$$	*Solution.* 1. Square both sides: $4x^2 + 4x + 1 = 2x^2 + 1$, *or* $$2x^2 + 4x = 0;$$ $$2x(x + 2) = 0.$$
2. Solutions: $x = 0; x = -2$.	2. Solutions: $x = 0; x = -2$.
Test. Substitute $x = 0$ in (*a*); Does $0 + 1 = 1$? *Yes.*	*Test.* Substitute $x = 0$ in (*b*): Does $0 + 1 = -1$? *No.*
Substitute $x = -2$ in (*a*): Does $-4 + 1 = \sqrt{9}$? *No.*	Substitute $x = -2$ in (*b*): Does $-4 + 1 = -\sqrt{9}$? *Yes.*
$x = 0$ *is* a solution. $x = -2$ is *not* a solution.	$x = 0$ is *not* a solution. $x = -2$ *is* a solution.

In Example 1, the extraneous solution $x = 0$ was obtained in solving (*b*), and the extraneous solution $x = -2$ in solving (*a*). The necessity for the tests in Example 1 is shown by the fact that, although (*a*) and (*b*) are different equations, all distinction between them is lost due to squaring.

EXAMPLE 2.   Solve: $(x + 2)^{\frac{1}{2}} + \sqrt{5 - 2x} = 3.$ (3)

Solution.   1. Transpose the most complicated radical to one member, and all remaining terms to the other side; then square both sides:

$$\sqrt{5 - 2x} = 3 - \sqrt{x + 2};$$

$$5 - 2x = x + 2 - 6\sqrt{x + 2} + 9, \quad or \quad 6\sqrt{x + 2} = 3x + 6.$$

2. Simplify, and square both sides:

$$4(x + 2) = (x + 2)^2, \quad or \quad x^2 = 4; \quad hence \quad x = \pm 2.$$

Test. Substitution in (3) shows that $x = 2$ and $x = -2$ are solutions of (3).

*Also true if both sides are raised to any positive integral power.

EXERCISE 11

Obtain all real solutions of the equation.

1. $\sqrt{x-1} = 3$.   **2.** $\sqrt{2-x} = 2$.   **3.** $\sqrt{1-4x} = -3$.

4. $2\sqrt{3} = 4x$.   **5.** $3\sqrt{x} = 9-2x$.   **6.** $\sqrt{z} = 6-z$.

7. $\sqrt{2x+5} - 1 = \sqrt{2x}$.   **8.** $\sqrt{x-2} - \sqrt{2x+3} = 2$.

9. $\sqrt{3+6x} + 3\sqrt{2x-1} = 6$.   **10.** $\sqrt{x+4} = \sqrt{x} + 1$.

11. $\sqrt{y-1} + \sqrt{y+6} = 7$.   **12.** $\sqrt{3x+1} = \sqrt{x} + 1$.

13. $\sqrt{3-2x} - \sqrt{2+2x} = 3$.   **14.** $\sqrt{x} - \sqrt{x-4} = 2$.

15. $\sqrt{3+x} - \sqrt{3-x} = \sqrt{x}$.

18. INEQUALITIES IN ONE VARIABLE

If A and B involve at least one variable, then $A < B$ is called an **open inequality.** Hereafter, unless otherwise implied, *"inequality"* will mean *"open inequality."* In the terminology for equations on page 34, if we merely change the word *"equation"* to *"inequality,"* corresponding terminology is obtained for inequalities. Thus, two inequalities in x are said to be *equivalent* if they have the same solution set. An inequality is called an *absolute* or *identical inequality* if it is true for all values of the variables involved.

ILLUSTRATION 1.   The solution set of the inequality $x < 2$ consists of all numbers x located to the left of 2 on a number scale. This illustrates the fact that an inequality may have infinitely many solutions. The inequality $|x| < 0$ is inconsistent because $|x|$ is positive or zero for all values of x. The inequality $x^2 \geqq 0$ is an absolute inequality because it is true for all values of x.

On a number scale, the *graph* of a given set, M, of numbers is defined as the set of points on the scale representing the numbers of M. The *graph of an inequality* in a single variable x is the set of points on the scale representing the solutions of the inequality. That is, its graph is the graph of the *solution set* of the inequality.

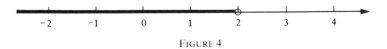

FIGURE 4

ILLUSTRATION 2.   The graph of $x < 2$ is the thick part of the scale in Figure 4, where the point 2 is circled to show that it is not included.

Recall that, if b and c are any real numbers, then

$$b < c \quad \text{means that} \quad (c - b) \text{ is positive.} \tag{1}$$

THEOREM I. Suppose that H, K, R, P, and N are number symbols involving the same variables, with P > 0 and N < 0 for all values of the variables. Then, the inequality H < K is equivalent to (has the same solutions as) each of the following inequalities:

$$H + R < K + R; \quad PH < PK; \quad NH > NK. \tag{2}$$

Proof of equivalence to NH > NK. By (1),

$$\text{"} H < K \text{"} \quad \text{means that} \quad {}'K - H > 0; \tag{3}$$

$$\text{"} NH > NK \text{"} \quad \text{means that} \quad N(H - K) > 0. \tag{4}$$

Since N is *negative,* (4) means that $(H - K)$ is *negative,* or $(K - H)$ is *positive,* which is the same as (3). Hence, with $N < 0$, "$H < K$" and "$NH > NK$" have the *same meaning* and thus the *same solutions.* Equivalence to each of the other inequalities in (2) is proved similarly.

Theorem I remains valid if "$<$" is changed to "$\leq$," and "$>$" to "$\geq$" throughout. This is true because the results with "$=$" are consequences of well known properties of equations.

In solving inequalities, frequently we shall proceed from a given inequality through successive equivalent inequalities leading to the final solution. In such a procedure, Theorem I justifies addition of the same number on both sides of an inequality, multiplication of both sides by a positive number, and multiplication of both sides by a negative number with *simultaneous reversal of the sign of inequality.*

If H and K are polynomials in a variable x, the inequality $H < K$ is said to be *linear* in x in case $H < K$ is equivalent to an inequality

$$cx < d, \tag{5}$$

(or, equally well, $cx > d$), where c and d are constants and $c \neq 0$. *To solve* a linear inequality in x means to obtain a simple description of the solution set for the inequality, in the form

$$x < k \quad or \quad k < x. \tag{6}$$

EXAMPLE 1.   Solve: $$\frac{5x}{3} - 1 < 13 - \frac{3x}{2}. \tag{7}$$

Solution.   1. Multiply both sides by 6 to clear of fractions:

$$10x - 6 < 78 - 9x. \tag{8}$$

2. Add $(9x + 6)$ to both sides: $19x < 84.$ \hfill (9)

3. Divide by 19: $$x < \frac{84}{19}. \tag{10}$$

Because of (2), inequality (7) is equivalent to (8), then to (9), and finally to (10). Hence, the solution set of (7) consists of all numbers less than 84/19.

THEOREM II.   *If $U < V$ and $V < W$, then $U < W$.*

Proof. Let $V - U = h$ and $W - V = k$. Then, by (1), h and k are positive. We obtain

$$W - U = (W - V) + (V - U) = h + k,$$

which is *positive.* Hence, by (1), we have $U < W$.

On account of Theorem II, it is said that the "*less than*" relationship is **transitive.**

19. SYSTEMS OF INEQUALITIES IN A SINGLE VARIABLE

Any type of assertion about the equality or inequality of numbers may be called a **numerical statement,** or simply a *statement* when the context is clear. Such a statement might be a single equation or inequality, or a system of equations or inequalities. The terminology relating to solutions for a single equation on page 34 extends to numerical statements of any variety if the word *equation* on page 34 is changed throughout to *numerical statement.*

ILLUSTRATION 1.   The statement (a system of equations)

$$2x - y = 3 \quad and \quad x + y = 9$$

has a single solution ($x = 4, y = 5$), obtained as in elementary algebra.

A statement such as

$$U < V < W \qquad means \ that \qquad U < V \ and \ V < W, \tag{1}$$

and thus abbreviates a *system of two inequalities.*

ILLUSTRATION 2.   The statement

$$-2 < x < 3 \qquad means \ that \qquad \{-2 < x \quad and \quad x < 3\} \tag{2}$$

Hence, on a number scale, the solution set of (2) consists of all numbers, x, to the *right* of -2 *and* to the *left* of 3, that is, all numbers *between* -2 and 3. The graph of (2) in Figure 5 is the interval with -2 and 3 as endpoints, where these endpoints are not part of the graph.

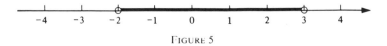

FIGURE 5

An *interval* of numbers, shown on a number scale, is called a **closed interval** if it includes its endpoints; an **open interval** if neither endpoint is included; a **half-open** or **half-closed interval** if just one endpoint is included. If *b* and *c* are unequal numbers,

> *the graph of* $b < x < c$ *consists of all numbers between b and c, forming an open interval with b and c as endpoints.* (3)

The graph of $b \leqq x \leqq c$ is the *closed* interval with *b* and *c* as endpoints.

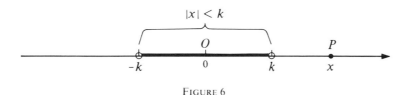

FIGURE 6

Suppose that $|x| < k$, where we recall that $|x|$ is the distance between the origin and $P:(x)$ on a number scale, as in Figure 6. The points on the scale at distance *k* from $O:(0)$ are $\pm k$. Hence, "$|x| < k$" is equivalent to stating that $P:(x)$ lies *between* $-k$ *and* $+k$ *on the scale*. Or,

$$|x| < k \qquad \text{is equivalent to} \qquad -k < x < k. \qquad (4)$$

If $|x| > k$, the distance between $O:(0)$ and $P:(x)$ on the scale is *greater* than *k*. That is, *P* lies beyond *k* to the *right*, or beyond $-k$ to the *left,* and

$$|x| > k \qquad \text{is equivalent to} \qquad \{x < -k \quad \text{OR} \quad k < x\}. \qquad (5)$$

For brevity, we may refer to "*the interval* $\{b < x \leqq c\}$" or "*the interval* $\{|x| < k\}$," when we mean the *solution set* of the given statement about *x*, or the *graph* of the statement. A number *h* is said to be an *interior point* of an interval of numbers if *h* is *not* an endpoint of the interval. Thus, $x = 3$ is an interior point of the interval $\{2 < x < 7\}$.

In any reference to an *interval,* we shall mean one whose length is *not zero.* The set of "*all numbers**x > *k*," where *k* is fixed, may be described as the *infinite interval* $\{k < x < +\infty\}$, where $+\infty$ (read *plus infinity*) is used to show that the interval is unbounded on the right. The set of "*all numbers*† $x < h$" forms the *infinite interval* $\{-\infty < x < h\}$, where $-\infty$ indicates that the interval is unbounded on the left. The set of *all real numbers* forms the *infinite interval* $\{-\infty < x < +\infty\}$. An interval of numbers sometimes may be called a *finite interval* if it is of the type in (3), possibly closed at one or both endpoints.

---

*Or $x \geqq k$, and thus closed at the *left.*
†Or $x \leqq h$, and thus closed at the *right.*

EXAMPLE 1.   Solve: $6 - x < 3x - 2 \leqq x + 6$ (6)

Solution.   1. Statement (6) is equivalent to

(a) $6 - x < 3x - 2$ *and*   (b) $3x - 2 \leqq x + 6.$ (7)

2. *Solution of* (a): We obtain $8 < 4x,$ *or* $2 < x.$

3. *Solution of* (b): We obtain $2x \leqq 8,$ *or* $x \leqq 4.$

4. Hence, (7) is equivalent to

$\{2 < x$ *and* $x \leqq 4\},$ *or* $\{2 < x \leqq 4\}.$ (8)

The solution set of (7) is the interval of numbers from 2 to 4, open at 2 and closed at 4.

EXAMPLE 2.   Solve: $|x - 3| < 2$ (9)

Solution.   1. With $(x - 3)$ replacing x in (4),

$|x - 3| < 2$ *is equivalent to* $\{-2 < x - 3 < 2\}.$ (10)

2. From (10), we obtain $1 < x < 5$ by adding 3 to each member, or the solution set of (9) consists of all numbers on the open interval with endpoints 1 and 5, as in Figure 7.

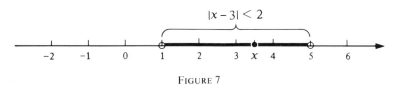

FIGURE 7

Comment. From (5) on page 5, $|x - 3|$ is the distance on a number scale between the points x and 3. Hence, (9) states that this distance is less than 2. Or, x is on the open interval of length 4 with center at 3.

EXAMPLE 3.   Solve: $|x - 3| > 2.$ (11)

Solution.   1. With $(x - 3)$ replacing x in (5),

$|x - 3| > 2$ *is equivalent to* $\{x - 3 < -2$ OR $2 < x - 3\}.$ (12)

2. From (12), the solution set of (11) consists of

all x such that $\{x < 1$ OR $5 < x\}.$ (13)

The graph of (13) consists of the infinite open intervals from $-\infty$ to 1, and from 5 to $+\infty$, as seen in Figure 7.

EXAMPLE 4.   Solve: $x^2 - 9 < 0.$ (14)

Solution. From (14), $x^2 < 9.$ (15)

Notice that $x^2 = |x|^2$. Hence, from (15),

$$|x|^2 < 9. \tag{16}$$

Take the positive square root of each side of (16):

$$\sqrt{|x|^2} < \sqrt{9}, \quad or \quad |x| < 3. \tag{17}$$

From (17), the solution set of (14) is the interval $\{-3 < x < 3\}$.

EXERCISE 12

Solve the inequality.

1. $4x - 20 < 0.$ **2.** $3x + 8 > 17.$ **3.** $\dfrac{3}{2}x - 5 < 2.$

4. $7 - 3x < 0.$ **5.** $10 - 3x > 2x.$ **6.** $-2x > 9.$

7. $3x - 2 < 5x + 9.$ **8.** $2x - \dfrac{3}{2} > \dfrac{7}{3}x - 2.$

9. $2x - \dfrac{4}{3} > \dfrac{3}{5}x - 1.$ **10.** $\dfrac{1}{4}x - 5 \leqq 3x + 2.$

11. $\dfrac{1}{4}x - \dfrac{3}{5} \leqq \dfrac{2}{3}x - \dfrac{11}{5}.$ **12.** $\dfrac{5}{3} - \dfrac{2}{5}x \leqq -\dfrac{3}{2} + \dfrac{7}{3}x.$

By use of inequalities, express the given fact about x.
13. x lies between 3 and 9; between -4 and 3.
14. x lies on the half closed interval from -3 to 8, closed at -3.
15. x lies on the closed interval with -2 and 6 as endpoints.

Solve. Then graph the statement on a number scale. The statement may be inconsistent.
16. $5 < 4x - 3 \leqq 13.$ **17.** $3 - 2x < 5x - 2 \leqq 2x + 4.$
18. $3 \leqq 5 + 2x < 8.$ **19.** $8 \leqq 3x - 1 \leqq 3 - x.$

20. $\dfrac{5x}{3} - 3 \leqq 2x - \dfrac{3}{2} < x + \dfrac{1}{4}.$ **21.** $\dfrac{x}{5} + 1 \leqq \dfrac{3x}{2} - \dfrac{2}{3} < \dfrac{x}{2} + 5.$

22. $\dfrac{x}{4} < \dfrac{7x}{4} + 8 < \dfrac{32}{3} - \dfrac{x}{4}.$ **23.** $\dfrac{5 + x}{2} \leqq 5x - 2 \leqq 16 - x.$

24. $\dfrac{1}{2} + 3x < 2x - 4 < -3 + \dfrac{7x}{3}.$ **25.** $\dfrac{3}{2} + 3x < \dfrac{4}{3} + \dfrac{3x}{2} < \dfrac{5}{2} - \dfrac{x}{3}.$

Write an equivalent statement without use of any absolute value. Then solve the statement, and also graph it on a number scale.
26. $|x - 3| \leqq 5.$ **27.** $|x + 2| < 1.$ **28.** $|x - 3| \geqq 2.$
29. $|x - a| < 2.$ **30.** $|x - a| < d.$ **31.** $|x - a| > d.$

Solve the inequality.

32. $4 - x^2 > 0$.   **33.** $25 - 4x^2 > 0$.   **34.** $36 - x^2 > 0$.

★*In the following problems, use* (1) *on page* 43 *or properties already proved for inequalities.*

35. If $A < B$ and $R < S$, prove that $A + R < B + S$.

36. If $0 < R < S$ and $0 < A < B$, prove that $AR < BS$. (First obtain $RA < RB$.)

37. If $R < S$ and $R + U < S + V$, prove that it is *not* always true that $U < V$.

Hint. Prove the result by obtaining a single *counterexample*, that is, particular values of the variables for which the hypotheses are true and $U < V$ is false.

38. If $0 < R, 0 < S$, and $R < S$, prove that $R^2 < S^2$; $R^3 < S^3$.

39. If $1 < u$, prove that $u < u^2$; $u < u^3$.   If $0 < u < 1$, prove that $u^2 < u$; $u^3 < u$.

Comment. If n is any positive integer, in Chapter 12 the student may prove that, with data as in Problem 38, $R^n < S^n$; with $1 < u$ as in Problem 39, then $u < u^n$; with $0 < u < 1$, then $u^n < u$.

Note 1. Sometimes an inequality can be proved to be an *absolute inequality* by obtaining from it a more simple equivalent inequality which can be recognized easily as an absolute inequality.

40. Prove that $a^2 + b^2 \geq 2ab$ for all values of a and b.

41. If $u + v > 0$ and $u \neq v$, prove that $\dfrac{u + v}{2} > \dfrac{2uv}{u + v}$.

42. Prove that, for any real numbers x and y,

$$|x + y| \leq |x| + |y|. \tag{1}$$

Hint. First prove that, for *nonnegative* x, and then for *negative* x, $-|x| \leq x \leq |x|$, and similarly for y. Finally use Problem 35 to show that

$$-(|x| + |y|) \leq x + y \leq |x| + |y|.$$

Then use (4) on page 45.

20. ELEMENTARY OPERATIONS ON SETS

In the following discussion, suppose that all sets are subsets of a certain set T, called the **universal set**, or the **universe.**

DEFINITION I.   *If T is the universe, then the **complement**, K', of any set K is the set of elements of T which are not in K.*

Although the elements of T are thought of at present as *any* objects, visualize them as points in a plane. Also, let T be all points of the plane inside some closed curve C, in Figure 8. Then, let K be the set of points inside C in Figure 8 where K is not ruled. With this representation, the complement of K is the set K' which is ruled. The interpretation of the elements of a set as points in a plane is extremely useful. A corresponding figure like Figure 8 is called a **Venn diagram,** in honor of the mathematician John Venn (1834–1923) who initiated use of diagrams to illustrate relations between sets.

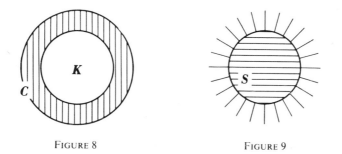

FIGURE 8   FIGURE 9

ILLUSTRATION 1.   If S is the set of points covered by horizontal rulings in Figure 9, and T is the whole plane, then S' is indicated by radial lines.

DEFINITION II.   The **union** *of any* number of sets is the set consisting of all elements which are in one or more of the sets.*

The "*union of sets A and B*" is represented by "$A \cup B$." The language of Definition II indicates that $A \cup B = B \cup A$, or the *order* in which the sets are shown is immaterial. Hence, it is said that the operation of forming the union of sets A and B is *commutative*. The symbol "$\cup$" can be read "*union*" wherever met. The union of three sets (A, B, C) is represented by $A \cup B \cup C$, with the sets in any order. The language of Definition II, where the order of the sets is immaterial, shows that

$$A \cup B \cup C = A \cup (B \cup C) = B \cup (A \cup C) = etc. \qquad (1)$$

Thus, in (1), any two of the sets may be associated as desired. Hence, it is said that the operation of *union* of sets is *associative*.

---

*For simplicity we use "*any number,*" instead of just "*two*" sets in Definition II and, again, in Definition III. Moreover, these definitions are in a form useful in later mathematics, where the union and intersection of infinite numbers of sets must be considered.

ILLUSTRATION 2.   Let T be the set of all points in the plane in Figure 10, and let A and B be the sets of points shown by vertical and horizontal rulings, respectively. Then $A \cup B$ consists of all ruled points, including some in both sets. $A \cup B$ consists of all elements in A alone, or in B alone, or in both A and B.

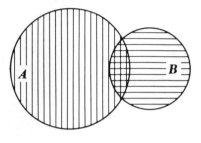

FIGURE 10

DEFINITION III.   The **intersection** *of any number of given sets is the set of elements belonging to all of the given sets.*

If A and B are sets, the "*intersection of A and B*" is represented by "$A \cap B$," where "$\cap$" may be read "*intersection*" wherever met. In Figure 10, $A \cap B$ consists of all points in the doubly ruled region. To denote the intersection of A, B, and C, we write "$A \cap B \cap C$," where the order of the sets is immaterial. Just as in the case of the *union* of sets, the operation of *intersection* of sets is seen to be *commutative* and *associative*. Thus, $A \cap B = B \cap A$ and $A \cap B \cap C = A \cap (B \cap C)$.

ILLUSTRATION 3.   In Figure 11, let A be the set of points on the large circle, and let B represent the set of points on the small circle. Then $A \cap B$ consists of the points P and Q where the circles intersect. Thus, the concept of *set intersection* is consistent with the meaning of *intersection* for geometric figures.

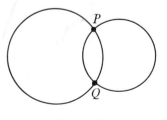

FIGURE 11

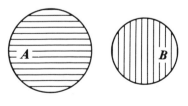

FIGURE 12

DEFINITION IV. To say that certain sets A, B, C, $\cdots$ are **mutually exclusive** *or* **disjoint,** *means that the intersection of any two of the sets is the empty set,* ϕ.

ILLUSTRATION 4. In Figure 12, A and B are mutually exclusive sets of points.

ILLUSTRATION 5. In a group, T, of people, let H be the subset consisting of those with blue eyes, and K be the subset with brown eyes. Then, $H \cap K = \phi$, or H and K are mutually exclusive.

ILLUSTRATION 6. Let $S = \{1, 2, 3, 4, 5, 6, 7\}$ and $T = \{5, 6, 7, 8, 9, 10\}$. Then $S \cap T = \{5, 6, 7\}$, and $S \cup T = \{1, 2, 3, 4, 5, 6, 7, 8, 9, 10\}$.

A set, T, of elements may be described by introducing a variable, say r, whose domain is T, and specifying a condition on r which is satisfied if and only if r is in T.

ILLUSTRATION 7. The set T of all positive real numbers can be represented by

$$T = \{x \mid x > 0\}. \tag{2}$$

In (2), the vertical rule is read "*such that.*" Then, (2) is read "*T is the set of all numbers x such that* $x > 0$."

A symbol as used on the right in (2) is said to be in "*set builder*" notation.

21. UNIONS AND INTERSECTIONS OF INTERVALS

On page 45 we agreed to represent an interval of the number scale by inserting in braces a statement, involving inequalities, whose graph is the given interval. Thus, the interval between -3 and 4 on the scale is represented by $\{-3 < x < 4\}$, to be read "*the interval* $-3 < x < 4$." Observe that this symbol can be thought of as a simplification of set builder notation. In it, we would write $\{x \mid -3 < x < 4\}$. Our agreed symbol merely omits "$x \mid$." We shall continue to use our abbreviated type of symbol.

The concepts of the union and the intersection of sets can be applied with considerable advantage when the sets are intervals of the number scale, as in the following examples.

ILLUSTRATION 1. The statement

$$-2 < x \leq 3 \quad means\ that \quad \{-2 < x \quad and \quad x \leq 3\}.$$

Hence, $\{-2 < x \leq 3\}$ is the *intersection* of $\{-2 < x\}$ *and* $\{x \leq 3\}$, or

$$\{-2 < x \leq 3\} = \{-2 < x\} \cap \{x \leq 3\}. \tag{1}$$

Thus, in Figure 13, the interval on the left in (1) is the part of the number scale where the intervals $\{-2 < x\}$ and $\{x \leqq 3\}$ overlap.

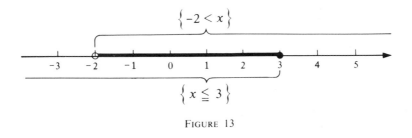

FIGURE 13

ILLUSTRATION 2.   Let A and B be the sets, or intervals of numbers

$$A = \{-4 < x \leqq 2\} \quad and \quad B = \{-2 < x \leqq 5\},$$

as in Figure 14. Observe that

$$A \cap B = \{-2 < x \leqq 2\} \quad and \quad A \cup B = \{-4 < x \leqq 5\}.$$

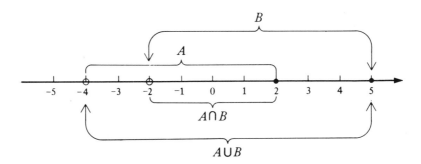

FIGURE 14

ILLUSTRATION 3.   The inequality $|x| > 4$ is equivalent to

$$x < -4 \quad OR \quad 4 < x.$$

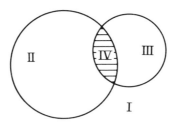

FIGURE 15

Let $A = \{\,|x| > 4\}$, $G = \{x < -4\}$ and $H = \{4 < x\}$. Then, A is the union of G and H:

$$A = G \cup H, \qquad or \qquad \{\,|x| > 4\} = \{x < -4\} \cup \{4 < x\}.$$

ILLUSTRATION 4.   Let the universe U be all points in the plane, R be the set of points inside the large circle and S be the set inside the small circle in Figure 15. Various regions in the figure are numbered (I)–(IV), respectively, where (IV) is ruled. For simplicity, we omit specifying where boundary points of the regions belong in this classification. Then (I) $= R' \cap S'$; (II) $= R \cap S'$; (III) $= S \cap R'$; (IV) $= R \cap S$.

EXERCISE 13

In Problems 1 and 2, the universe is the set $U = \{all\ positive\ integers\}$.
1. If $R = \{2, 3, 4, 5, 6, 7, 8\}$, $S = \{4, 5, 6, 7, 8, 9, 10, 11\}$, and $T = \{6, 7, 8, 9, 10, 11, 12, 13, 14, 15\}$, describe by the roster method or otherwise each of the following sets: S'; $R \cap S$; $R \cup S$; $R \cup S \cup T$; $R \cap S \cap T$; $(R \cap S) \cup T$; $R \cap (S \cup T)$; $T \cap R'$; $S' \cap (R \cup T)$; $R' \cap S'$; $R \cap S'$.
2. If $A = \{all\ positive\ integers\ divisible\ by\ 3\}$, describe A'.
3. By the roster method, describe all proper subsets of men which can be formed from {John, Harry, Bill}.
4. In a group of one hundred men, let A be the set of men where each has a college degree, and B be the set where each man has an annual income of at least $15,000. There are sixty men in A, forty men where each has a college degree and annual income of at least $15,000, and twenty-five men who do not have a college degree and earn less than $15,000 per year. Draw a Venn diagram showing the data, and find how many of the men do not have a college education and earn at least $15,000 per year.
5. In a group of ninety people, forty-five have the antigen A in their blood; some have the antigen B; twenty-five have both of the antigens A and B and thus are of blood type AB; fifteen have neither A nor B. Draw a Venn diagram for the data, and find how many people of the group have the antigen B.

On a number scale, let S be the graph of the first inequality and T be the graph of the second inequality. Show $S \cap T$ on the scale. Use an open circle to show exclusion of any endpoint and a dot to show its inclusion. Then, write a statement of the form $A < B < C$ whose graph is $S \cap T$.
6. $-2 < x$; $x < 5$.   7. $-4 < x$; $x < 0$.   8. $x < 5$; $-4 < x$.

In remaining problems, the universe is the set of all real numbers.

9. If $T = \{x < 5\}$, describe T'.   **10.** If $A = \{-2 < x\}$, describe A'.

Obtain the graph, H, of the statement on a number scale. Also, by use of inequalities, define two intervals A and B so that $H = A \cap B$.

11. $-4 < x \leq 2$.   **12.** $-5 \leq x \leq 0$.   **13.** $|x| < 2$.   **14.** $|x - 2| < 4$.

15. If $S = \{x < -3\}$ and $T = \{x > 3\}$, indicate $S \cup T$ on a number scale. Also, define $S \cup T$ by use of a single inequality involving an absolute value.

Show the graph, T, of the inequality on a number scale. Also, define two intervals A and B by use of inequalities so that $T = A \cup B$.

16. $|x| > 5$.   **17.** $|x| > 1$.   **18.** $|x - 3| > 2$.

19. Describe two intervals S and T by use of inequalities so that $A = S \cap T$, if $A = \{-1 \leq x < 4\}$.

Show intervals A and B on a number scale. Then indicate $A \cap B$ and $A \cup B$ on the figure, if possible.

20. $A = \{-3 \leq x < 2\}$; $B = \{-1 < x \leq 4\}$.

21. $A = \{-6 < x \leq 2\}$; $B = \{0 < x \leq 5\}$.

22. $A = \{x \leq 3\}$; $B = \{1 \leq x\}$.

23. $A = \{x < -4\}$; $B = \{4 < x\}$.

24. If $A = \{|x| > 5\}$, describe A' by use of inequalities.

chapter 3

Introduction to Functions and Graphs

22. COORDINATES IN A PLANE

In a given plane, draw two perpendicular lines OX and OY intersecting at point O. Each of these lines will be referred to as a *coordinate axis*, where one is horizontal and the other is vertical in the typical case, as in Figure 16. On OX, and also on OY, choose arbitrarily a unit for distance as the basis for number scales on OX and OY, with point O representing zero on each scale. The units for distances on OX and OY need *not* be the same. Positive numbers will be to the right on OX and upward on OY, with negative numbers to the left on OX and downward on OY. Any horizontal line segment in the plane will be a *directed segment,* with positive and negative directions as on OX, and will be measured in terms of the scale unit on OX. Vertical line segments will be *directed,* with positive and negative directions as on OY, and will be measured in terms of the scale unit on OY. Let P be any point in the plane of OX and OY. Then, we define an ordered pair of coordinates for P as follows:

> The **horizontal coordinate,*** or the *x-coordinate of P is the directed distance, x, measured parallel to OX from the vertical axis OY to P. The* **vertical coordinate,**† *or the y-coordinate of P is the directed distance, y, measured parallel to OY from the horizontal axis OX to P.*

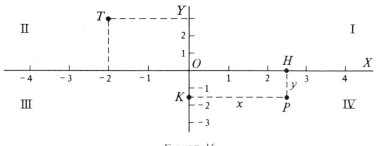

FIGURE 16

*Sometimes called the *abscissa* of P.
†Frequently called the *ordinate* of P in later discussion.

55

Coordinates as just described are called *rectangular coordinates* because the coordinate axes are perpendicular.* The intersection, O, of the axes is called the **origin** of coordinates. The plane in which the axes OX and OY lie is referred to as the *coordinate plane,* or the *xy-plane* when x and y are used for the coordinates of an arbitrary point P. We shall use "$P:(x,y)$" to mean "P *with coordinates* (x,y)." We may read "$P:(x,y)$" simply as "P,x,y." For each point P there is just one set of coordinates. For each pair of numbers (x,y), there is just one point P having (x,y) as the coordinates.

ILLUSTRATION 1.   To plot $T:(-2, 3)$, as in Figure 16, we may erect a perpendicular to OX at $x = -2$, and go 3 vertical units upward to reach P. Or, we may erect a perpendicular to OY at $y = 3$ and go 2 horizontal units to the left to reach T.

ILLUSTRATION 2.   With $(x = a, y = b)$ given as the coordinates of a point, we may refer to it as "*the point* (a, b)."

The *projection* of a point P on a line L is defined as the foot of the perpendicular from P to L. For any point $P:(x,y)$ in the coordinate plane, the projection of P on OX is the point $H:(x, 0)$, and on OY is the point $K:(0, y)$, as in Figure 16.

The coordinate axes divide the plane into four *quadrants,* numbered I, II, III, and IV, counterclockwise, as in Figure 16. A point on a coordinate axis is not said to be in any quadrant.

Note 1.   If P and Q are any two points, "PQ" will refer to the *segment* PQ until otherwise mentioned.

ILLUSTRATION 3.   The points in an *xy*-plane for which the horizontal coordinate is $x = 6$ form the line perpendicular to the *x*-axis where $x = 6$.

23. LENGTHS OF LINE SEGMENTS IN AN *xy*-PLANE

Hereafter, unless otherwise mentioned, the scale units on the coordinate axes will be *equal* in any *xy*-plane. Also, in such a case, this unit will be used in measuring distance in any direction in the plane. If some result is such that it remains valid when the scale units on the axes are unequal, this fact will be emphasized.

If a line segment P_1P_2 is parallel to OX, as in Figure 17, then P_1 and P_2 have the same *y*-coordinate. Thus, consider $P_1:(x_1, c)$ and $P_2:(x_2, c)$ where P_1P_2 is horizontal. The projections H_1 and H_2 of P_1

*Coordinates can be defined where the axes are not perpendicular.

and P_2, respectively, on OX are $H_1:(x_1, 0)$ and $H_2:(x_2, 0)$. From page 5, $\overline{H_1 H_2} = x_2 - x_1$ and hence

$$\overline{P_1 P_2} = x_2 - x_1; \qquad |\overline{P_1 P_2}| = |x_2 - x_1|. \qquad (1)$$

In the same fashion, with $Q_1:(d, y_1)$ and $Q_2:(d, y_2)$, we have $Q_1 Q_2$ parallel to OY and

$$\overline{Q_1 Q_2} = y_2 - y_1; \qquad |\overline{Q_1 Q_2}| = |y_2 - y_1|. \qquad (2)$$

We observe that (1) and (2) are true even if the scale units on the coordinate axes are unequal. In such a case, $\overline{P_1 P_2}$ is measured in terms of the unit for OX, and $\overline{Q_1 Q_2}$ in terms of the unit for OY.

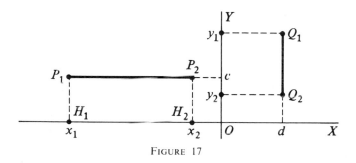

FIGURE 17

ILLUSTRATION 1. By use of (1), with $P_1:(4, -3)$ and $P_2:(7, -3)$, we have $\overline{P_1 P_2} = 7 - 4 = 3$ and $\overline{P_2 P_1} = 4 - 7 = -3$; $|\overline{P_1 P_2}| = |\overline{P_2 P_1}| = 3$.

If P_1 and P_2 are any points in an xy-plane, and if $P_1 P_2$ is not vertical or horizontal, usually we shall *not* consider $P_1 P_2$ to be a directed line segment. Hence, we shall use $\overline{P_1 P_2}$ for the *length* (nonnegative) of $P_1 P_2$. However, we may also use $|\overline{P_1 P_2}|$ for this length in any case where there is a possibility that $P_1 P_2$ *might* be horizontal or vertical.

THEOREM I. *If the scale units on the coordinate axes are equal, the distance, d, between $P_1:(x_1, y_1)$ and $P_2:(x_2, y_2)$ is*

$$d = |\overline{P_1 P_2}| = \sqrt{(x_2 - x_1)^2 + (y_2 - y_1)^2}. \qquad (3)$$

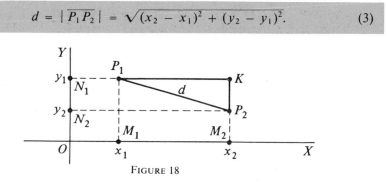

FIGURE 18

Proof.   1. In Figure 18, let K be the intersection of the lines perpendicular to OY through P_1 and perpendicular to OX through P_2. The Pythagorean theorem can be used for $\Delta P_1 K P_2$ because a single unit of length applies to line segments in any direction in the plane. Hence, $\overline{P_1 P_2}^2$ is equal to the sum of the squares of the lengths of the perpendicular sides of $\Delta P_1 K P_2$, or

$$d^2 = \overline{P_1 P_2}^2 = \overline{P_1 K}^2 + \overline{P_2 K}^2. \tag{4}$$

2. From (1) and (2), in Figure 18 we obtain

$$|\overline{P_1 K}| = |\overline{M_1 M_2}| = |x_2 - x_1|; \quad |\overline{P_2 K}| = |\overline{N_1 N_2}| = |y_2 - y_1|.$$

Since $|x_2 - x_1|^2 = (x_2 - x_1)^2$ and $|y_2 - y_1|^2 = (y_2 - y_1)^2$, we have

$$d^2 = (x_2 - x_1)^2 + (y_2 - y_1)^2. \tag{5}$$

On extracting square roots in (5) we obtain (3).

We note that $|\overline{P_1 P_2}|$ may be written simply $\overline{P_1 P_2}$ when $P_1 P_2$ is not vertical or horizontal, because then we have agreed that $\overline{P_1 P_2} \geqq 0$. In using (3), the *order* in which we label two given points as P_1 and P_2 is immaterial because *squares* of differences of coordinates appear in (3).

Illustration 2.   The distance between $P:(-3, 5)$ and $Q:(2, -3)$ is

$$\overline{PQ} = \sqrt{[2 - (-3)]^2 + (-3 - 5)^2} = \sqrt{25 + 64} = \sqrt{89}.$$

From (3), the distance of $P:(x, y)$ from the origin $O:(0,0)$ is

$$d = \sqrt{x^2 + y^2}. \tag{6}$$

Example 1.   Find the point on OX equidistant from the points $R:(-2, 3)$ and $S:(2, 5)$.

Solution.   Let $P:(x,0)$ be the unknown point. Then $|\overline{PR}| = |\overline{PS}|$, or $\overline{PR}^2 = \overline{PS}^2$. Hence, from (3),

$$(x + 2)^2 + 3^2 = (x - 2)^2 + 5^2;$$

$$x^2 + 4x + 4 + 9 = x^2 - 4x + 4 + 25; \; or$$

$$8x = 16; \quad x = 2.$$

Thus, the desired point is $P:(2, 0)$.

By use of $P_1:(x_1, c)$ and $P_2:(x_2, c)$ in (3), we obtain

$$|\overline{P_1 P_2}| = \sqrt{(x_2 - x_1)^2} = |x_2 - x_1|,$$

as in (1). Similarly, with $Q_1:(d, y_1)$ and $Q_2:(d, y_2)$, from (3) we obtain $|\overline{Q_1 Q_2}| = |y_2 - y_1|$, as in (2). However, the student is advised to use (1) and (2) when they apply instead of referring to (3).

EXERCISE 14

1. In which quadrant is $P:(x, y)$ located if $x < 0$ and $y > 0$?

If P represents the first point and Q the second point, find $\overline{PQ}$ and the length of PQ.

2. $(0, 3); (0, 8).$ **3.** $(-2, 0); (-4, 0).$
4. $(5, 2); (1, 2).$ **5.** $(3, -2); (3, -9).$
6. $(4, -3); (4, 0).$ **7.** $(-1, 3); (-5, 3).$
8. $(5, 3); (5, 7).$ **9.** $(-2, -1); (0, -1).$

Find the distance between the points. Table I may be used for square roots.

10. $(3, 7); (-6, 7).$ **11.** $(7, 2); (2, 14).$
12. $(2, -2); (5, 2).$ **13.** $(-5, 0); (-9, -4).$
14. $(1, 3); (5, 0).$ **15.** $(-3, -1); (1, 6).$
16. $(c, d); (3, -2).$ **17.** $(-2, 5); (-2, -3).$

Prove that the triangle with the given vertices is isosceles (two sides of equal lengths).

18. $(-4, 9); (-3, 2); (1, 4).$ **19.** $(-2, 3); (-5, 1); (-4, 6).$

Prove that the triangle with the given vertices is equilateral.

20. $(0, -2); (0, 8); (5\sqrt{3}, 3).$ **21.** $(2, 0); (-6, 0); (-2, 4\sqrt{3}).$

If the sum of the squares of the lengths of two sides of a triangle is equal to the square of the length of the third side, then the triangle is a right triangle. (This is the converse of the Pythagorean theorem.) Prove that the triangle with the given vertices is a right triangle.

22. $(1, 0); (3, 1); (0, 7).$ **23.** $(3, 2); (5, 3); (0, 8).$
24. Find x if $(x, -3)$ is equidistant from $(2, -3)$ and $(6, 5).$
25. Find the point on the y-axis equidistant from $(2, 1)$ and $(4, 5).$

24. GRAPHS OF RELATIONS AND EQUATIONS IN TWO VARIABLES

Let x and y be variables. Then, a nonempty set of pairs of corresponding values of x and y is called a **relation** between them.

ILLUSTRATION 1.   Let T be the following set of pairs of corresponding values of x and y. Then T is a *relation* between x and y.

$x =$	3	3	-1	-3	0	4	5	2	1
$y =$	-2	4	-2	-1	3	1	2	3	4

The *graph of a relation T* between two variables x and y, whose domains consist of real numbers, is defined as the set of points in an xy-plane whose coordinates (x, y) are number pairs in T.

ILLUSTRATION 2.   The graph of the relation T in Illustration 1 would consist of the nine points $(3, -2), (3, 4), \cdots$ given by the table.

A *solution* of an equation in x and y is *a pair of corresponding values of x and y* which satisfy the equation.   Then, the *solution set* of a consistent equation is a set of ordered pairs of numbers (x, y), and hence is a *relation* between the variables.   It is said that the variables x and y are *related by the equation.*

ILLUSTRATION 3.   In $2x - 3y = 6$, if $y = 2$ then $x = 6$ to satisfy the equation, so that $(x = 6, y = 2)$ is one of its solutions.

DEFINITION I.   The **graph of an equation** *in two variables x and y is the set of points in an xy-plane whose coordinates are solutions of the equation.*

Thus, the graph of an equation in x and y is the *graph of the solution set* of the equation.   To obtain the graph, when no special method is available, calculate representative solutions and draw the graph through the corresponding points.   The scale units need not be equal in the coordinate plane.

ILLUSTRATION 4.   To graph $2x - 3y = 6$, we compute the following solutions. The graph, in Figure 19, is seen to be a straight line, L. This illustrates the fact, to be proved later, that *the graph of a linear equation in x and y is a line.*

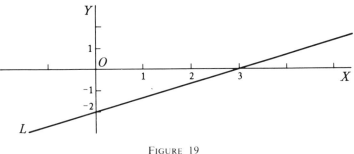

FIGURE 19

$x =$	-1	0	3	4
$y =$	$-8/3$	-2	0	$2/3$

The **x-intercepts** of a set, T, of points in an xy-plane are the values of x at the points (if any) where T intersects the x-axis. The **y-intercepts** are the values of y where T intersects the y-axis. We obtain the intercepts of the graph of an equation in x and y as follows.

To find the **x-intercepts,** *place y* = 0 *in the equation and solve for x.*

To find the **y-intercepts,** *place x* = 0 *in the equation and solve for y.*

ILLUSTRATION 5.   To graph $2x - 3y = 6$, place $x = 0$ and find $y = -2$, the y-intercept. The x-intercept is $x = 3$. The graph is given in Figure 19.

ILLUSTRATION 6.   To graph the equation $y - x^2 - 4x - 2 = 0$, first we solve for y:

$$y = x^2 + 4x + 2. \tag{1}$$

With assigned values for x, the following solutions (x, y) of (1) are obtained. The graph in Figure 20 was drawn through the corresponding points. The graph is called a **parabola.** It is desirable sometimes to refer to an equation by giving it the name of its graph. Thus, in Figure 20 we observe *the parabola $y = x^2 + 4x + 2$.*

$y =$	7	2	-1	-2	-1	2	7
$x =$	-5	-4	-3	-2	-1	0	1

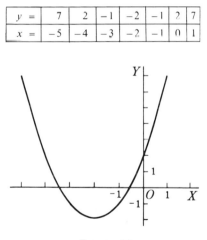

FIGURE 20

An equation for a set, T, of points in an *xy*-plane is an equation in *x* and *y* where *T* is the graph of the equation. A set *T* may be such that no equation can be found for it. If one equation for *T* is obtained, then infinitely many equivalent equations can be written for *T*. The particular one of these with which we deal may be called **THE** equation of *T*.

ILLUSTRATION 7.   The graph of $2x - 3y = 6$ is the line L in Figure 19 on page 60. This line also is the graph of the equivalent equation $-4x + 6y = -12$.

Either *the equation of a set T* of points in an *xy*-plane, or *the graph T of an equation in x and y* is determined by the following conditions.

If P:(x, y) is in T, the coordinates of P satisfy the equation of T.

If P:(x, y) is not in T, then (x, y) is not a solution of the equation of T.

Suppose that a certain equation in x and y has the graph T in an xy-plane. Then, sometimes, it is said that the equation *represents T.* Thus, in Illustration 7, the equation $2x - 3y = 6$ *represents the line L* in Figure 19.

ILLUSTRATION 8.   Consider the equation $x - 4 = 0$ as a linear equation in x and y where the coefficient of y is zero. If we place $x = 0$ in $x - 4 = 0$, we obtain $-4 = 0$, which is a contradiction. Hence, there is *no y-intercept.* The graph of $x - 4 = 0$ in the xy-plane is the line perpendicular to the x-axis where $x = 4$.

25. SPECIAL CASES IN GRAPHING EQUATIONS

Recall that a product is zero when and only when one of the factors is zero. Suppose that a polynomial P in the variables x and y is a product of polynomial factors. Then, the solution set of $P = 0$ is the union of the solution sets of the new equations obtained on setting each factor of P separately equal to zero. Hence, the graph of $P = 0$ is the *union* of the graphs of the new equations.

EXAMPLE 1.   Graph: $(3x - 2y + 6)(x - 2y + 4) = 0.$ (1)

Solution.   1. A pair of numbers (x, y) is a solution of (1) when and only when

(a) $3x - 2y + 6 = 0,$ OR   (b) $x - 2y + 4 = 0.$ (2)

2. The graph of (a) is L, and of (b) is M, in Figure 21. The solution set of (1) is the *union* of the solution sets of (a) and (b). Let T be the graph of (1). Then T is the *union of L and M,* that is, $T = L \cup M.$ Or, the graph of (1) consists of the *two lines L and M.*

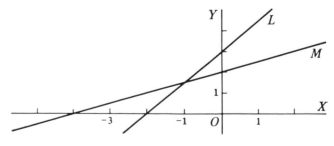

FIGURE 21

ILLUSTRATION 1.   To graph $2x^2 - 5xy - 3y^2 = 0$, first factor:

$$(x - 3y)(2x + y) = 0. \tag{3}$$

Equation (3) is equivalent to the statement

$$x - 3y = 0 \qquad \text{OR} \qquad 2x + y = 0. \tag{4}$$

Hence, the graph of (3) is the union of the graphs of the equations in (4). Or, the graph of (3) consists of the lines $L:\{x - 3y = 0\}$ and $M:\{2x + y = 0\}$.

If an equation in x and y has *no real solution,* then the equation has *no graph.* In such a case, it can be said that the graph is the *empty set,* ϕ.

ILLUSTRATION 2.   The equation $4x^2 + y^2 = -2$ has no real solution because the left-hand side is positive or zero for all real values of x and y. Hence, the equation has no solution, or the graph is the empty set, ϕ.

The graph of an equation may be a single point, or a collection of isolated points.

ILLUSTRATION 3.   The only solution of $x^2 + 9y^2 = 0$ is $(x = 0, y = 0)$. Hence, the graph of the given equation is the single point $O:(0,0)$.

ILLUSTRATION 4.   The equation

$$(x^2 + y^2)[(x - 2)^2 + (y + 3)^2] = 0 \tag{5}$$

is equivalent to the statement

$$\text{(a)} \quad x^2 + y^2 = 0 \qquad \text{OR} \qquad \text{(b)} \quad (x - 2)^2 + (y + 3)^2 = 0. \tag{6}$$

In (6), equation (a) has a single solution $(x = 0, y = 0)$, and (b) has a single solution $(x = 2, y = -3)$. Hence, the graph of (5) consists of just the two points $(0, 0)$ and $(2, -3)$.

EXAMPLE 2.   Graph the equation: $|x - y| = 4.$ \qquad (7)

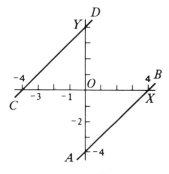

FIGURE 22

Solution.   Equation (7) is equivalent to the statement.

$$(a)\quad x - y = 4 \qquad OR \qquad (b)\quad x - y = -4. \tag{8}$$

The student may verify that the graph of (*a*) is line AB, and of (*b*) is line CD in Figure 22. The graph of (7) is the *union* of the graphs of (*a*) and (*b*) of (8). Or, the graph of (7) consists of the two lines AB and CD.

EXERCISE 15

Graph the equation in an xy-plane.

1. $5x + 3y = 15$.
2. $4y - 3x = 12$.
3. $2x - 9 = 0$.
4. $y + 7 = 0$.
5. $2x - y = 0$.
6. $3y = 2x$.
7. $2x - 5y + 10 = 0$.
8. $2y - 3x + 6 = 0$.
9. $y = 0$.

Graph the equation with $x = -2$ used in the table of solutions.

10. $y = x^2 + 4x - 3$.
11. $y = 2 - 4x - x^2$.
12. $y = 5 - 8x - 2x^2$.

Graph the equation with $y = 2$ used in the table of values.

13. $x = y^2 - 4y + 7$.
14. $x = 5 + 4y - y^2$.
15. $x = 2y^2 - 8y + 3$.

Write an equation for the line satisfying the condition.

16. Parallel to the *y*-axis with *x*-intercept 5.
17. Parallel to the *x*-axis with *y*-intercept -6.

In an xy-plane, graph the equation if it has a graph.

18. $(2x - 3y)(x + y) = 0$.
19. $(x - 3)(x - y + 5) = 0$.
20. $x^2 - 4y^2 = 0$.
21. $9x^2 - y^2 = 0$.
22. $x^2 + 4y^2 = 0$.
23. $2x^2 + 3y^2 = -2$.
24. $|x + 2y| = 3$.
25. $|x - 4| = 3$.
26. $(x - 3)(x + 2) = 0$.
27. $y^2 - y - 2 = 0$.
28. $2x^2 + 7xy - 4y^2 = 0$.
29. $3x^2 + 5xy - 2y^2 = 0$.

26. FUNCTION CONCEPT

Frequently in mathematics we meet situations involving two related variables x and y where, for each value of one variable, say x, a *single* value of y is given by some rule. Then, it is natural to say that "*y is a function of x*," where the name *function* is used with a semicolloquial meaning. We shall make the meaning precise.

DEFINITION II.   *Suppose that, for each value of x in a set, D, of numbers, some rule specifies a single corresponding value of y, and let R be set of all of these values of y. Then, the resulting set of ordered pairs of numbers (x, y) is called a* **function,** *F, whose* **domain** *is D and* **range** *is R.*

In Definition II, each value of y in R is called a *value* of the function. We refer to x as the **independent variable** and to y as the **dependent variable.**

We call F "*a function of x*," to indicate that x is to be used as the symbol for the independent variable.

ILLUSTRATION 1.   Let $D = \{-3,-2,-1,0,1,2,3\}$. For each number x in D, let $y = x^2$; the corresponding values for y are $\{9,4,1,0,1,4,9\}$. Then the given rule defines the set of ordered pairs

$$F = \{(-3, 9), (-2, 4), (-1, 1), (0, 0), (1, 1), (2, 4), (3, 9)\}. \qquad (1)$$

This *set F* is a *function* whose *domain* is D and *range* is $R = \{0, 1, 4, 9\}$.

In Definition II, the *rule* involved specifies a *correspondence* between the values of x in D and the values of y in R. Thus, a function may be thought of as a *rule of correspondence* between the domain and the range. This *rule itself** sometimes is referred to *as if it were the function*. With this terminology in Illustration 1, we might refer to "*the function x^2*," meaning that this formula gives the value of y corresponding to any assigned value of x.

ILLUSTRATION 2.   Let $D = \{2 \leq x \leq 5\}$ and let $y = 3x - 1$. If $x - 2$ then $y - 5$; if $x - 3$, then $y - 8$; if $x = 4$, then $y = 11$. If x varies from 2 to 5, we find that y varies from 5 to 14. Thus, with x in D, the corresponding values of y form the interval $R = \{5 \leq y \leq 14\}$. This correspondence is a function F with domain D and range R; F *maps D onto R*, as indicated by representative arrows in Figure 23. F is defined as a function of x by the equation $y = 3x - 1$. We might call F "*the function $(3x - 1)$*."

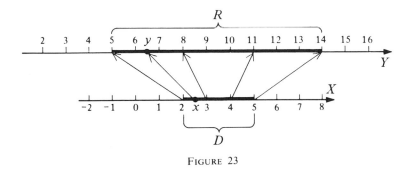

FIGURE 23

Any formula, in a variable x, which specifies a single number corresponding to each value of x, thus defines a function whose values are assigned by the formula. However, a function may be defined merely by

*This convenient elliptical phraseology for functions is met at all levels of mathematics. For related comments, see (18) on page 35, Volume I, of *A Programmed Course in Calculus;* prepared by the Committee on Educational Media of the Mathematical Association of America (New York: W. A. Benjamin, Inc., 1968.).

specifying the corresponding ordered pairs of numbers (x, y), for instance in a table, without use of a formula. If a function is described by use of a formula for its values, the function may be named corresponding to the nature of its formula. Thus, in previous courses in mathematics, the student has met *algebraic, exponential, logarithmic,* and *trigonometric* functions, which are called the *elementary functions* of mathematical analysis. They will be studied in this text.

ILLUSTRATION 3.   In Definition II, if $y = h$, a constant, at all values for x, then F is called a *constant function*.

DEFINITION III.   In an xy-plane, the **graph** *of a function of a variable x is the set of points whose coordinates (x, y) form pairs of corresponding values of x and the function.*

The following procedure is a consequence of Definition III. To obtain the graph of a function of x defined by a formula in x, *place y equal to the formula and graph the resulting equation.* Usually, in graphing a function, values of the independent variable are plotted on the horizontal axis, but this is not essential.

ILLUSTRATION 4.   To graph the function $(x^2 + 4x + 2)$, we let $y = x^2 + 4x + 2$ and graph this equation. The graph is in Figure 20 on page 61.

In the description of a function, the letters used for the variables are immaterial. Thus, suppose that a function F is defined by stating that, for every number x in the domain $D = \{-2 < x < 6\}$, the value of F is $y = -2x + 5$. The same function is described by stating that, for each number u in the domain $D = \{-2 < u < 6\}$, the value of F is $v = -2u + 5$.

If F is a function of x, we may represent the value of F corresponding to any value of x by $F(x)$, to be read "F *of x*," or "F *at x.*" Thus,

$$F(x) \text{ is the } \textbf{value of } F \text{ corresponding to the number } x. \qquad (2)$$

In (2), $F(x)$ is referred to as a symbol in *functional notation,* with x as the **argument** in the symbol. Symbols like $F, G, H, f, \cdots$ are used for functions.

ILLUSTRATION 5.   If we let $F(x) = 3x^2 - x$, this assigns F as a symbol for a function, and gives a formula, $(3x^2 - x)$, for computing the value of F at any value for x. Thus, the value of F at $x = -2$ is

$$F(-2) = 3(-2)^2 - (-2) = 14.$$

ILLUSTRATION 6.   If $F(x) = 3x^3 + 4$ and $G(y) = 2y^2$, then

$$F(G(y)) = 3(2y^2)^3 + 4 = 3(8y^6) + 4 = 24y^6 + 4.$$

In functional notation, Definition III yields the following result.

$$\left.\begin{array}{l}\textit{The graph of a function } F(x) \textit{ in an xy-plane is the}\\ \textit{graph of the equation } y = F(x).\end{array}\right\} \tag{3}$$

Frequently hereafter, we shall use the terminology

$$y \textit{ is a function of } x, \tag{4}$$

to mean that there exists some function F where we shall let $y = F(x)$.

ILLUSTRATION 7.   With $f(x) = 2x^3 - 3x^2 - 12x + 5$, a graph of f is a graph of the equation $y = 2x^3 - 3x^2 - 12x + 5$. Coordinates of points $(x, y = f(x))$ on the graph are obtained by substituting values for x, as given in the following table, which is the basis for the graph of f in Figure 24.

$y = f(x) =$	-15	1	12	5	-8	-15	-4	12
$x =$	-2.5	-2	-1	0	1	2	3	3.5

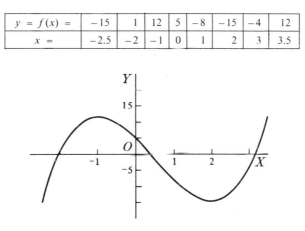

FIGURE 24

A function F is called a **polynomial function** of degree $n \geq 0$ in x in case $F(x)$ is a polynomial of degree n in x, or*

$$F(x) = a_0 + a_1 x + a_2 x^2 + \cdots + a_n x^n, \tag{5}$$

where $(a_0, a_1, a_2, \cdots, a_n)$ are constants with $a_n \neq 0$.   Polynomial functions of degrees 1, 2, 3, and 4 are called *linear, quadratic, cubic,* and *quartic* functions, respectively.   In Illustration 7, f is a cubic function.

ILLUSTRATION 8.   Any linear function $f(x)$ is of the form $f(x) = mx + b$, where $\{m, b\}$ are constants and $m \neq 0$. Any quadratic function $f(x)$ is of the form $f(x) = ax^2 + bx + c$, where $\{a, b, c\}$ are constants with $a \neq 0$.

---

*In (5), if $n = 0$ and $a_0 = 0$, then $F(x)$ is *not* called a polynomial of degree zero, but may be referred to as the *zero polynomial.*

If $P(x)$ and $Q(x)$ are polynomials in x, and $f(x) = P(x)/Q(x)$, then f is called a **rational function** of x. Since $P(x) = P(x)/1$, any polynomial function $P(x)$ is a rational function. Graphs of rational functions will be considered in a later chapter.

ILLUSTRATION 9.   If $f(x) = \dfrac{3x^2 - 5}{2x + 7}$, then f is a rational function of x.

The operation represented by $\sqrt[n]{A}$ may be called the *extraction of an nth root.* Then addition, subtraction, multiplication, division, raising to a power, and extraction of a root are called the operations of algebra. If $f(x)$ is defined by operations of algebra applied to the number x, then f is called an **algebraic* function** of x. If f is an algebraic function of this variety which is *not* a rational function, then f is called an **irrational function** of x. In this case, the symbol for $f(x)$ involves at least one radical, or an equivalent power, and cannot be expressed in the form $P(x)/Q(x)$ where $P(x)$ and $Q(x)$ are polynomials.

ILLUSTRATION 10.   If $f(x) = \sqrt{x^2 + 5}$, then f is an irrational function of x.

ILLUSTRATION 11.   A function may be defined by different formulas on different subsets of its domain. Thus, let a function f be defined as follows:

$$\text{When } x < 2: \qquad f(x) = -2.$$
$$\text{When } x = 2: \qquad f(x) = 0.$$
$$\text{When } x > 2: \qquad f(x) = -2x + 6.$$

The graph of $y = f(x)$ in Figure 25 consists of that part of the horizontal line $y = -2$ where $x < 2$; the isolated point $(2,0)$; that part of the line $y = -2x + 6$ where $x > 2$. The circles at $(2, -2)$ and $(2,2)$ indicate that these points are not in the graph.

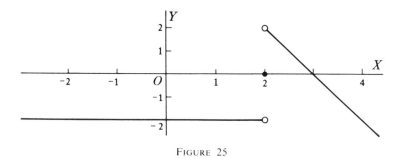

FIGURE 25

*A more general definition of an algebraic function is met in calculus.

On page 59, a relation between two variables x and y was defined as a set, T, of ordered pairs (x, y) of values of the variables. From Definition II, *any function is a relation* between the independent and dependent variables, x and y, with the special characteristic that, for any value $x = x_0$, there is *just one* ordered pair (x, y) in the function-relation with $x = x_0$. Thus, every function of one variable is a relation between two variables, but a relation is *not* a function unless the relation has the special characteristic just mentioned.

Note 1. If the word *number* is changed to *element* in Definition II, it describes a function whose domain is a set D of elements of *any* variety, and range is a corresponding set R of elements of any sort. Thus, in the foundation of the theory of probability, D would consist of all *subsets* of a given sample space of arbitrary elements, and R would be a set of real numbers on the interval $\{0 \leq p \leq 1\}$, where each value of p is the *probability* of some subset belonging to D.

27. FUNCTIONS OF MORE THAN ONE VARIABLE

Let D be a set of pairs of numbers (x, y). We may visualize D as a set of *points* in an xy-plane. For each point (x, y) in D, suppose that some rule specifies a single number z, and let R be the set of all values of z. Then, this correspondence between points in D and numbers in R, or the resulting set of ordered *triples* of numbers (x, y, z), is called a *function F of the two variables x and y*. We refer to D as the domain and to R as the range of F, with (x, y) called the independent variables and z the dependent variable.

We write $z = F(x, y)$, meaning that both $F(x, y)$ and z may represent the value of F corresponding to assigned numbers (x, y) in D. Similarly, functions of three or more variables may be considered. Thus, $f(x, y, z)$ would represent the value of a function f of three independent variables (x, y, z). Hereafter, unless otherwise indicated, in any reference to a *function* we shall mean a function of a *single variable*.

ILLUSTRATION 1.   If $g(x, y) = x^2 + xy$, then $g(2, -3) = 4 - 6 = -2$.

EXERCISE 16

1. The table defines a relation T between x and y. Is T a function (*a*) with x as the independent variable, or (*b*) with y as the independent variable? Why?

$x =$	-1	0	1	2	3	4
$y =$	2	-1	2	4	-2	3

Tabulate the whole set of ordered pairs of numbers which form the function H. Describe the range, R, of H. Graph H in an xy-plane.

2. The domain of H is $D = \{-2,-1,0,1,2,3,4\}$. The value of H at any number x in D is $(3x - 4)$.

3. The domain of H is $D = \{-3,-2,-1,0,1,2,3\}$. The value of H at any number x in D is $(x^2 - 2)$.

Graph the function.

4. $2x + 5$.   **5.** $-x + 4$.   **6.** $x - 7$.   **7.** 8.

8. $(x^2 - 6x - 3)$, with $x = 3$ used in the table of ordered pairs.

9. $(-x^2 + 2x - 5)$, with $x = 1$ used in the table of ordered pairs.

10. A function G has the domain $D = \{1, 2, 3, 4, \cdots, 18\}$. The value, y, of G corresponding to any number x in D is the largest integral multiple of 4 which is at most equal to x. List the set of ordered pairs (x,y) which form G. What is the range of G? Construct a figure like Figure 23 on page 65 to show how the domain is mapped on the range.

If $f(x) = x - x^2$, find the value of the symbol.

11. $f(3)$.   **12.** $f(1)$.   **13.** $f(2)$.   **14.** $f(\tfrac{1}{3})$.   **15.** $[f(-2)]^2$.

If $h(u) = u^2 - u^4$, find the value of the symbol.

16. $h(-4)$.   **17.** $4h(2)$.   **18.** $h(-\tfrac{1}{3})$.   **19.** $h(\sqrt{a + b})$.

If $G(x,y) = x^2 - xy - y^2$, find the value of the symbol.

20. $G(1, 3)$.   **21.** $G(-2, 2)$.   **22.** $G(c, d^2)$.   **23.** $G(3, h)$.

24. If $f(v) = 2v + 1$ and $g(w) = 2 - w^2$, find $f(1)g(3)$; $f(2)/g(4)$; $f(g(x))$; $g(f(t))$.

Graph the function.

25. $g(x) = |x|$.   **26.** $h(x) = |x + 2|$.   **27.** $f(x) = 9 - x^2$.

28. The function $f(x)$ defined as follows for all real numbers x:

$$f(x) = 2 \text{ when } x < 0; \quad f(0) = 3; \quad f(x) = 2x + 1 \text{ when } x > 0.$$

29. The function $F(x)$ defined as follows:

$$F(x) = x \text{ if } x \leqq -1; F(x) = 0 \text{ if } -1 < x < 0; F(x) = x^2 + 1 \text{ if } x \geqq 0.$$

30. Graph the function $[x]$, where $[x]$ represents the greatest integer at most equal to x.

31. The domain of a function F is $D = \{1, 2, 4, 9, 16\}$. For any number x in D, the value of F is $\sqrt{x}$. Obtain a graph of F.

32. The domain of a function F is $D = \{0, .5, 1, 2, 3\}$. For any number x in D, the value of F is x^2. Obtain a graph of F.

28. FUNCTIONS DEFINED BY EQUATIONS

Any equation in two variables x and y can be written in the form

$$H(x, y) = 0. \tag{1}$$

Assume that there exists a function $f(x)$ with domain M so that (1) is satisfied when $y = f(x)$ for all values of x in M, or $H(x, f(x)) = 0$. We shall refer to f as a **solution function** of (1) for y as a function of x. It may happen that a single solution function $f(x)$ exists so that all solutions of (1) are given by $(x, y = f(x))$ with x in M. Then, (1) is equivalent to the equation $y = f(x)$, and it is said that (1) *defines y as a function of x*. In this case, the graph of (1) is the graph of the solution function $f(x)$, or of the equation $y = f(x)$; thus the graph, C, of (1) has *just one point* $(x_0, y_0 = f(x_0))$ corresponding to each value $x = x_0$. Or, C is met in *just one point* by a perpendicular to the x-axis at any point $x = x_0$ on the domain for x. Under these circumstances, it may be possible to obtain $f(x)$ by solving (1) for y in terms of x.

ILLUSTRATION 1.   The equation $y - x^2 - 4x - 2 = 0$ defines y as a quadratic function of x because, on solving for y, we obtain $y = x^2 + 4x + 2$, which gives the single solution function $(x^2 + 4x + 2)$ for y as a function of x. The graph, C, of this function, or of the given equation, is the parabola in Figure 20 on page 61. Notice that C is met in just one point by any perpendicular to the x-axis.

Possibly more than one solution function for y as a function of x may be needed in order to specify all solutions of an equation (1). Thus, (1) might be equivalent to the statement

$$y = f(x) \qquad or \qquad y = g(x), \tag{2}$$

involving two solution functions f and g. In such a case, (1) does *not* define y as a function of x. Under favorable circumstances, as in examples below, the equivalent solution functions are obtained by solving (1) for y in terms of x.

The preceding discussion can be repeated with the roles of x and y interchanged. Thus, a function $k(y)$ will be called a *solution function for x as a function of y* in case (1) is satisfied by $x = k(y)$ for all values of y on the domain of k. Also, (1) is said to define x as a function of y if all solutions of (1) are given by $(x = k(y), y)$ for y on the domain of k, that is, in case (1) is equivalent to the equation $x = k(y)$. It may not be possible to find a single solution function $k(y)$ so that all solutions (x, y) of (1) are given by $(x = k(y), y)$. To obtain solution functions for x as a function of y, we attempt to solve (1) for x in terms of y.

ILLUSTRATION 2.   From $2x + 3y = 6$, on solving for y in terms of x, we find $y = -\frac{2}{3}x + 2$. Thus, we obtain the linear solution function $\left(-\frac{2}{3}x + 2\right)$ for y as a function of x, and $2x + 3y = 6$ is equivalent to the equation $y = -\frac{2}{3}x + 2$. Similarly, on solving for x in terms of y, we find that the given equation also defines x as a function of y, and is equivalent to the equation $x = -\frac{3}{2}y + 3$. The graph of either of the functions, thus obtained, in an xy-plane is the line which is the graph of $2x + 3y = 6$.

ILLUSTRATION 3.   The graph of $x = y^2$ is the parabola in Figure 26. The equation $x = y^2$ defines x as a function of y, because just one value of x corresponds to each value of y. Thus, the parabola in Figure 26 is met in just one point by any line perpendicular to the y-axis. From $x = y^2$,

$$y = +\sqrt{x} \qquad OR \qquad y = -\sqrt{x}.$$

Hence, "$x = y^2$" does NOT define y as a function of x, because the equation has the two solution functions $+\sqrt{x}$ and $-\sqrt{x}$ for y. The graph of $y = \sqrt{x}$ is the part of the parabola given as an unbroken curve *above* the x-axis in Figure 26. The graph of $y = -\sqrt{x}$ is the broken curve *below* the x-axis. The graph of $x = y^2$ is the *union* of the graphs of $y = \sqrt{x}$ and $y = -\sqrt{x}$.

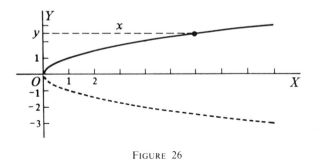

FIGURE 26

An equation such as $y = f(x)$ is said to define y *explicitly* as a function of x, because the equation describes the solution function f without added investigation. Similarly, an equation $x = g(y)$ defines x *explicitly* as a function of y. On the other hand, any consistent equation $H(x,y) = 0$, under favorable circumstances, is said to define either x or y *implicitly* as a function of the other variable, because no solution function is given explicitly. We use this phraseology even when it may prove impossible or inconvenient to obtain from $H(x,y) = 0$ one or more equivalent equations defining y, or x, *explicitly* as a function of the other variable. In calculus, implicit functions are met in important situations.

Illustration 4.   The equation $x = y^2$ defines x *explicitly* as a function of y, and defines y *implicitly* as a function of x.

EXERCISE 17

By solving for y in terms of x, obtain the function of x defined by the equation. Also, find the function of y defined by the equation.

1. $2x - 3y = 8$.   **2.** $4x + 5y = 7$.   **3.** $6x - 3y = 4$.
4. Find the function of x defined by the equation $3y + 3x^2 - 12x = 5$. Then graph it with $x = 2$ used in the table of solutions (x,y).
5. Find the function of y defined by $3x - 3y^2 + 18y = 8$. Then graph it with $y = 3$ used in the table of solutions (x,y).
6. Find the function of y defined by the equation $2x - 4y^2 + 8y = 7$. Then graph it with $y = 1$ used in the table of solutions (x,y).
7. Find the function of x defined by the equation $y - 2x^3 + 3x^2 + 12x = 8$. Then graph it, with $x = 2$ and $x = -1$ used in the table of solutions (x,y).

Obtain equations defining two solution functions for y as a function of x. Then graph the given equation by use of the solution functions.

8. $x^2 = 4y^2$.   **9.** $9x^2 - 25y^2 = 0$.   **10.** $4y^2 = 9x^2$.
11. Graph the equation $y = 4x^2$. Does it define y as a function of x; x as a function of y?
12. Graph the equation $x = y^3$. Does it define y as a function of x; x as a function of y?

29. SLOPE OF A LINE

In an xy-plane where the scale units on the coordinate axes may be unequal, consider a line L which is not vertical. Let $P_1:(x_1,y_1)$ and

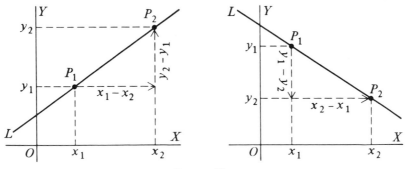

Figure 27

$P_2:(x_2, y_2)$ be any two distinct points on L. If we move on L from P_1 to P_2 as in Figure 27, the ordinate changes from y_1 to y_2, by an amount $(y_2 - y_1)$. Meanwhile, the horizontal coordinate changes from x_1 to x_2, by an amount $(x_2 - x_1)$. The property which distinguishes a line from all other sets of points in the xy-plane is the fact that the ratio $(y_2 - y_1)/(x_2 - x_1)$ is a constant, for all choices of P_1 and P_2. Then, this ratio is given a name, as follows, and is illustrated in Figure 27.

*DEFINITION IV.   The **slope**, m, of a nonvertical line L in an xy-plane is the ratio of the change $(y_2 - y_1)$ in the ordinate to the change $(x_2 - x_1)$ in the abscissa as we move from any point $P_1:(x_1, y_1)$ to a distinct second point $P_2:(x_2, y_2)$ on L. That is,*

$$\text{\textbf{slope}} = m = \frac{y_2 - y_1}{x_2 - x_1}. \tag{1}$$

ILLUSTRATION 1.   To find the slope of the line through $H:(-3, 4)$ and $K:(2, -5)$, we may use (1) with H as P_1 and K as P_2, or with K as P_1 and H as P_2, because the alteration in notation merely multiplies both numerator and denominator in (1) by -1. Thus,

$$m = \frac{-5 - 4}{2 - (-3)} = -\frac{9}{5}; \quad or \quad m = \frac{4 - (-5)}{-3 - 2} = -\frac{9}{5}.$$

The slope of a nonvertical line L is *positive* if, colloquially, L "*slopes upward*" to the right, and is *negative* if L "*slopes downward*" to the right, as illustrated in Figure 27. Slope is not defined for a vertical line, or a vertical line is said to have *no slope*. In the terminology of surveying,

$$(slope\ of\ a\ nonvertical\ line\ in\ the\ xy\text{-}plane) = \frac{\textbf{rise}}{\textbf{run}},$$

where the rise, $(y_2 - y_1)$, and the run, $(x_2 - x_1)$, are measured from any first point $P_1:(x_1, y_1)$ to any distinct second point $P_2:(x_2, y_2)$ on the line.

We accept the fact that two lines L_1 and L_2 in an xy-plane are parallel when and only when L_1 and L_2 have the same slope.

Note 1.   Hereafter, if A and B are given points, "AB" will represent the whole line through A and B, except when the context shows that AB means the line segment from A to B.

EXAMPLE 1.   With $R:(2, -3)$, $S:(5, -6)$, $U:(4, 7)$, and $V:(7, 4)$, determine whether or not RS is parallel to UV.

Solution.   The slopes m_1 and m_2 of RS and UV, respectively, are

$$m_1 = \frac{-6 - (-3)}{5 - 2} = -1; \quad m_2 = \frac{7 - 4}{4 - 7} = -1.$$

Since $m_1 = m_2$, RS and UV are parallel.

To state that certain points are **collinear** means that the points lie on a line. Suppose, now, that three distinct points R, S, and T are not on a vertical line. Then, the points are *collinear* when and only when the line through RS is the same as the line through ST, and thus RS and ST have the same slope.

ILLUSTRATION 2. To prove that $A:(-2, -4)$, $B:(-1, -1)$, and $C:(2, 8)$ are collinear, we compute

$$(slope\ of\ AB) = \frac{-1 + 4}{-1 + 2} = 3; \qquad (slope\ of\ BC) = \frac{8 + 1}{2 + 1} = 3.$$

Hence, A, B, and C are collinear.

We emphasize the fact that all content of this section applies in an xy-plane where the scale units on the axes may be *unequal*.

30. PERPENDICULARITY OF LINES

In an xy-plane, when angles between lines are being considered, it is agreed that *equal scale units* will be used on the coordinate axes.

Let L be any nonvertical line with slope m through the origin, and let $P:(x,y)$ be any point on L other than the origin. Then, we shall need the fact that

$$m = \frac{y}{x}. \tag{1}$$

Proof of (1). The points $O:(0,0)$ and $P(x,y)$ are on L. Hence, by use of (1) on page 74, we have

$$m = \frac{y - 0}{x - 0} \qquad or \qquad m = \frac{y}{x}.$$

THEOREM II. *In an xy-plane, let L_1 and L_2 be lines with slopes m_1 and m_2, respectively. Then, to say that L_1 and L_2 are perpendicular is equivalent to stating that*

$$m_1 m_2 = -1, \quad or \quad m_1 = -\frac{1}{m_2}, \quad or \quad m_2 = -\frac{1}{m_1}. \tag{2}$$

Proof. 1. If L_1 and L_2 do not intersect at the origin, lines respectively parallel to L_1 and L_2 can be drawn through the origin whose slopes and angles of intersection are the same as those of L_1 and L_2. Hence, for convenience, we shall assume that L_1 and L_2 intersect at the origin, as in Figure 28. Suppose that L_1 and L_2 have slopes m_1 and m_2, respectively, and are perpendicular. Then, neither L_1 nor L_2 is *horizontal* or *vertical;* one line, say L_1, has *positive* slope and the other line, L_2, has *negative* slope, as in Figure 28.

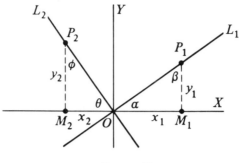

FIGURE 28

2. On L_1, choose $P_1:(x_1,y_1)$ with $x_1 = 1$. On L_2, choose $P_2:(x_2y_2)$ such that $y_2 = 1$. Then, by (1),

$$m_1 = \frac{y_1}{x_1} = \frac{y_1}{1} = y_1 \quad and \quad m_2 = \frac{y_2}{x_2} = \frac{1}{x_2}. \tag{3}$$

3. Construct right Δ's OM_1P_1 and OM_2P_2. Then, a statement that L_1 and L_2 are perpendicular is equivalent to saying that, in these triangles,*

$$\alpha = \phi \quad and \quad \beta = \theta. \tag{4}$$

4. In Δ's OP_1M_1 and P_2OM_2, we have equalities (4) and $\overline{OM_1} = \overline{M_2P_2} = 1$. Hence the Δ's are congruent, with corresponding sides of equal lengths (positive). Therefore

$$|\overline{M_1P_1}| = |\overline{OM_2}|, \quad or \quad y_1 = -x_2 \tag{5}$$

because the x-coordinate of P_2 is *negative*. Then $x_2 = -y_1$ and, from (3),

$$m_2 = \frac{1}{x_2} = -\frac{1}{y_1} = -\frac{1}{m_1},$$

which proves (2).

ILLUSTRATION 1.   If a line L_1 has the slope $m_1 = \frac{2}{3}$ and L_2 is perpendicular to L_1, then the slope of L_2 is $m_2 = -1 \div \frac{2}{3} = -\frac{3}{2}$.

EXAMPLE 1.   Prove that the line AB through $A:(-1,3)$ and $B:(2,-4)$ is perpendicular to the line CD through $C:(2,1)$ and $D:(9,4)$.

Solution.   Let the slopes of AB and CD be m_1 and m_2, respectively. From (1) on page 74,

$$m_1 = \frac{-4-3}{2+1} = -\frac{7}{3}; \quad m_2 = \frac{4-1}{9-2} = \frac{3}{7}.$$

Since $\frac{3}{7} = -1 \div \left(-\frac{7}{3}\right) = 1 \cdot \frac{3}{7}$, AB and CD are perpendicular.

*Read $\{\alpha,\beta,\phi,\theta\}$ as $\{alpha, beta, phi, theta\}$.

EXERCISE 18

Obtain the slope of the line L through the points, and the slope of a line perpendicular to L.

1. $(-1, 3); (4, 2)$.   **2.** $(-3, -2); (-1, 5)$.   **3.** $(4, 1); (-2, 0)$.
4. $(-1, 0); (3, -2)$.   **5.** $(4, 3); (-1, -2)$.   **6.** $(6, -4); (-2, 1)$.

Prove that the lines UV and RS are parallel, or that they are perpendicular.

7. $R:(2, 4); S:(0, 0); U:(-1, 2); V:(0, 4)$.
8. $R:(-2, -3); S:(-3, -5); U:(3, -2); V:(1, -1)$.
9. $R:(2, 6); S:(2, 3); U:(1, -5); V:(2, -5)$.

Prove that the points are collinear.

10. $(1, 2); (2, 4); (0, 0)$.   **11.** $(-1, 2); (1, 3); (5, 5)$.
12. $(1, 4); (2, 7); (0, 1)$.   **13.** $(-4, -6); (0, -7); (-8, -5)$.

Find u, or v, if the points are collinear.

14. $(2, 3); (1, 5); (u, 2)$.   **15.** $(4, -1); (2, -2); (3, v)$.
16. Prove that $(-3, 2), (3, 5), (6, -1)$, and $(0, -4)$ are vertices of a square.
17. Prove that $(3, 2), (0, 8)$, and $(1, 1)$ are vertices of a right triangle.
18. Prove that $(-1, 2)$, $(0, -1)$, $(3, 3)$, and $(4, 0)$ are vertices of a parallelogram.

31. STANDARD EQUATIONS FOR LINES IN AN xy-PLANE

In an xy-plane, if a vertical line L has the x-intercept a, the equation of L is $x = a$. If a horizontal line L has the y-intercept b, the equation of L is $y = b$. We state these results as corresponding standard forms (1) and (2) below:

Equation of line parallel to x-axis: $y = b$.   (1)

Equation of line parallel to y-axis: $x = a$.   (2)

The line L through $P_1:(x_1, y_1)$ with slope m has the equation

(point-slope form) $y - y_1 = m(x - x_1)$.   (3)

Proof.   1.   Equation (3) is satisfied when $(x = x_1, y = y_1)$, and hence P_1 is on the graph of (3).

2. In Figure 29, let $P:(x, y)$ be any point other than P_1. By (1) on page 74, the slope of $P_1 P$ is $(y - y_1)/(x - x_1)$. Line L is the set of points consisting of P_1 and all other points P for which m is the slope of $P_1 P$, or for which

$$m = \frac{y - y_1}{x - x_1}, \quad or \quad y - y_1 = m(x - x_1).$$

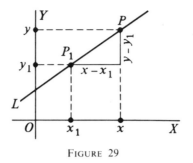

FIGURE 29

Hence, (3) is the equation of L because (3) is satisfied by the coordinates of P_1, also of all other points on L, and of no other points.

ILLUSTRATION 1.   The equation of the line with slope 2 through $(-2, 3)$ is

$$y - 3 = 2[x - (-2)], \quad or \quad y = 2x + 7.$$

The line L with slope m and y-intercept b has the equation

(**slope-intercept form**) $y = mx + b.$ (4)

Proof. L meets the y-axis at the point with coordinates $(0, b)$. From (3) with $x_1 = 0$ and $y_1 = b$, the equation of L is

$$y - b = m(x - 0), \quad or \quad y = mx + b.$$

ILLUSTRATION 2.   The line with slope -2 and y-intercept 5 is $y = -2x + 5$.

With an equation $Ax + By = C$, where $B \neq 0$, we may solve for y and change the equation to the form (4). Thus, we show that the graph of the given equation is a *line*, whose slope is the coefficient obtained for x, and y-intercept is the value for b in the equation of type (4) which is obtained.

EXAMPLE 1.   Show that $2x - 3y = 6$ is the equation of a line, and find the y-intercept and the slope of this line.

Solution.   Solve for y in terms of x:

$$3y = 2x - 6; \quad y = \frac{2}{3}x - 2.$$ (5)

On comparing (5) with (4), we see that $2x - 3y = 6$ is the equation of a line with slope $\frac{2}{3}$ and y-intercept -2.

EXAMPLE 2.   Find the line L through the points $(1, 2)$ and $(-4, 4)$.

Solution.   *"To find a line,"* means *"to find its equation."* The slope, m, of the line is

$$m = \frac{4 - 2}{-4 - 1} = -\frac{2}{5}.$$

By use of (3) with $P_1:(1,2)$, the equation of L is

$$y - 2 = -\frac{2}{5}(x - 1), \qquad or \qquad 2x + 5y = 12.$$

ILLUSTRATION 3.   To find the line L through $(-2,5)$ and $(-2,7)$, we cannot use (3) because L is vertical and hence has no slope, but (2) applies. The equation of L is $x = -2$.

EXAMPLE 3.   Obtain the line L through $(-2,3)$, if the y-intercept of L is 4.

Solution.   The point $(0,4)$ is on L. By use of $(-2,3)$ and $(0,4)$, the slope of L is

$$m = \frac{4 - 3}{0 + 2} = \frac{1}{2}.$$

Hence, by use of (4), the equation of L is $y = \frac{1}{2}x + 4$.

EXAMPLE 4.   Find the equation of the line L through $U:(-2,5)$ which is perpendicular to the line through $V:(3,2)$ and $W:(-2,6)$.

Solution.   The slope of the line VW is $(6 - 2)/(-5)$, or $-4/5$. Hence, the slope, m, of L is the negative reciprocal of $-4/5$, or $m = 5/4$. By use of the point-slope form, the equation of L is

$$y - 5 = \frac{5}{4}(x + 2), \qquad or \qquad 4y = 5x + 30.$$

If $F(x) = b$, a *constant,* for all values of x, the graph of the constant function F is the line $y = b$, whose slope is zero and y-intercept is b. If F is a *linear* function of x, then

$$F(x) = mx + b, \qquad where \ m \neq 0,$$

and the graph of F is the line $y = mx + b$ with slope $m \neq 0$ and y-intercept b. Thus, if $F(x) = mx + b$, *the graph of the linear or constant function F is the line $y = mx + b$ with slope m and y-intercept b.*

EXAMPLE 5.   Find the equation of the line L through $(2,-3)$ perpendicular to the line $H:\{3x + 2y = 6\}$.

Solution.   1.   Write H in the slope-intercept form to obtain its slope:

$$2y = 6 - 3x; \qquad y = -\frac{3}{2}x + 3.$$

Hence, the slope of H is $-3/2$; the slope of L is $2/3$.

2.   By the point-slope form, the equation of L is

$$y + 3 = \frac{2}{3}(x - 2), \qquad or \qquad 3y = 2x - 13.$$

32. GENERAL LINEAR EQUATION IN x AND y

Any linear equation in x and y is equivalent to an equation $Ax + By + C = 0$ where A and B are not both zero.

THEOREM III.   *The graph of any linear equation in x and y is a line.*

Proof.   1. In $Ax + By + C = 0$, if $B \neq 0$ we may solve for y and obtain

$$By = -Ax - C, \qquad or \qquad y = -\frac{A}{B}x - \frac{C}{A}, \tag{1}$$

which is the equation of a *line* with slope $-A/B$ and y-intercept $-C/A$.

2. If $B = 0$, then $A \neq 0$ and the linear equation becomes $Ax + C = 0$, or $x = -C/A$. This is the equation of a vertical line with x-intercept $-C/A$. Hence, in all cases, the graph of $Ax + By + C = 0$ is a *line*.

To graph a particular line $L:\{Ax + By + C = 0\}$, it is not desirable to change to any one of the standard forms of Section 31. To obtain L, find two points on it (three if a check is desired), and draw the graph.

EXERCISE 19

Write an equation for the line with given slope m, or satisfying other specified conditions.

1. $m = -2$; y-intercept 5.
2. $m = 3$; y-intercept -4.
3. Horizontal; y-intercept 6.
4. Vertical; x-intercept 1.
5. Through $(2,-4)$; $m = 5$.
6. Through $(-1,-1)$; $m = 4$.
7. Through $(-1,2)$; $m = -1$.
8. Through $(0,0)$; $m = -2$.
9. Through $(5,0)$; $m = 2$.
10. Through $(-6,1)$; $m = 0$.
11. Vertical; x-intercept -2.
12. Horizontal; y-intercept -3.
13. Slope $-2/3$; x-intercept 3.
14. Slope 5; x-intercept -2.
15. Through $(4,3)$ and $(-2,1)$.
16. Through $(-4,0)$ and $(3,5)$.
17. Through $(0,0)$ and $(-4,5)$.
18. Through $(0,0)$ and $(3,6)$.
19. Through $(1,2)$ and $(-3,5)$.
20. Through $(0,3)$ and $(-2,6)$.
21. Through $(2,0)$ and $(-1,-5)$.
22. Through $(-2,1)$ and $(-2,4)$.
23. x-intercept 4; y-intercept 2.
24. x-intercept -2; y-intercept 6.

Show that the points lie on a line and find its equation.

25. $(-2,1)$; $(3,0)$; $(-7,2)$.
26. $(-1,-3)$; $(2,-4)$; $(-4,-2)$.

Find the lines through W which are, respectively, parallel to and perpendicular to UV.

27. $U:(-2,3)$; $V:(0,4)$; $W:(1,-2)$.
28. $U:(1,-3)$; $V:(2,0)$; $W:(3,5)$.

Write the line in the slope-intercept form to find the slope and y-intercept.

29. $3x + 2y = 5$. **30.** $2y - 4x = 7$. **31.** $x - 3y = 5$.

Find the lines through the point which are, respectively, parallel to and perpendicular to the given line.

32. $3y - 2x = 5; (-2,4)$. **33.** $x + 4y = 6; (-1,-3)$.
34. $-2x + 6y = 7; (-1,2)$. **35.** $2x - 5y = 1; (-2,-5)$.
36. $y - 5 = 0; (3,-1)$. **37.** $2x - 6 = 0; (0,-3)$.

38. By use of (3) on page 77, show that the line L through distinct points $P_1:(x_1,y_1)$ and $P_2:(x_2,y_2)$, not on a vertical line, has the following equation, called the **two-point form** for L.

$$y - y_1 = \frac{y_2 - y_1}{x_2 - x_1}(x - x_1). \tag{1}$$

Then solve Problems 19 and 20 by use of (1).

39. A line L has the x-intercept $a \neq 0$ and y-intercept $b \neq 0$. Show that L has the following equation, called the **intercept form** for L:

$$\frac{x}{a} + \frac{y}{b} = 1. \tag{2}$$

Then, solve Problems 23 and 24 by use of (2).

Note 1. The *point-slope* and the *slope-intercept* forms for equations of lines are sufficient and convenient for all applications in calculus. Hence, (1) and (2) are not recommended for memorization.

40. Graph each linear function: $(3x - 5); -6x$.
41. Prove that the lines intersect to form a parallelogram.

$$4x - 6y = 7; \quad 8x + 10y = 5; \quad 3y - 2x = 3; \quad 4x + 5y = 6.$$

33. SYSTEMS OF LINEAR EQUATIONS

Let f, g, h, and k represent functions of the variables x and y, and consider the system of equations

$$f(x,y) = g(x,y) \quad and \quad h(x,y) = k(x,y). \tag{1}$$

A *solution* of (1) is a pair of numbers (x,y) which is a solution of both equations. In solving (1), we say that we are solving the equations *simultaneously*. If (1) has at least one solution, then (1) is said to be *consistent*. Otherwise (1) is called *inconsistent*.

EXAMPLE 1. Solve algebraically: $\begin{cases} x - y = 6, & (2) \\ x + 2y = 3. & (3) \end{cases}$

Solution.   On subtracting corresponding sides of the equations, in the order (2) from (3), we find $3y = -3$ or $y = -1$. From (2) with $y = -1$, we obtain $x = 5$. Hence, the solution of the system is ($x = 5$, $y = -1$).

To find the real solutions of any system (1) graphically, graph both of the equations on the same xy-plane. Then, the coordinates (x,y) of each point of intersection of the graphs is a solution of system (1).

ILLUSTRATION 1.   To solve system [(2),(3)] graphically, we graph (2) and (3) in Figure 30. The lines which are the graphs intersect at $P:(x = 5, y = -1)$, which is the solution found in Example 1.

EXAMPLE 2.   Discuss the system:

$$\begin{cases} 3x + 5y = -7, & (4) \\ 6x + 10y = 15. & (5) \end{cases}$$

Solution.   1. *To solve algebraically,* multiply by 2 in (4):

$$6x + 10y = -14. \qquad (6)$$

Then, subtract each side in (6) from the corresponding side in (5). This gives $0 = 29$. Hence, if (4) and (5) have a solution (x,y), we obtain the *contradictory* result $0 = 29$. Therefore, system [(4),(5)] has no solution, or the equations are inconsistent.

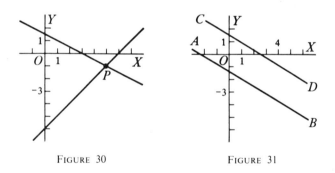

FIGURE 30   FIGURE 31

2. *Graphical solution.* The graph of (4) is line AB, and of (5) is line CD in Figure 31. These lines appear to be parallel, but this must be checked. The slope-intercept forms of (4) and (5) are, respectively,

$$y = -\frac{3}{5}x - \frac{7}{5} \quad and \quad y = -\frac{3}{5}x + \frac{3}{2}.$$

where the slopes are the same, but the y-intercepts are different. Hence, the graphs AB and CD of (4) and (5) are distinct* parallel lines which thus

---

*If the graphs of two linear equations in x and y are the *same line*, we may refer to the graphs as *coincident parallel lines.*

do not intersect. Therefore, system [(4),(5)] has no solution, or is inconsistent.

If each line of a set of lines in an xy-plane passes through a given point P, the lines are called **concurrent.** Three lines are concurrent when and only when their equations form a consistent system. To investigate its consistency, solve any two of the equations simultaneously; then test their solution, if one exists, by substitution in the third equation. If it is satisfied, the lines are concurrent.

EXAMPLE 3.   Find the intersection of the following lines, or prove that they are not concurrent:

$$\begin{cases} 2y - x = 4. & (7) \\ 3x + y = 9. & (8) \\ x + 2y = 7. & (9) \end{cases}$$

Solution.   1. On solving the system [(7),(8)], we obtain the solution $(x = 2, y = 3)$.

2. Substitution of $(x = 2, y = 3)$ in (9) gives $2 + 6 = 7$, or $8 = 7$. Hence, (9) is not satisfied; the given equations have no solution, and thus their graphs are not concurrent.

EXAMPLE 4.   Find the value of the constant h in case the following lines are perpendicular:

$$3x + y = 5 \quad and \quad 2x + hy = 6. \qquad (10)$$

Solution.   In the slope-intercept form, equations (10) become, respectively,

$$y = -3x + 5 \quad and \quad y = -\frac{2}{h}x + \frac{6}{h}. \qquad (11)$$

Hence, the slopes are -3 and $-2/h$. The lines are perpendicular when

$$-\frac{2}{h} = -\frac{1}{-3}, \quad or \quad -\frac{2}{h} = \frac{1}{3}. \qquad (12)$$

From (12), $h = -6$.

Suppose that a linear equation in x and y involves a third variable k, with domain D, so that one and only one line L_k is obtained as the graph of the equation for each number k in D. Then we may refer to the set of lines $\{L_k\}$ as *a family of lines,* with k as the *parameter.*

ILLUSTRATION 2.   The family of nonvertical lines through the point $(3,2)$ has the equation $y - 3 = k(x - 2)$. The value of the parameter k for any line L_k of the family is the slope of L_k.

EXERCISE 20

Solve graphically and algebraically.

1. $\begin{cases} 2x + y = 2, \\ 2y + 3x = 5. \end{cases}$
2. $\begin{cases} y - x = 3, \\ 3y + 4x = 2. \end{cases}$
3. $\begin{cases} 4x - y = 4, \\ 2x + 3y = 9. \end{cases}$

4. $\begin{cases} 2x + 5y = 10, \\ 10y + 2x = 8. \end{cases}$
5. $\begin{cases} 2x + 5 = 0, \\ 4x - y = -6. \end{cases}$
6. $\begin{cases} x + 3y = 6, \\ 2x + 6y = 12. \end{cases}$

7. $\begin{cases} 3x + 2y - 5 = 0, \\ 6x = 3 - 4y. \end{cases}$
8. $\begin{cases} 2x + 3y = 5, \\ 2y - 3 = 0. \end{cases}$
9. $\begin{cases} 3y - 5 = 0, \\ 2x + 7 = 0. \end{cases}$

10. Graph each of the statements and describe the graph:

$$2x + y = 5 \text{ or } x - y = 3; \qquad 2x + 5y = 5 \text{ and } x - y = 3.$$

Prove that the lines with the given equations are concurrent, or that the lines are not concurrent.

11. $\begin{cases} 4x + y = 4. \\ y + 3x = 5. \\ 2x - y = 3. \end{cases}$
12. $\begin{cases} 2y - x = 3. \\ 3y + 2x = 1. \\ 2x - 5y = -7. \end{cases}$

With k as a parameter, graph the lines of the given family for two values of k. State a characteristic geometric property of the family of lines.

13. $y = kx + 3.$ 14. $y = 2x + k.$ 15. $y + 3 = k(x - 2).$

16. Find the line of the family in Problem 13 passing through the point $(x = 2, y = 6)$.

★17. By considering slopes and y-intercepts, prove that, if lines

$$L_1: \{A_1x + B_1y + C_1 = 0\} \qquad and \qquad L_2: \{A_2x + B_2y + C_2 = 0\}$$

coincide, there exists a constant k such that the equation of L_2 can be obtained by multiplying both sides of the equation of L_1 by k. Conversely, if the preceding fact is true, prove that L_1 and L_2 coincide.

★*Note 1.* Consider two nonparallel lines L_1 and L_2:

$$L_1: \{a_1x + b_1y + c_1 = 0\}; \qquad L_2: \{a_2x + b_2y + c_2 = 0\}. \tag{1}$$

Let W_1 and W_2 represent the left-hand members of the equations of L_1 and L_2, respectively. Then, if k is any number, the equation $W_1 + kW_2 = 0$ is linear in x and y, and represents a line H through the intersection of L_1 and L_2. Conversely, any line H through the intersection of L_1 and L_2 (except L_2 itself) can be obtained from $W_1 + kW_2 = 0$ by using the proper value for k. Hence $W_1 + kW_2 = 0$ is called an equation for the *family* of lines through the intersection* of L_1 and L_2.

★*Solve by use of Note 1, without first finding the intersection of the given lines.*

18. Find the line through the point $(3, -2)$ and the intersection of

$$L_1:\{2x - 3y + 4 = 0\} \qquad and \qquad L_2:\{4x + y - 2 = 0\}.$$

Hint. Substitute $(x = 3, y = -2)$ in $W_1 + kW_2 = 0$, as in Note 1.

19. Find the line with slope 1 through the intersection of L_1 and L_2 of Problem 18.

20. In the triangle formed by the following lines, find the equations of the altitudes from the vertices perpendicular to the opposite sides, without finding the vertices.

$$2x + 3y = 5; \qquad x + 2y = 7; \qquad 5x + 2y = 2.$$

Then prove that the altitudes meet in a common point.

21. For any triangle, prove the property of the altitudes which was verified in Problem 20 for a particular triangle.

Hint. Let the vertices be $(-a, 0)$, $(0, b)$, and $(c, 0)$.

*In a plane, the set of all lines through a given point, or of all lines parallel to a given line, sometimes is called a *pencil* of lines.

chapter 4

Equations of Common Curves

34. PLANE SECTIONS OF A CONE

At the center of a circle, T, erect a line M perpendicular to the plane in which T lies. Select any point V on M, where V is not in the plane of T, as seen in Figure 32. From any point Q on T, draw a line L through V. Then, the set of points swept out by L as Q moves through all points on T is a surface of infinite extent called a complete *right circular cone* whose vertex is V and axis is M. Each position of L is called a *ruling* of the cone. V divides the cone into two parts, each called a *nappe*. In Figure 33, each nappe is cut off above and below V by a plane perpendicular to the cone's axis.

If a plane cuts the cone, the intersection of the plane and the cone is called a **conic section,** or simply a **conic.** First, suppose that the plane does not pass through V. Then, if the plane cuts just one nappe and is not parallel to a ruling of the cone, the conic section AB, as in Figure 33, is called an **ellipse.** *The ellipse is a circle if the plane is perpendicular to the* axis of the cone. If the plane cuts just one nappe and is parallel to a ruling, the conic CDE, as in Figure 33, is called a **parabola.** If the plane cuts both nappes, the conic consists of two separated branches ABC and DEF, as in Figure 34, and is called a **hyperbola.**

If the cone is cut by a plane through the vertex V, the only point of

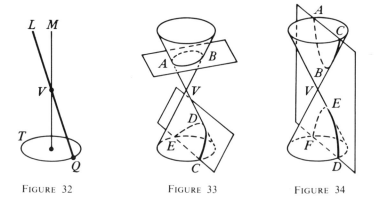

FIGURE 32   FIGURE 33   FIGURE 34

intersection may be V itself, so that the conic section is just this point. Or, the plane through V may be tangent to the cone along a ruling, and then the conic is just this line (thought of as two coincident lines because of the tangency). Or, the plane through V may cut the conic along two rulings, so that the conic then is two lines intersecting at V. These plane sections through V are called **degenerate conics.**

We shall call ellipses, hyperbolas, and parabolas the **nondegenerate conic sections.** We classify as degenerate conics any sets of points of the following natures: a single point; two lines in the same plane, where the lines may intersect in just one point, or be coincident, or be parallel and distinct. Two distinct parallel lines cannot be obtained as plane sections of a cone, but are included as degenerate conics for a certain algebraic reason which we shall not discuss. Hereafter, if we refer to a *conic section,* or simply to a *conic,* we shall mean a *nondegenerate conic,* that is, an ellipse, hyperbola, or parabola.

By use of the definitions of the conics as plane sections of a cone, certain geometric properties can be demonstrated for each variety of conic. The corresponding proofs are beyond the scope of this text. On the basis of the specified geometric properties, standard equations can be obtained for each type of conic, located in an *xy*-plane. The derivations of these equations will not be introduced until late in this chapter, except for the special case of a circle. At first, we shall accept (or, define) each type of conic as the graph of a certain variety of equation.

ILLUSTRATION 1.   A hyperbola consists of two separated parts, called *branches,* as seen in Figure 34 on page 86. Each branch is of *"infinite extent,"* because each nappe of the cone in Figure 34 extends without bound upward and downward. A limited part of a hyperbola is shown as branches AB and CD in Figure 35. Corresponding to any hyperbola, in its plane there exist two characteristic lines, shown as broken lines in Figure 35. If a point P on either branch of the hyperbola in Figure 35 recedes upward or downward beyond all limits, then the

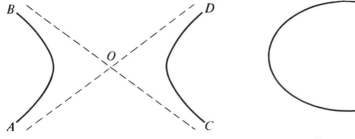

FIGURE 35   FIGURE 36

distance between P and a corresponding broken line approaches zero.*
A line with such a property is called an **asymptote** of the curve. Thus, a
hyperbola has two asymptotes, and each branch of the hyperbola approaches each asymptote. An ellipse is shown in Figure 36. A parabola
was met in Illustration 6 on page 61.

35. EQUATIONS FOR A CIRCLE

In this section and its later applications, it will be implied that the
scale units on the axes in the coordinate plane are equal, and that this unit
is used in measuring distance in any direction in the plane. With this
agreement, the distance formula of page 57 can be used.

A *circle* is defined as the set of points in a plane at a fixed distance
$r \geq 0$ from a specified point C in the plane, where r is called the radius
and C is called the center of the circle.

THEOREM I.   *In an xy-plane, the circle W with center C:(h,k) and
radius r has the equation*

$$(x - h)^2 + (y - k)^2 = r^2. \tag{1}$$

Proof.   1. In Figure 37, if $P:(x,y)$ is any point in the plane, by use
of the distance formula (3) on page 57 we obtain

$$|\overline{CP}|^2 = (x - h)^2 + (y - k)^2. \tag{2}$$

2. P is on the circle W when and only when $|\overline{CP}| = r$, or $\overline{CP}^2 = r^2$,
which gives

$$(x - h)^2 + (y - k)^2 = r^2$$

as the equation of W.

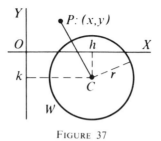

FIGURE 37

*Also, as proved in calculus, the slope of the tangent line at P approaches the slope
of the broken line.

We shall call (1) the **center-radius** form for the equation of a circle. If its center is (0,0) and radius is r, the equation of the circle, from (1), is

$$x^2 + y^2 = r^2. \tag{3}$$

ILLUSTRATION 1.   The equation of the circle with center $(-2,3)$ and radius 4 is

$$[x - (-2)]^2 + (y - 3)^2 = 16, \textit{ or}$$
$$x^2 + 4x + y^2 - 6y = 3.$$

If $r = 0$ in (1) or (3), the graph of the equation is a single point, the center of the corresponding circle, with radius $r = 0$. In such a case, we refer to the graph as a *point-circle*. Hereafter, if we refer to a graph simply as a circle, we allow for the possibility that it is a point-circle.

ILLUSTRATION 2.   The equation

$$(x - 1)^2 + (y - 2)^2 = -4 \tag{4}$$

is not satisfied by any real numbers (x,y), because the squares on the left are never negative. Hence, (4) has no graph, or the graph of (4) is the empty set. However, (4) can be looked upon as being in the form (1) with $r^2 = -4$, which would imply that $r = \sqrt{-4} = 2i$, which is imaginary. For this reason, if r^2 is negative in an equation of the form (1), it is said to represent an *imaginary circle*.

On expanding in (1), we obtain

$$x^2 + y^2 - 2hx - 2ky + (h^2 + k^2 - r^2) = 0. \tag{5}$$

In (5), let

$$D = -2h; \quad E = -2k; \quad F = h^2 + k^2 - r^2.$$

Then, from (5),

$$x^2 + y^2 + Dx + Ey + F = 0. \tag{6}$$

Hence, every circle has an equation of the form (6). Conversely, for any particular values of $\{D,E,F\}$, by completing a square with the terms in x, and separately with the terms in y, we can change any equation (6) to the form (1). Hence, for any values of $\{D,E,F\}$, (6) is the equation of a circle, real or imaginary. We refer to (6) as the **general form** for the equation of a circle.

EXAMPLE 1.   Obtain the center and radius of the circle.

$$x^2 + y^2 - 6x + 4y - 23 = 0. \tag{7}$$

Solution.   Recall that, to complete a square with $(x^2 + bx)$, we add $(\frac{1}{2}b)^2$. From (7),

$$(x^2 - 6x \quad) + (y^2 + 4y \quad) = 23. \tag{8}$$

To complete squares in (8), we add $[\frac{1}{2}(6)]^2$ or 9 with the terms in x, and $[\frac{1}{2}(4)]^2$ or 4 with the terms in y, and add both 4 and 9 on the right, to obtain

$$(x^2 - 6x + 9) + (y^2 + 4y + 4) = 36, \text{ or}$$
$$(x - 3)^2 + (y + 2)^2 = 36, \text{ or}$$
$$(x - 3)^2 + [y - (-2)]^2 = 36. \tag{9}$$

On comparing (9) with (1), we see that (7) represents a circle with center $(3, -2)$ and radius $r = \sqrt{36}$, or $r = 6$.

If $\{a, b, c, d\}$ are constants with $a \neq 0$, an equation

$$ax^2 + ay^2 + bx + cy + d = 0 \tag{10}$$

can be changed to the form (6) by dividing both sides in (10) by a. Hence, if an equation of the second degree in x and y is of the form (10) where the coefficients of x^2 and y^2 are *equal,* and no term in xy is present, then the graph of the equation is a circle, real or imaginary.

ILLUSTRATION 3.   The graph of

$$9x^2 - 18x + 9y^2 + 36y + 29 = 0 \tag{11}$$

is a circle. To obtain its center-radius form, divide by 9 in (11) and complete squares on the left:

$$x^2 - 2x + y^2 + 4y = -\frac{29}{9};$$

$$(x^2 - 2x + 1)^2 + (y^2 + 4y + 4) = 5 - \frac{29}{9};$$

$$(x - 1)^2 + (y + 2)^2 = \frac{16}{9}.$$

Hence, the graph of (11) is the circle with center $(1, -2)$ and radius $r = \sqrt{\frac{16}{9}}$, or $r = \frac{4}{3}$.

Either the center-radius form (1), or the general form (6) for a circle involves three constants. Hence, to obtain the equation of a circle, we must have enough data to write three equations involving the constants, if none of them is given.

Tangents to curves of general types are discussed in calculus. In this text, we shall refer to tangents on an intuitional basis. We accept the fact that the tangent to a circle *T* at any point *P* on *T* is the line through *P* perpendicular to the radius drawn from the center of *T* to *P*.

EXAMPLE 2.   Obtain an equation for the circle, *T*, with center *C*:(4,3) which is tangent to the *y*-axis.

Solution.   In Figure 38 we observe that the radius of *T* from *C* perpendicular to *OY* meets *OY* at (0,3). Hence, the radius of *T* is *r* = 4. From the center-radius form (1) with *h* = 4, *k* = 3, and *r* = 4, an equation for *T* is

$$(x - 4)^2 + (y - 3)^2 = 16.$$

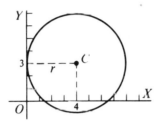

FIGURE 38

EXERCISE 21

Write an equation for the circle with center C and radius r.
1. $C:(3,4); r = 2.$ 2. $C:(-2,5); r = 3.$ 3. $C:(3,-2); r = 4.$
4. $C:(-2,-4); r = 3.$ 5. $C:(0,0); r = 4.$ 6. $C:(0,3); r = 3.$
7. $C:(-2,0); r = 0.$ 8. $C:(a,0); r = a.$ 9. $C:(0,b); r = b.$

If the graph is a real circle, find its center and radius.
10. $x^2 + 2x + y^2 - 4y = 4.$ 11. $x^2 - 6x + y^2 - 4y = 3.$
12. $x^2 + y^2 + 4x + 2y = -1.$ 13. $x^2 + y^2 + 6x + 3 = 4y.$
14. $x^2 + y^2 + 8y + 19 = 0.$ 15. $x^2 + y^2 - 6x = 11.$
16. $x^2 + y^2 + 5x + \dfrac{9}{4} = 6y.$ 17. $x^2 + y^2 + 6x + 39 = 4y.$
18. $4x^2 + 4y^2 + 16x - 8y = 5.$ 19. $9x^2 + 9y^2 + 18x + 29 = 36y.$

Obtain the equation of the circle satisfying the given conditions.
20. Center (3,0); tangent to the *y*-axis.
21. Center (0,-2); tangent to the *x*-axis.
22. Tangent to the line $y = 4$; center (2,7).
23. Tangent to the line $x = -3$; center (-5,4).
24. Tangent to the line $y = -2$; center (4,1).
25. Center (0,0) and passing through (4,-3).
26. Center (2,-3) and passing through (3,-1).

27. With (2,1) and (2,7) as the endpoints of a chord and radius 5.

28. With the x-intercepts -2 and 6, with radius 5.

29. Tangent to the x-axis at (2,0) and passing through (4,4).

★30. Find the intersections of the following line and circle by solving the two equations simultaneously. Check by graphing.

$$x^2 + 2x + y^2 + 4y = 15; \qquad x + y - 3 = 0.$$

Partial solution.   From the linear equation, $y = 3 - x$. On using this in the quadratic equation to eliminate y, we obtain

$$x^2 + 2x + (3 - x)^2 + 4(3 - x) = 15. \tag{1}$$

After solving (1) for x, corresponding values of y are obtained from $y = 3 - x$.

★31. Suppose that the equations

$$x^2 + y^2 + D_1 x + E_1 y + F_1 = 0 \qquad and$$

$$x^2 + y^2 + D_2 x + E_2 y + F_2 = 0$$

represent circles. Prove that the circles (*a*) are concentric if and only if $D_1 = D_2$ and $E_1 = E_2$; (*b*) are the same circle if and only if $D_1 = D_2, E_1 = E_2$ and $F_1 = F_2$.

★Note 1.   Consider any two nonconcentric circles T_1 and T_2:

$$T_1 : \{x^2 + y^2 + D_1 x + E_1 y + F_1 = 0\}; \tag{2}$$

$$T_2 : \{x^2 + y^2 + D_2 x + E_2 y + F_2 = 0\}. \tag{3}$$

Let W_1 and W_2 represent the left-hand sides of the equations of T_1 and T_2, respectively. Then, for any number $h \neq -1$, the equation

$$W_1 + hW_2 = 0 \tag{4}$$

is of the form (10) on page 90, and hence is the equation of some circle.* Let h be a parameter (a variable) whose domain is all real numbers except $h = -1$. Then, the student may prove that (4) is an equation for the family of all circles except T_2 through the intersections, if any, of T_1 and T_2. If $h = -1$ in (4), the equation is linear and hence represents a line, L, through the intersections, if any, of T_1 and T_2. Line L is called the **radical axis** of T_1 and T_2, even when they do not intersect. The equation of L arises naturally if (2) and (3) are solved simultaneously to find the intersections of T_1 and T_2.

    ★*Refer to Note 1 in solving the following problems.*

32. Find the intersections of the following circles algebraically, by first obtaining the equation of their radical axis, and then proceeding as in

---

*Possibly an imaginary circle, whenever a circle is mentioned.

Problem 30:

$$x^2 + y^2 + 20x - 12y + 111 = 0; \tag{5}$$

$$x^2 + y^2 + 12x - 16y + 75 = 0. \tag{6}$$

Then draw the circles and their radical axis.
33. Find the equation of the circle through $P:(6,11)$ and the intersections of (5) and (6) by use of (4), without using the results of Problem 32. Then, on the coordinate system for Problem 32, construct the circle obtained through P.

36. DOMAINS FOR THE VARIABLES AND INTERCEPTS

In graphing an equation $H(x,y) = 0$, it is necessary to exclude any values of x, or of y, for which the equation specifies that the values of the other variable would be imaginary or undefined. When possible and convenient, we solve $H(x, y) = 0$ for y in terms of x, and for x in terms of y. Then it may be possible to determine any excluded values of x or of y by inspection of $H(x,y)$ itself, and observation of the results obtained on solving $H(x,y) = 0$ for x and y. With experience, and in favorable cases, this process of solution may be merely *visualized*, to gain speed. In graphing $H(x,y) = 0$, the domains for x and for y consist of all real numbers except those which are excluded. Also, the x-intercepts and y-intercepts should be obtained at an early stage because frequently this information can be obtained easily.

EXAMPLE 1.   Graph the equation

$$4x^2 + 9y^2 = 36. \tag{1}$$

Solution.   1. *The intercepts.* If $x = 0$ in (1) then $y^2 = 4$ or $y = \pm 2$, the y-intercepts. Hence, the points $(0,2)$ and $(0,-2)$ are on the graph. If $y = 0$ then $x^2 = 9$ or $x = \pm 3$, the x-intercepts, and we have the points $(3,0)$ and $(-3,0)$ on the graph.

2. *Domains of the variables.* Solve (1) for y and for x:

$$9y^2 = 36 - 4x^2; \qquad y^2 = \frac{4}{9}(9 - x^2); \qquad y = \pm\sqrt{\frac{4}{9}(9 - x^2)}.$$

Hence, $$y = \frac{2}{3}\sqrt{9 - x^2} \quad or \quad y = -\frac{2}{3}\sqrt{9 - x^2}. \tag{2}$$

Similarly, $$x = \frac{3}{2}\sqrt{4 - y^2} \quad or \quad x = -\frac{3}{2}\sqrt{4 - y^2}. \tag{3}$$

To obtain real values for y in (2), we must have $9 - x^2 \geq 0$, or

$$x^2 \leq 9, \quad or \quad |x| \leq 3. \tag{4}$$

To obtain real values for x in (3), we must have $|y| \leq 2$. Therefore, in graphing (1), the domains for the variables are $\{-3 \leq x \leq 3\}$ and $\{-2 \leq y \leq 2\}$.

3. In addition to the intercept points, coordinates for other points on the graph were obtained by substituting values for x in (2), as given in the following table, which was the basis for the graph of (1) in Figure 39. The graph is an *ellipse*.

$x =$	-3	-2	-1	0	1	2	3
$y = \dfrac{2}{3}\sqrt{9 - x^2}$	0	1.5	1.9	2	1.9	1.5	0
$y = -\dfrac{2}{3}\sqrt{9 - x^2}$	0	-1.5	-1.9	-2	-1.9	-1.5	0

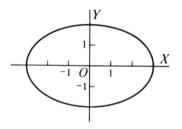

FIGURE 39

Comment.   In (2), we see that (1) has the irrational solution function $\frac{2}{3}\sqrt{9 - x^2}$ for y as a function of x; the graph of this function is the *upper half* of the ellipse in Figure 39. The *lower half* of the ellipse is the graph of the solution function $-\frac{2}{3}\sqrt{9 - x^2}$ for y as a function of x. Similarly, the graphs of the solution functions for x as a function of y in (3) are, respectively, the right-hand and left-hand halves of the ellipse.

In the preceding solution, the details in (2), (3), and (4) illustrate a routine for obtaining the domains of the variables in graphing an equation $H(x,y) = 0$. However, it may be inconvenient or impossible to solve for x or for y as in (2) and (3). Even when this can be done, it is desirable to be alert to the possibility of omitting the routine and obtaining the results more easily by observing special features of the given equation.

ILLUSTRATION 1. Consider obtaining the graph of $|x| + |y| = 3$. Without formality, observe that the x-intercepts are the solutions of $|x| = 3$, or $x = \pm 3$; the y-intercepts are $y = \pm 3$. Also, 3 is the *largest* possible value for $|x|$, so that the domain for x is $\{-3 \leqq x \leqq 3\}$. The student will obtain the graph in the next exercise.

In applications of graphs of equations in later courses in mathematics and in applied mathematics, the purpose for which a graph is drawn dictates the manner of its preparation. Sometimes it is essential to obtain the graph of an equation $H(x,y) = 0$ with great accuracy. In such a case, not only the general features of the graph should be shown clearly, but also numerous solutions (x,y) of the equation should be obtained in order to locate many points on the curve correctly. On some other occasion, it may be satisfactory to produce a graph which is merely *qualitatively* correct, in the sense that general features are shown, but only certain essential points on the curve are located accurately. In this course, the student should develop the ability to prepare graphs in either one of the two ways just mentioned. The solution of Example 1 illustrates the preparation of a meticulously accurate graph. If a graph of (1) were desired on an enlarged scale, (2) and (3) would enable us to calculate as many accurate points as desired. In succeeding sections of the text, emphasis will be placed frequently on methods for obtaining qualitatively accurate graphs quickly. In the present section, the emphasis is placed on obtaining meticulously accurate graphs.

37. SYMMETRY OF GRAPHS

Two points P and Q are said to be symmetric with respect to a line L in case L is the perpendicular bisector of the segment PQ, as in Figure 40.

In a plane, let us refer to an arbitrary set of points (of the variety to be considered) as a *curve, T*. Then, T is said to be symmetric with respect to a *line L* in case, for each point P in T, there is a point Q in T such that P and Q are symmetric to L. Then L is the perpendicular bisector of PQ. In such a case, L is called an *axis of symmetry* for T.

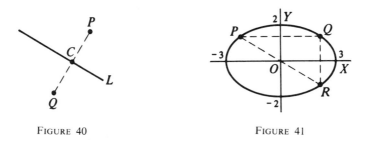

FIGURE 40   FIGURE 41

ILLUSTRATION 1. A circle, T, is symmetric with respect to any diameter. In Figure 41, P and Q are symmetric with respect to OY; Q

and R are symmetric with respect to OX; the ellipse has OX and OY as axes of symmetry.

Two points P and R are said to be symmetric with respect to a *point C* in case C is the midpoint of *PR*. Thus, in Figure 41, P and R are symmetric with respect to the origin. A curve T is said to be symmetric with respect to a *point C*, called a *center of symmetry*, in case, for each point P in T there is a point Q in T which is symmetric to P with respect to C. Then, C is the midpoint of segment PQ.

A *chord* of a curve T is defined as a line segment joining two distinct points P and Q in T. Then, to state that a point C is a center of symmetry for T means that every chord PQ of T through C is bisected by C.

The following facts are illustrated in Figure 42.

   I. *Points (x,y) and $(-x,y)$ are symmetric with respect to OY.*
  II. *Points (x,y) and $(x,-y)$ are symmetric with respect to OX.*
 III. *Points (x,y) and $(-x,-y)$ are symmetric with respect to the origin.*

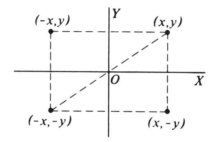

FIGURE 42

Usually, in any reference to symmetry, it will be inferred that symmetry to OX, OY, or the origin is involved. Each of the following tests for symmetry involves showing that, if $P:(x,y)$ is on the graph of an equation $H(x,y) = 0$, then the coordinates of the point symmetric to P also form a solution of $H(x,y) = 0$.

Tests for symmetry. *The graph of $H(x,y) = 0$ has the indicated symmetry when and only when an equivalent equation is obtained by making the specified substitution.*

Symmetry to the x-axis: *replace y by* $-y$.
Symmetry to the y-axis: *replace x by* $-x$.
Symmetry to the origin: *replace x by* $-x$, *and y by* $-y$.

   ILLUSTRATION 2.   Recall the graph of

$$4x^2 + 9y^2 = 36 \tag{1}$$

in Figure 39 on page 94. Equation (1) is unaltered if we replace x by $-x$, or y by $-y$, or if *both* replacements are made. Hence, the graph of (1), as in Figure 39, is symmetric to OX, to OY, and to the *origin* as a center of symmetry, which is called the *center of the ellipse.*

In Illustration 2 we met a special case of the fact that the graph of an equation $H(x,y) = 0$ is symmetric to OX if $H(x,y)$ involves y only with *even* exponents; to OY if $H(x,y)$ involves x only with *even* exponents; and to the *origin* if both x and y are involved only with *even* exponents. Also, it was seen that, if a graph is symmetric to both OX and OY, then the graph also is symmetric to the origin. Now, think of OX and OY as if they were *any* two perpendicular axes of symmetry. Then, we reach the following conclusions.

> When a curve T in a plane has two perpendicular axes
> of symmetry, their intersection is a center of symmetry   (2)
> for T.

Consider any ellipse as in Figure 43 in a given plane. At the moment, it is immaterial whether or not any coordinate system will be imposed on the plane. The ellipse has two axes of symmetry, hereafter to be called simply the *axes* of the ellipse. Suppose that it is intersected by one of its axes in points P and Q, and by the other axis in points R and S. Then, the *longer* of the segments PQ and RS is called the **major axis** of the ellipse. The *shorter* of the segments is called the **minor axis** of the ellipse. If the two segments PQ and RS are of *equal lengths,* the ellipse is a *circle*. In Figure 43, the major axis is RS and the minor axis is PQ. In Figure 44, the major axis of the ellipse is PQ and the minor axis is RS.

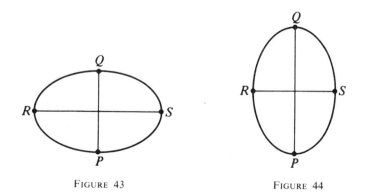

FIGURE 43   FIGURE 44

If an ellipse is located in an xy-plane, the appearance of the ellipse determines which of its axes is the major axis and which is the minor axis. The measures of these axes in terms of the scale units (possibly unequal) on the coordinate axes are immaterial.

Note 1.   The tests for symmetry are useful in checking the graph of an equation after the graph has been drawn.   Also, sometimes symmetry may be of aid in simplifying construction of a table of coordinates of points on the graph.

EXAMPLE 1.   Graph: $\qquad\qquad xy = 6.$ $\qquad\qquad$ (3)

Solution.   1. *Intercepts, and domains for x and y.*   If $x = 0$ in (3) then $0 = 6$, which is contradictory.   Hence, there is no point $P:(x,y)$ on the graph, T, of (3) with $x = 0$.   Or, there is no y-intercept and the y-axis does not intersect T.   Similarly there is no x-intercept for T.   From (3), $y = 6/x$; hence the domain for x is all real numbers except $x = 0$. Similarly, the domain for y is all real numbers except $y = 0$.

2. On substituting values for x in $y = 6/x$, we obtain the coordinates of points in the following table, as a basis for the graph in Figure 45.   The graph is a *hyperbola*.   With "$\rightarrow$" meaning "*approaches,*" in the table we read "$|y|$ *approaches plus infinity as* $|x| \rightarrow 0$," which means that $|y|$ grows large without bound as $x \rightarrow 0$ through positive values, or through

$y =$	$-.1$	-1	-6	-60	$\|y\| \rightarrow +\infty$	60	6	1	$.1$
$x =$	-60	-6	-1	$.1$	$\|x\| \rightarrow 0$	$.1$	1	6	60

negative values. These results correspond to the fact that the graph has the y-axis as an *asymptote*.   It is approached *upward* from the right, where $y \rightarrow +\infty$ as $x \rightarrow 0$ through positive values, and *downward* from the left, where $y \rightarrow -\infty$ as $x \rightarrow 0$ through negative values.   Similarly, from $x = 6/y$, we see that the x-axis also is an asymptote.

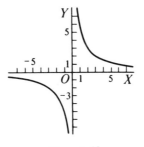

FIGURE 45

3. *Symmetry.*   If we replace x by $-x$ and y by $-y$, then equation (3) is unaltered, because $(-x)(-y) = xy$.   Therefore, if (x,y) is a solution of (3), then $(-x, -y)$ also is a solution.   Or, if $P:(x,y)$ is on the graph, then $Q:(-x, -y)$ also is on the graph, and it is symmetric to the origin as a center of symmetry.

EXERCISE 22

1. For each given point, plot the symmetric point with respect to each coordinate axis, and the origin: (3,2); (−4,3); (−2,−2); (3,−5).
2. Test each equation for the symmetry of its graph with respect to the coordinate axes and the origin. Do not graph the equation.

$$x^2 + 5y^2 = 9; \qquad 3xy + 4 = 0; \qquad x^2 = 7y.$$

Test the equation for the various types of symmetry of its graph, if it has one. Find any x-intercepts and y-intercepts of the graph. Calculate a table of representative solutions of the equation and draw its graph.

3. $xy = -4$ 4. $y = 2x^2$.   5. $x^2 + 4y^2 = 16$.
6. $x = y^2$.   7. $y = -\sqrt{4 - x^2}$.   8. $2xy = 5$.
9. $2x^2 + y^2 = -4$.   10. $x = \sqrt{9 - y^2}$.   11. $|x| + |y| = 3$.

Hint for Problem 11. Consider four parts of the graph separately. Thus, find solutions when $\{x \geq 0, y \geq 0\}$; when $\{x \leq 0, y \geq 0\}$; etc.

12. $|x| - |y| = 2$.   13.* $[x] = [y]$.   14. $y = |x - 1|$.

38. CENTRAL CONICS IN SIMPLE POSITIONS

Consider a consistent quadratic equation

$$Ax^2 + By^2 = C, \tag{1}$$

where $\{A,B,C\}$ are constants and $\{A,B\}$ are not both zero. Then, as in Illustration 2 on page 96, the tests for symmetry show that the graph, T, of (1) is symmetric to OX, OY, and the origin. Thus, T has the origin as a *center of symmetry,* which will be called the *center* of T. At this point in the chapter, we accept without discussion the facts about the graph of (1) which are mentioned in the following Cases I, II, and III. The remarks about a circle in Case I are a consequence of results in Section 35. Except for the statements about asymptotes in Case II, the facts about ellipses and hyperbolas in Cases I and II will be proved later in the chapter. Special instances of Case III were met in Exercise 15 on page 64. Except for the remarks about a circle in Case I, all of the facts in Cases I, II, III, and IV are true in an *xy*-plane where the scale units on the coordinate axes are *equal* or *unequal.*

Case I. Ellipse. *If $\{A,B,C\}$ are all positive or all negative, the graph of $Ax^2 + By^2 = C$ is an ellipse whose axes of symmetry are the coordinate axes, and center is the origin. This ellipse is a circle whose center is the origin if $A = B$ and the scale units are equal on the coordinate axes.*

---

*Recall that $[x]$ means *the greatest integer at most equal to x.*

Case II.   Hyperbola. *If all of* $\{A,B,C\}$ *are different from zero, with A and B of opposite signs, the graph of* $Ax^2 + By^2 = C$ *is a hyperbola whose axes of symmetry are the coordinate axes and center is the origin. An equation for the asymptotes of the hyperbola is obtained on replacing C by* 0 *in* (1):

Asymptotes: $Ax^2 + By^2 = 0.$ (2)

Case III.   Degenerate conics, or empty set ϕ **(no graph).** *If neither one of the conditions in I or II is satisfied, the graph of* $Ax^2 + By^2 = C$ *is a single point, the origin; or two lines parallel* to a coordinate axis, and possibly coincident on an axis; or two nonparallel lines intersecting at the origin; or the empty set* ϕ.

The graphs mentioned in Cases I, II, and III are called the **central conics** because, in each case, the conic has a center of symmetry. The conics obtained as graphs of (1) are said to be in their most simple positions in the *xy*-plane.

EXAMPLE 1.   Obtain a qualitatively accurate graph of

$$4x^2 + 9y^2 = 36.$$ (3)

Solution.   1. By Case I, the graph will be an ellipse.

2. If $x = 0$ then $y = \pm 2$, the *y*-intercepts. If $y = 0$ then $x = \pm 3$, the *x*-intercepts. This gives the four intercept points $(0,2)$, $(0,-2)$, $(3,0)$, and $(-3,0)$. Then, with appreciation of the shape of an ellipse, the graph of (3) is drawn through the intercept points, with a well rounded shape at the ends of the major axis. The graph is in Figure 39 on page 94.

EXAMPLE 2.   Obtain a qualitatively accurate graph of

$$4y^2 - 9x^2 = 36.$$ (4)

Solution.   1. By Case II, the graph will be a hyperbola.

2. *Intercepts.* If $x = 0$ then $y^2 = 9$ or $y = \pm 3$, and the *y*-intercept points are $(0,3)$ and $(0,-3)$. If $y = 0$ in (4) then $x^2 = -4$, or $x = \pm\sqrt{-4} = \pm 2i$. Thus, there are *no* *x*-intercepts, or we say that the *x*-intercepts are *imaginary*.

3. *The asymptotes.* By (2), an equation for the asymptotes is

$$4y^2 - 9x^2 = 0, \quad or \quad (2y - 3x)(2y + 3x) = 0,$$ (5)

which is equivalent to

$$2y - 3x = 0 \quad or \quad 2y + 3x = 0.$$ (6)

---

*In Case III, if the graph is two parallel lines, then each point on the line parallel to the preceding lines and midway between them is a center of symmetry. Thus, this type of degenerate conic has a *line of centers.*

The graph of $2y - 3x = 0$ is the broken line with positive slope in Figure 46, obtained by use of the solutions ($x = 2$, $y = 3$) and ($x = 0$, $y = 0$). The graph of $2y + 3x = 0$ is the broken line with negative slope in Figure 46. Notice that the asymptotes are the diagonals of the rectangle with vertices ($\pm2, \pm3$) as shown in Figure 46.

4. Each branch of the hyperbola was drawn through a y-intercept point, $(0,3)$ or $(0,-3)$, to approach the asymptotes gracefully in Figure 46.

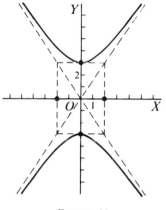

FIGURE 46

Comment.   To obtain a meticulously accurate graph of (4), as many accurate points of the graph as desired could be obtained by substituting values for x in (4) and computing y. Thus, if $x = \pm2$ in (4), then $4y^2 = 36 + 36 = 72$;

$$y = \pm\sqrt{18} = \pm\sqrt{9(2)} = \pm3\sqrt{2} = \pm4.2. \tag{7}$$

Hence, the four points ($\pm2, \pm4.2$) are on the graph.

Case IV.   *If $h \neq 0$, the graph of $xy = h$ is a hyperbola whose asymptotes are the coordinate axes.*

We met a special instance of Case IV in Example 1 on page 98.

EXAMPLE 3.   Obtain a qualitatively accurate graph of

$$xy = -6. \tag{8}$$

Solution.   1. By Case IV, the graph is a hyperbola with OX and OY as asymptotes. The branches of the hyperbola are in quadrants II and IV because xy is negative in (8), and hence x and y have opposite signs in any solution (x,y) of (8).

2. The following brief table of coordinates is a basis for the graph in Figure 47. Each branch was drawn through two accurate points to approach the asymptotes gracefully. The axes of symmetry of any hyperbola are the lines bisecting the angles made by the asymptotes. These axes are shown as broken lines in Figure 47.

$y =$	2	3	$\|y\| \to +\infty$	-3	-2
$x =$	-3	-2	$\|x\| \to 0$	2	3

When Case I applies with (1), and it thus represents an *ellipse,* the equation can be changed to the equivalent *"intercept form"*

$$\frac{x^2}{a^2} + \frac{y^2}{b^2} = 1, \tag{9}$$

where we agree to take a and b as the *positive* square roots of the *positive* denominators. The x-intercepts of (9) are $\pm a$; the y-intercepts are $\pm b$; the lengths of the axes of the ellipse are $2a$ and $2b$, respectively, as indicated in a graph of (9) in Figure 48.

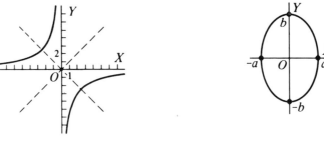

FIGURE 47   FIGURE 48

ILLUSTRATION 1.   To alter $9x^2 + 4y^2 = 36$ to the intercept form, divide both sides by 36, to obtain

$$\frac{x^2}{4} + \frac{y^2}{9} = 1. \tag{10}$$

The ellipse in Figure 48 may be considered as the graph of (10) with $a = 2$ and $b = 3$. The lengths of the axes of the ellipse (10) are 6 and 4, respectively. The major axis for (10) is vertical in Figure 48.

When Case II applies with (1), and it thus represents a *hyperbola,* the equation can be changed to one of the *"intercept forms"*

$$\frac{x^2}{a^2} - \frac{y^2}{b^2} = 1, \textit{ or} \tag{11}$$

$$\frac{y^2}{a^2} - \frac{x^2}{b^2} = 1, \tag{12}$$

where we agree to take $a > 0$ and $b > 0$ as the *positive* square roots of the *positive* denominators. The x-intercepts of (11) are $\pm a$ and the y-intercepts are imaginary. The y-intercepts of (12) are $\pm a$ and the x-intercepts are imaginary.

A graph of (11) is in Figure 49; the asymptotes have the equation

$$\frac{x^2}{a^2} - \frac{y^2}{b^2} = 0, \quad or \quad \left(\frac{x}{a} - \frac{y}{b}\right)\left(\frac{x}{a} + \frac{y}{b}\right) = 0. \tag{13}$$

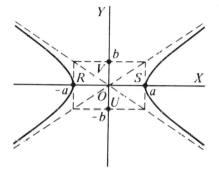

FIGURE 49

From (13), the asymptotes L_1 and L_2 are as follows:

$$L_1 : \left\{\frac{x}{a} - \frac{y}{b} = 0\right\}; \quad L_2 : \left\{\frac{x}{a} + \frac{y}{b} = 0\right\}. \tag{14}$$

The origin and the point $(x = a, y = b)$ determine L_1. The origin and the point $(x = a, y = -b)$ determine L_2. Hence, in Figure 49, the asymptotes are the diagonals of the rectangle shown with vertices $(\pm a, \pm b)$. This rectangle is called the *fundamental rectangle* for the hyperbola (11). One axis of symmetry for a hyperbola intersects it in two points. The line segment with these points as endpoints is called the **transverse axis** of the hyperbola. In Figure 49, the transverse axis is RS, and $\overline{RS} = 2a$. Notice that the transverse axis RS is a diameter of the fundamental rectangle. The diameter UV of the rectangle perpendicular to RS is called the **conjugate axis** of the hyperbola. The length of this axis in Figure 49 is $2b$. The transverse and conjugate axes lie on the axes of symmetry (the coordinate axes) of the hyperbola.

ILLUSTRATION 2.   To alter $4x^2 - 9y^2 = 36$ to its intercept form, divide both sides by 36:

$$\frac{x^2}{9} - \frac{y^2}{4} = 1, \tag{15}$$

where $a = 3$ and $b = 2$ in the form (11). From (15), the x-intercepts are ± 3, in Figure 49 if $a = 3$ and $b = 2$. The transverse axis is horizontal and is 6 units long; the conjugate axis is vertical and is 4 units long. The fundamental rectangle and its diagonals, the asymptotes, were drawn. Then each branch of the hyperbola was constructed through an x-intercept point to approach the asymptotes gracefully in Figure 49.

EXERCISE 23

By inspection, name the graph of the equation. Then draw its graph. Use equal scale units on the coordinate axes except where indicated otherwise. Specify the lengths of the major and minor axes of each ellipse. For each hyperbola, give the equations of its asymptotes and draw them. Use rapid methods where not directed otherwise, if permitted by the instructor, but then obtain a qualitatively accurate and well shaped graph.

Obtain the graph of the equation as specified.

1. $4x^2 + 25y^2 = 100$, by use of at least eight accurate points.

2. $x^2 - 4y^2 = 16$, by use of at least ten accurate points.

Obtain a qualitatively accurate graph, quickly.

3. $xy = -3$.   **4.** $xy = 8$.   **5.** $4x^2 - y^2 = 0$.
6. $x^2 - 4 = 0$.   **7.** $25y^2 - 4x^2 = 100$.   **8.** $9x^2 + 25y^2 = 225$.
9. $3x^2 + 9y^2 = 0$.   **10.** $2x^2 + 5y^2 = -8$.   **11.** $y^2 = 0$.
12. $9x^2 - y^2 = 36$.   **13.** $y^2 - 9x^2 = 36$.   **14.** $2x^2 + 1 = 0$.

Hint. Draw the graphs for Problems 12 and 13 on the same xy-plane. The graphs are called **conjugate hyperbolas** because they have the same asymptotes.

15. $x^2 - y^2 = 25$.   **16.** $4y^2 + x^2 = 16$.   **17.** $4x^2 + y^2 = 16$.

Note 1. The graph in Problem 15 is called an **equilateral hyperbola** because its transverse and conjugate axes have the same length.

18. Graph $x^2 + y^2 = 4$ with (*a*) equal scale units on the coordinate axes; (*b*) the scale unit on OX twice as long as the scale unit on OY; (*c*) the scale unit on OY twice as long as the scale unit on OX. In each of (*b*) and (*c*), state whether the major axis is horizontal or vertical.

Note 2. Problem 18 emphasizes the necessity of the statements about scale units in Case I of Section 38.

Write an equation for the ellipse with the given intercepts whose center is the origin and axes are on the coordinate axes. Also, construct the ellipse rapidly but with a good appearance.

19. x-intercepts ± 3; y-intercepts ± 4.

20. x-intercepts ± 1; y-intercepts ± 2.

Write an equation for the hyperbola whose center is the origin and axes are on the coordinate axes, with the specified lengths and locations for the transverse and conjugate axes. Then draw the hyperbola, with its fundamental rectangle and asymptotes.

21. Vertical transverse axis with length 4; conjugate axis with length 2.

22. Horizontal transverse axis with length 6; conjugate axis with length 8.

23. Axes on the coordinate axes; one axis horizontal with length 4; the other axis vertical with length 6. (There are two answers. The hyperbolas are conjugate hyperbolas.)

24. Axes on the coordinate axes; transverse and conjugate axis each with length 8. The two results will be *conjugate* hyperbolas, which are also *equilateral* hyperbolas if the scale units on the coordinate axes are equal.

39. QUADRATIC FUNCTIONS AND QUADRATIC INEQUALITIES

A polynomial function of the second degree in a variable x is called a **quadratic function** of x. Thus, any quadratic function $f(x)$ is of the form

$$f(x) = ax^2 + bx + c, \tag{1}$$

where $\{a, b, c\}$ are constants and $a \neq 0$. Without proof here, we accept the fact that the graph of (1) is a *parabola*.

Example 1. Graph the function $(x^2 - 2x - 2)$.

Solution. 1. Let $y = x^2 - 2x - 2$. We assign values to x and then compute y, to obtain the solutions (x, y) in the following table. The parabola through these points is the graph of the given function.

2. The lowest point, V, of the parabola is called its *vertex*, where $y = -3$, which is the smallest or *minimum value* of the function. Hence, V is called the *minimum point* of the graph. The broken vertical line through V is referred to as the *axis* of the parabola. Later it will be proved that the parabola is *symmetric to its axis*. In Figure 50, the equation of the axis is $x = 1$. The parabola is said to be *concave upward* because the curve bends *counterclockwise* if we travel on it from left to right.

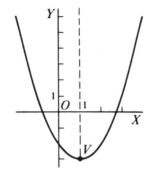

FIGURE 50

$y =$	6	1	-2	-3	-2	1	6
$x =$	-2	-1	0	1	2	3	4

ILLUSTRATION 1.   The graph of $y = -2x^2 + 3x - 5$ would be a parabola concave *downward*, and its vertex V would be called the *maximum point* of the curve. The value of y at V would be the *maximum value* of the function $(-2x^2 + 3x - 5)$.

The following results will be used immediately. Proofs of most of them are available later in the chapter. The facts are true in an xy-plane where the scale units on the axes may be equal or unequal.

Summary.   *Concerning the graph of a quadratic function ($ax^2 + bx + c$), or of the equation*

$$y = ax^2 + bx + c. \tag{2}$$

 I. *The graph is a parabola with its axis perpendicular to the x-axis. The parabola is concave upward if $a > 0$, and downward if $a < 0$.*

 II. *At the vertex of the parabola, $x = -b/2a$, which gives the function its minimum or its maximum value according as $a > 0$ or $a < 0$.*

III. *The equation of the axis is $x = -b/2a$.*

ILLUSTRATION 2.   To graph the equation $y = x^2 - 2x - 2$: By the Summary, at the vertex, $x = -(-2)/2$ or $x = 1$; then $y = -3$. The vertex is $V:(1, -3)$, as in Figure 50. The equation of the axis is $x = 1$.

The roles of x and y in the Summary can be interchanged to give information about the graph of

$$x = ay^2 + by + c, \tag{3}$$

where $a \neq 0$. Thus, the graph of (3) is a parabola whose axis is perpendicular to the y-axis, with the equation $y = -b/2a$. The parabola is concave in the direction of the positive x-axis, or to the *right*, if $a > 0$, and to the *left* if $a < 0$.

ILLUSTRATION 3.   To graph $x = 2y^2 - 8y - 1$:   At the vertex, $y = -(-8/4)$ or $y = 2$; then $x = -9$. The vertex is $V:(-9,2)$. The axis is the line $y = 2$. The parabola, in Figure 51, is concave to the right.

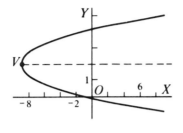

FIGURE 51

The most simple forms for (2) and (3) are, respectively,

$$y = Hx^2 \quad and \quad x = Ky^2, \tag{4}$$

where $H \neq 0$ and $K \neq 0$. Illustrations of graphs of (4) are seen in Figures 52–54.

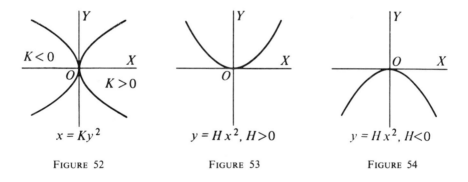

$x = Ky^2$ $y = Hx^2, H>0$ $y = Hx^2, H<0$

FIGURE 52   FIGURE 53   FIGURE 54

Consider a quadratic equation

$$Ax^2 + By^2 + Dx + Ey + F = 0, \tag{5}$$

where not both of A and B are zero. If $B = 0$ and $E \neq 0$, then (5) is linear in y and we can solve (5) for y in terms of x. The result will be of the form $y = ax^2 + bx + c$, where $a \neq 0$, or (5) defines y as a quadratic function of x. If $A = 0$ and $D \neq 0$ in (5), then (5) is linear in x, and we can solve for x in terms of y. The result will be of the form $x = ay^2 + by + c$, where $a \neq 0$. Thus, in either of the cases just mentioned, the graph of (5) is a *parabola*.

ILLUSTRATION 4.   To graph $2x^2 + 3y - 2x + 7 = 0,$ (6)

solve for y: $$y = -\frac{2}{3}x^2 + \frac{2}{3}x - \frac{7}{3}.$$

where y is a quadratic function of x. Hence, the graph of (6) is a parabola with a vertical axis.

ILLUSTRATION 5.   To graph $3x + y^2 - 4y = 5,$ (7)

solve for x: $$x = -\frac{1}{3}y^2 + \frac{4}{3}y + \frac{5}{3},$$

where x is a quadratic function of y. Hence, the graph of (7) is a parabola with a horizontal-axis.

For any function $f(x)$, if c is a number such that $f(c) = 0$, then c is called a **zero** of $f(x)$. Thus, a *zero* of $f(x)$ is a *solution* of the equation $f(x) = 0$. Also, if c is a real number and $f(c) = 0$, then $x = c$ is an *x-intercept of the graph of* $y = f(x)$. The preceding remarks lead to the following statements, which apply to any function $f(x)$, but will be used in this text only when $f(x)$ is a polynomial and, at present, a polynomial of the second degree. Any zero mentioned below will be a real number, because the domain for x in $f(x)$ consists of real numbers when a graph of f is involved.

Applications of the graph, T, of y = f(x).

The zeros of $f(x)$, or the solutions of $f(x) = 0$, are the x-intercepts of T.

The solution set of $f(x) < 0$ is the set of values of x for which T is below the x-axis.

The solution set of $f(x) > 0$ is the set of values of x for which T is above the x-axis.

The preceding statements show that, by use of the graph of $y = f(x)$, a graphical solution can be obtained for the equation $f(x) = 0$, and the inequalities $f(x) < 0$ and $f(x) > 0$.

EXAMPLE 2.   Solve the following equation and inequalities graphically:

$$x^2 - 4x - 1 = 0; \quad x^2 - 4x - 1 \leq 0; \quad x^2 - 4x > 1.$$ (8)

Solution.   1. Let $f(x) = x^2 - 4x - 1$. Then (8) becomes

$$f(x) = 0; \quad f(x) \leq 0; \quad f(x) > 0.$$

2. In Figure 55, the parabola, T, is the graph of $y = f(x)$. The x-intercepts of T, or the solutions of $f(x) = 0$, are $x = 4.2$ and $x = -.2$, approximately.

3. T is *at or below* the x-axis, or $f(x) \leq 0$, when $-.2 \leq x \leq 4.2$. Or, the solution set of $f(x) \leq 0$ is the interval $\{-.2 \leq x \leq 4.2\}$.

4. T is *above* the x-axis, or $f(x) > 0$, when $x < -.2$ and when $4.2 < x$. Hence, the solution set, W, of $f(x) > 0$ is the following union of intervals:

$$W = \{x < -.2\} \cup \{4.2 < x\}.$$

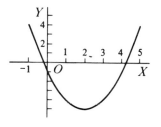

FIGURE 55

If the zeros of a quadratic polynomial $f(x)$ are rational numbers, so that $f(x)$ can be factored conveniently, the solution set of $f(x) < 0$, or of $f(x) > 0$, can be obtained easily without drawing a graph of $y = f(x)$. For this purpose, we emphasize that

$(x - a)$ *is* **negative**, *or* $\quad x - a < 0 \quad$ *when* $\quad x < a;$ $\qquad$ (9)

$(x - a)$ *is* **positive**, *or* $\quad x - a > 0 \quad$ *when* $\quad x > a.$ $\qquad$ (10)

EXAMPLE 3. Solve without graphing:

$$2x^2 + x - 6 < 0. \qquad (11)$$

Solution. 1. From (11), $\qquad (2x - 3)(x + 2) < 0, or$

$$2[x - (-2)](x - \tfrac{3}{2}) < 0, \qquad (12)$$

because $(2x - 3) = 2(x - \tfrac{3}{2})$.

2. In Figure 56, the points $x = -2$ and $x = 3/2$ divide the number scale into the following open intervals:

$$\{x < -2\}; \quad \{-2 < x < \tfrac{3}{2}\}; \quad \{\tfrac{3}{2} < x\}. \qquad (13)$$

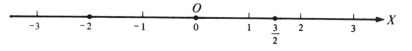

FIGURE 56

By use of (9) and (10), we determine the nature, positive or negative, of each linear factor in (12) on each interval in (13). Then we pick the intervals belonging to the solution set of (11), or (12).

On $\{x < -2\}$: $x - (-2) < 0$ *and* $x - \frac{3}{2} < 0.$ (14)

On $\{-2 < x < \frac{3}{2}\}$: $x - (-2) > 0$ *and* $x - \frac{3}{2} < 0.$ (15)

On $\{\frac{3}{2} < x\}$: $x - (-2) > 0$ *and* $x - \frac{3}{2} > 0.$ (16)

Hence, the solution set for (12) is the interval $\{-2 < x < \frac{3}{2}\}$, where one linear factor is positive and one is negative in (12).

Comment.   The solution set of $2x^2 + x - 6 > 0$ is the *union* of the intervals in (14) and (16) where both linear factors in (12) are positive, or both are negative. Or, the solution set is $\{x < -2\} \cup \{\frac{3}{2} < x\}$.

EXERCISE 24

Obtain a graceful graph of the function or equation by use of a minimal number of accurately located points. Use unequal scale units on the axes if desirable in order to prevent the parabola from being too slender.
1. $-x^2$.   2. $y = 3x^2$.   3. $x = 2y^2$.   4. $x = -3y^2$.
5. $x - 2x + 5$.   6. $-2x^2 + 8x - 3$.   7. $x = y^2 - 4y + 3$.
8. $x = -y^2 + 2y - 7$.   9. $y = -2x^2 + 4x - 5$.   10. $y = 2x^2 - 4x + 3$.
11. $2x^2 - 8x - 2y + 3 = 0$.   12. $4y^2 - 2x - 8y - 7 = 0$.

Graph each of the equations in the problem on the same xy-plane:
13. $(x - 2)(x + 3) = 0$; $y = (x - 2)(x + 3)$.
14. $x^2 - 6x + 8 = 0$; $y = x^2 - 6x + 8$.

Solve the equation graphically. Also, by use of the graph involved, obtain the solution set of each inequality.
15. $x^2 - 2x - 1 = 0$; $x^2 - 2x - 1 < 0$; $x^2 - 2x - 1 \geq 0$.
16. $x^2 + 2x - 4 = 0$; $x^2 + 2x - 4 \leq 0$; $x^2 + 2x - 4 > 0$.

Find the solution set of each inequality without drawing a graph. First factor the polynomial if necessary.
17. $(x - 2)(x + 3) < 0$; $(x - 2)(x + 3) \geq 0$.
18. $2x^2 - x - 3 \leq 0$; $2x^2 - x - 3 > 0$.
19. $2x^2 - x - 15 < 0$; $2x^2 - x - 15 \geq 0$.

Find an equation for the parabola with its vertex at the origin in an xy-plane, and with a coordinate axis as the axis of the parabola, if it passes through the specified point. Construct each parabola obtained.
20. $(2,4)$.   21. $(-3,1)$.   22. $(-3,-2)$.

Hint. The equation may be of the form $y = Hx^2$ or $x = Ky^2$. Find H and K.

40. TRANSFORMATION BY TRANSLATION OF AXES

Consider two systems of coordinates superimposed on the same plane, an xy-system which we shall call the *original system,* and a new $x'y'$-system. Then, each point P in the plane will have two sets of coordinates, (x,y) and (x',y'). Any curve having an equation $f(x,y) = 0$ will have a related equation $F(x',y') = 0$ in the new system. The process of obtaining (x',y') corresponding to any given point $P:(x,y)$, or of finding the new equation $F(x',y') = 0$ for a curve whose equation is $f(x,y) = 0$, is spoken of as a *transformation of coordinates.*

In Figure 57, let OX and OY be the *original* coordinate axes. Let $O'X'$ and $O'Y'$ be new coordinate axes respectively parallel to OX and OY, with the same positive directions, the same scale unit on $O'X'$ as on OX, and the same scale unit on $O'Y'$ as on OY. A change of this nature from an xy-system to an $x'y'$-system is spoken of as a **translation of the axes** to a new origin O'. If O' has the old coordinates $(x = h, y - k)$, we shall prove that the coordinates (x,y) and (x',y') of any point P in the plane satisfy

$$x = x' + h \quad and \quad y = y' + k; or \tag{1}$$
$$x' = x - h \quad and \quad y' = y - k. \tag{2}$$

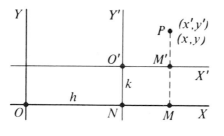

FIGURE 57

Proof. Let the projection of P on $O'X'$ be M', and on OX be M, and the projection of O' on OX be N. All line segments to be mentioned will be *directed.* For any position of P, as in Figure 57, we have

$$h = \overline{ON}; \quad x = \overline{OM}; \quad x' = \overline{O'M'} = \overline{NM};$$
$$x = \overline{OM} = \overline{ON} + \overline{NM} = h + x'.$$

Similarly, by projecting P and O' on OY, we would obtain $y = k + y'$. Thus, we have proved (1), and therefore (2).

ILLUSTRATION 1. Let the new origin be $O':(x = 2, y = -3)$. Then the point $P:(x = -1, y = 4)$ has new coordinates (x',y') as found from

(2) with (h = 2, k = -3):

$$x' = -1 - 2 = -3; \qquad y' = 4 - (-3) = 7.$$

A major objective in transformation of coordinates is the simplification of equations in terms of the new coordinates.

EXAMPLE 1.   Transform the following equation by translating axes to the new origin (x = 2, y = -3):

$$2x^2 - 8x + y^2 + 6y + 11 = 0. \qquad (3)$$

Solution.   1. From (1), substitute

$$x = x' + 2 \qquad and \qquad y = y' - 3 \qquad (4)$$

in (3), to obtain

$$2(x' + 2)^2 - 8(x' + 2) + (y' - 3)^2 + 6(y' - 3) + 11 = 0, \, or$$

$$2x'^2 + y'^2 = 6. \qquad (5)$$

2. In Figure 58, the xy-axes and the $x'y'$-axes are shown. The graph of (5) in the $x'y'$-system is an ellipse whose x'-intercepts are $\pm\sqrt{3}$ and y'-intercepts are $\pm\sqrt{6}$. With equal scale units on all axes, it is seen that the major axis of the ellipse is vertical, along $O'Y'$. The equations of the axes of symmetry of the ellipse are

$$y' = 0 \qquad and \qquad x' = 0,$$

or, from (4), $\qquad\qquad x - 2 = 0 \qquad and \qquad y + 3 = 0.$

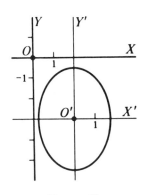

FIGURE 58

Observe that we use (1) to transform an equation *from* an xy-system *to* an $x'y'$-system. We use (2) for the inverse transformation *from* the $x'y'$-system *to* the xy-system. Thus, in (5), if we use $x' = x - 2$ and $y' = y + 3$, from (4), we thus would obtain the original equation (3).

Note 1.   Since (1) and (2) are *linear* in x, y, x', and y', the *degree* of an equation $f(x,y) = 0$ is not changed by a translation of axes.

EXERCISE 25

1. Find the new coordinates of the points $(3,7)$, $(-4,6)$, $(-1,-2)$, and $(0,0)$ if the axes are translated to the new origin $O':(2,4)$. Plot all points and show both sets of coordinate axes.

2. If the axes have been translated to $O':(x = -2, y = 3)$ in an xy-plane, find the old coordinates of the points whose new coordinates are $(-1,2)$, $(2,-3)$ $(0,0)$, and $(-4,3)$. Plot all points and show both sets of coordinate axes.

Transform the equation by translating axes to the specified new origin O'. Graph the final equation in the new coordinate system, showing both sets of coordinate axes. Draw any asymptotes possessed by the graph.

3. $O':(1,3)$; $x^2 - 2x + y^2 - 6y + 6 = 0$.
4. $O':(2,4)$; $y^2 - 8y - 4x + 24 = 0$.
5. $O':(-3,4)$; $xy + 3y - 4x - 18 = 0$.
6. $O':(1,2)$; $9(x - 1)^2 + 4(y - 2)^2 = 36$.   (Do not expand.)

7. $O':(3,-5)$, $\dfrac{(x - 3)^2}{9} - \dfrac{(y + 5)^2}{16} = 1$.   (Do not expand.)

8. $O':(-1,2)$; $4x^2 + y^2 + 8x - 4y = 8$.
9. $O':(2,-3)$; $x^2 - 4y^2 - 4x - 24y = 36$.

41. CENTRAL CONICS WITH AXES PARALLEL TO *OX* AND *OY*

Any polynomial equation of the second degree in x and y which does not involve the product xy is equivalent to an equation of the form

$$Ax^2 + By^2 + Dx + Ey + F = 0, \tag{1}$$

where $\{A,B\}$ are not both zero. If $B = 0$ and $E \neq 0$, in Section 39 it was found that (1) defines y as a quadratic function of x. If $A = 0$ and $D \neq 0$, then (1) defines x as a quadratic function of y. In either of the preceding cases, the graph of (1) is a *parabola,* which can be obtained as in Section 39. We proceed with the understanding that neither of the preceding conditions holds true for (1), so that (1) does *not* represent a parabola.*

In (1), we may complete a square with the terms in x if $A \neq 0$, and with the terms in y if $B \neq 0$. Then, by a translation of the axes to a new

*Hence, in (1), if $B = 0$ then $E = 0$ also; if $A = 0$ then $D = 0$ also.

origin, (1) can be changed to a new $x'y'$-form which will be recognized as the equation of a central conic in a simple position in the $x'y'$-plane. Finally, the graph of (1) can be obtained by drawing the graph of the $x'y'$-equation in the $x'y'$-system which has been superimposed on the xy-plane.

EXAMPLE 1.   Transform to a standard form in new coordinates and graph

$$2x^2 - 8x + y^2 + 6y + 11 = 0. \tag{2}$$

Solution.   1. Group the x-terms, and the y-terms:

$$2(x^2 - 4x + \quad) + (y^2 + 6y + \quad) = -11. \tag{3}$$

Add 4 within parentheses for the x-terms, add 9 for the y-terms, and therefore add $(8 + 9)$ to both sides of (3) to complete squares:

$$2(x^2 - 4x + 4) + (y^2 + 6y + 9) = -11 + 17, or$$
$$2(x - 2)^2 + (y + 3)^2 = 6. \tag{4}$$

2. Transform coordinates by use of

$$x' = x - 2 \quad and \quad y' = y + 3, \tag{5}$$

which translates the origin to $O':(x = 2, y = -3)$. By use of (5), equation (4) becomes

$$2x'^2 + y'^2 = 6. \tag{6}$$

Thus the graph of (2) is an ellipse whose center is at $O':(x = 2, y = -3)$; the x'-intercepts are $\pm\sqrt{3}$; the y'-intercepts are $\pm\sqrt{6}$. The graph of (2), or (6), is given in Figure 58 on page 112.

3. In the xy-system, the graph of (2) may be described as an ellipse whose center is $O':(x = 2, y = -3)$ and whose axes of symmetry are the lines $x = 2$ and $y = -3$. The horizontal axis of the ellipse has the length $2\sqrt{3}$; the length of the vertical axis is $2\sqrt{6}$.

EXAMPLE 2.   Remove linear terms in the following equation by a translation of axes and graph the equation:

$$9x^2 + 54x - 16y^2 + 64y - 127 = 0. \tag{7}$$

Solution.   1. To complete squares in the x-terms, and in the y-terms, we proceed as follows:

$$9(x^2 + 6x + \quad) - 16(y^2 - 4y + \quad) = 127;$$
$$9(x^2 + 6x + 9) - 16(y^2 - 4x + 4) = 127 + 81 - 64 = 144;$$
$$9(x + 3)^2 - 16(y - 2)^2 = 144. \tag{8}$$

2. To simplify (8), transform by use of

$$x' = x + 3 = x - (-3), \quad and \quad y' = y - 2, \qquad (9)$$

which is a translation to the new origin $O':(x = -3, y = 2)$. Then (8) becomes

$$9x'^2 - 16y'^2 = 144. \qquad (10)$$

3. The graph of (7) is the hyperbola, T, which is the graph of (10) in the $x'y'$-system, as shown in Figure 59 with the origin O'. From (10), on dividing by 144, we obtain

$$\frac{x'^2}{16} - \frac{y'^2}{9} = 1. \qquad (11)$$

T was constructed on the $x'y'$-system by the method employed in Example 2 on page 100. The axes of symmetry for T are $x' = 0$ and $y' = 0$, or $x = -3$ and $y = 2$, and the center of symmetry is O'. A single equation for the asymptotes of T is $9x'^2 - 16y'^2 = 0$, from (10), and separate equations for the asymptotes are

$$3x' - 4y' = 0 \quad and \quad 3x' + 4y' = 0. \qquad (12)$$

We could obtain xy-equations for the asymptotes by using (9) in (12).

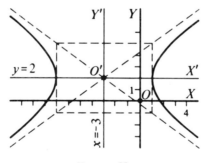

FIGURE 59

If the procedure employed in Examples 1 and 2 is used with any equation (1) which does *not* represent a parabola, the new $x'y'$-equation can be written in the form

$$Ax'^2 + By'^2 = C, \qquad (13)$$

where C is some constant, while A and B have the same values as in (1). This is true because Ax^2 in (1) produces Ax'^2, and By^2 in (1) produces By'^2 of (13) when the linear equations of the transformation by translation are used. Then, by reference to Cases I, II, and III on page 99, and 100 we conclude that (13), and hence (1), represents a *central conic*, which may be a degenerate conic. Therefore, any quadratic equation (1) represents *either a parabola or a central conic*.

   Note 1.   When (1) does not represent a parabola, we could consider various special cases, such as

$$A \text{ and } B \text{ both positive or both negative.} \tag{14}$$

Then, by reference to (13), we could reach corresponding conclusions as to the possible nature of the graph of (1). For instance, if (14) is true and $C \neq 0$ in (13), then (1) represents an ellipse if C has the same sign as A and B. There is negligible advantage in memorizing such facts. For any particular equation (1), by completing squares as in Examples 1 and 2, we quickly obtain the corresponding equation (13). Its form then gives information about the graph of (1), on the basis of preceding results in Cases I–III on pages 99 and 100.

EXERCISE 26

   Obtain a graph of the equation in the xy-plane, or prove that the graph is $\emptyset$. Show the axes of both the xy-system and any new x'y'-system of coordinates which is used.

1. $4x^2 + 9y^2 - 24x = 0$.
2. $4y^2 - 16y - x^2 = 0$.
3. $25x^2 - 100x - 4y^2 - 32y = 64$.
4. $y^2 + 4x^2 + 10y + 32x + 25 = 0$.
5. $3x^2 + 4y^2 - 12x + 8y + 19 = 0$.
6. $4x^2 - 25y^2 - 8x + 50y = 121$.
7. $x^2 - 4x - y^2 + 8y = 12$.
8. $4y^2 - 16y + x^2 - 6x + 25 = 0$.
9. $x^2 - 4x + 8 = 0$.
10. $y^2 + 6y = 0$.
11. $x^2 - 4x - 9y^2 + 72y = 140$.

   Only determine the nature of the graph of the equation.

12. $4x^2 - 8x + y^2 + 6y = -9$.
13. $2x^2 + 8x - y^2 + 2y = -3$.
14. $4x^2 - 24x + y^2 - 8y + 70 = 0$.

42. EQUATIONS FOR CONICS SATISFYING GIVEN CONDITIONS

   Recall the following facts from page 102 about equations of ellipses and hyperbolas having the coordinate axes in the *xy*-plane as axes of symmetry. Also, from (4) on page 107, recall the facts about equations for parabolas having the origin as the vertex and a coordinate axis in the *xy*-plane as an axis of symmetry.

   Ellipse: *x-intercepts* $\pm a$; $\left.\vphantom{\frac{x^2}{a^2}}\right\}$ $\dfrac{x^2}{a^2} + \dfrac{y^2}{b^2} = 1$.   (1)
   y-intercepts $\pm b$.

Hyperbola: $\left.\right\}$
x-intercepts $\pm a.$

$$\frac{x^2}{a^2} - \frac{y^2}{b^2} = 1. \tag{2}$$

Hyperbola: $\left.\right\}$
y-intercepts $\pm a.$

$$\frac{y^2}{a^2} - \frac{x^2}{b^2} = 1. \tag{3}$$

Parabola *with vertex* $(0,0)$ $\left.\right|$
and vertical axis, x $= 0.$

$$y = Hx^2, H \neq 0. \tag{4}$$

Parabola *with vertex* $(0,0)$ $\left.\right\}$
and horizontal axis, y $= 0.$

$$x = Ky^2, K \neq 0. \tag{5}$$

In each of the following examples, a point $(x = x_0, y = y_0)$, not the origin, will play the role occupied by the *origin* in (1)–(5). To obtain any desired equation, first the coordinates are transformed by a translation of axes to the new origin $O':(x = x_0, y = y_0)$. Then, an $x'y'$-equation is written by use of (1)–(5). Finally, an xy-equation is obtained by transforming the $x'y'$-equation.

EXAMPLE 1.   Obtain an equation for the ellipse in an xy-plane where the scale units on the coordinate axes are equal, if the ellipse satisfies the following conditions:

> *center* (4,3); *each axis parallel to a coordinate axis;* $\left.\right\}$
> *horizontal axis* 6 *units long and vertical axis* 4 *units* $\left.\right\}$ (6)
> *long.* $\left.\right\}$

Solution.   1. Let us introduce an $x'y'$-system of coordinates by translating the axes to the new origin $O':(x = 4, y = 3)$. The equations of transformation are

$$x' = x - 4 \quad and \quad y' = y - 3. \tag{7}$$

2. By use of (6), the ellipse, T, is drawn in Figure 60 with the center O', x'-intercepts ± 3, and y'-intercepts ± 2. By (1), an equation for T is

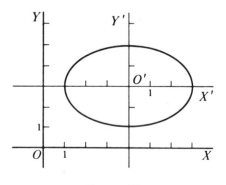

FIGURE 60

$$\frac{x'^2}{9} + \frac{y'^2}{4} = 1. \tag{8}$$

3. Use (7) in (8) to obtain the *xy*-equation of *T*:

$$\frac{(x-4)^2}{9} + \frac{(y-3)^2}{4} = 1, \quad or \quad 4(x-4)^2 + 9(y-3)^2 = 36.$$

ILLUSTRATION 1.   Suppose that the center of a hyperbola, *T*, is the point ($x = 4$, $y = 3$), as used in (7); transverse axis is vertical and is 6 units long; conjugate axis is 4 units long. In the *x'y'*-system of Figure 60, the *y'*-intercepts of *T* would be ±3, and the length of the semi-conjugate axis of *T* is 2. By use of (3) with $a = 3$ and $b = 2$, an *x'y'*-equation for *T* is

$$\frac{y'^2}{9} - \frac{x'^2}{4} = 1. \tag{9}$$

When (7) is used in (9), we obtain an *xy*-equation for *T*:

$$\frac{(y-3)^2}{9} - \frac{(x-4)^2}{4} = 1, \quad or \quad 4(y-3)^2 - 9(x-4)^2 = 36.$$

EXAMPLE 2.   (*a*) Obtain a general *xy*-equation for the parabola, *T*, whose vertex is the point $(4,-3)$ and axis of symmetry is horizontal. (*b*) Then, obtain the equation of the particular parabola of this type which passes through the point $P:(3,-2)$.

Solution.   (*a*) Introduce an *x'y'*-system of coordinates by translating the axes to the new origin $O':(x = 4, y = -3)$. The equations of transformation are

$$x' = x - 4 \quad and \quad y' = y - (-3). \tag{10}$$

A possible appearance for *T* is shown in Figure 61. From (5), an *x'y'*-equation for *T* is

$$x' = Ky'^2, \quad where\ K \neq 0. \tag{11}$$

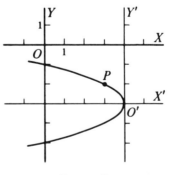

FIGURE 61

We use (10) in (11) to obtain an *xy*-equation for *T*:

$$x - 4 = K(y + 3)^2. \tag{12}$$

(b) To find an equation for *T* if it passes through ($x = 3$, $y = -2$), substitute these values in (12) to determine *K*:

$$3 - 4 = K(-2 + 3)^2, \quad or \quad K = -1.$$

Hence the desired equation is $x - 4 = -(y + 3)^2$. The axis of the parabola is the line $y + 3 = 0$.

By the method used in Example 2, the student may obtain the following results from (4) and (5).

Suppose that the vertex of a parabola in an xy-plane is the point (x_0, y_0). *Then, an equation for the parabola is as follows, where* $H \neq 0$ *and* $K \neq 0$:

(*Axis: the line* $x = x_0$)	$y - y_0 = H(x - x_0)^2.$	(13)
(*Axis: the line* $y = y_0$)	$x - x_0 = K(y - y_0)^2.$	(14)

EXERCISE 27

A conic in an xy-plane is described. First sketch the conic approximately, showing the fundamental rectangle and asymptotes of any hyperbola. Then write an equation for the conic in a new x'y'-system, and finally obtain an xy-equation for the conic. The scale units are assumed to be equal on the coordinate axes.

1. *Ellipse:* center ($x = 2$, $y = 6$); major axis horizontal with length 6; minor axis with length 4.
2. *Hyperbola:* center ($x = 5$, $y = 4$); transverse axis horizontal with length 2; conjugate axis with length 4.
3. *Hyperbola:* center ($x = -4$, $y = -3$); transverse axis vertical with length 6; conjugate axis with length 6.
4. *Ellipse:* center ($x = -3$, $y = 4$); major axis with length 4, parallel to *OY*; minor axis with length 2.
5. *Ellipse:* center ($x = -4$, $y = -6$); vertical axis with length 8; minor axis with length 6.
6. *Hyperbola:* center ($x = 6$, $y = -4$); transverse axis horizontal with length 8; conjugate axis with length 6.
7. *Ellipse:* center (x_0, y_0); major axis horizontal with length 2a; minor axis with length 2b.
8. *Hyperbola:* center ($x = x_0$, $y = y_0$); transverse axis vertical with length 2a; conjugate axis with length 2b.

Obtain an equation for the parabola which is described in an xy-plane. Equations (13) *and* (14) *on this page may be used.*

9. Vertex ($x = 2, y = 4$); axis vertical; passes through ($x = 3, y = 6$).
10. Vertex ($x = -1, y = 3$); axis vertical; passes through ($x = 0, y = 1$).
11. Vertex ($x = 2, y = -2$); axis horizontal; passes through ($x = 4, y = 0$).
12. Vertex ($x = 3, y = 5$); axis horizontal; passes through ($x = 5, y = 7$).

★43. NEW APPROACH TO NONDEGENERATE CONICS

Hereafter in this chapter, when we mention simply a *conic,* we shall mean one of the *nondegenerate conics,* that is, an ellipse, hyperbola, or parabola. Each of these will be defined as a set of points in a plane satisfying a certain geometric condition. By elaborate geometrical proofs, which we shall omit, it can be shown that, in each case, the definition is equivalent to the previous definition of the curve as a plane section of a right circular cone. In any *xy*-plane involved, we shall assume that the scale units on the coordinate axes are equal. Then, in the *xy*-plane, the distance formula of page 57 is available for use.

In each definition to be met, a point called a *focus* of the conic will be involved. The distance of any point *P* of the conic from a focus *F* of the conic will be called a **focal radius.**

Our objective in the next three sections will be to prove that each standard equation for a conic specified in Sections 38 and 39 was correctly described as an equation for the particular conic involved.

★44. STANDARD EQUATIONS FOR A PARABOLA

DEFINITION I.   A **parabola** *is the set of points* $\{P\}$ *such that the undirected distance between P and a fixed point F, called the* **focus,** *is equal to the undirected distance of P from a fixed line D, called the* **directrix,** *which does not pass through F.*

EXAMPLE 1.   Find an equation of a parabola whose focus is $F:(3,0)$ and directrix is the line $D:\{x = -3\}$.

Solution.   1. In Figure 62, let $P:(x,y)$ be any point in the *xy*-plane, where *P* is not *F* and is not on *D*. In Figure 62, PM is perpendicular to *D* and hence *M* is the point $(-3,y)$.   Let $h = |\overline{PF}|$ and $d = |\overline{PM}|$. From page 57,

$$h = \sqrt{(x - 3)^2 + y^2}; \qquad d = |-3 - x| = |3 + x|. \qquad (1)$$

2. *P* is on the parabola when and only when $h = d$, or $h^2 = d^2$, or

$$(x - 3)^2 + y^2 = (3 + x)^2, \qquad or \qquad y^2 = 12x, \qquad (2)$$

which is the equation of the parabola. It is shown in Figure 62.

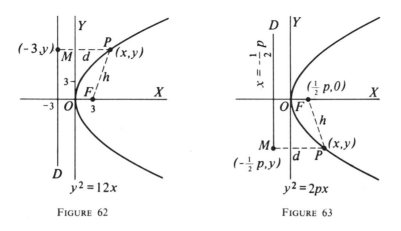

FIGURE 62   FIGURE 63

DERIVATION OF A STANDARD EQUATION FOR A PARABOLA

1. In an xy-plane, in Figure 63, let F be the focus and D be the directrix of a parabola where $p > 0$ is the undirected distance from F to D. Designate the x-axis as the line through F perpendicular to D, with the origin midway between F and D, and F in the positive x-direction from the origin. Then, as seen in Figure 63, F is the point $F:(\frac{1}{2}p,0)$ and D has the equation $x - -\frac{1}{2}p$.

2. Let $P:(x,y)$ be any point in the plane, with PM perpendicular to D, and PF as the segment joining P and F, in Figure 63. Then, we have the point $M:(-\frac{1}{2}p,y)$, and $\overline{MP} = x - (-\frac{1}{2}p)$. Hence,

$$| \overline{MP} | = | x + \tfrac{1}{2}p | \ and \ | \overline{PF} | = \sqrt{(x - \tfrac{1}{2}p)^2 + y^2}. \qquad (3)$$

3. $P:(x,y)$ is on the parabola if and only if $| \overline{PM} | = | \overline{PF} |$. Hence, an equation for the parabola is $\overline{PF}^2 = \overline{PM}^2$ or, from (3),

$$(x - \tfrac{1}{2}p)^2 + y^2 = (x + \tfrac{1}{2}p)^2, \ or \qquad (4)$$

$$x^2 - px + \tfrac{1}{4}p^2 + y^2 = x^2 + px + \tfrac{1}{4}p^2, or$$

$$y^2 = 2px, \qquad (5)$$

whose graph is shown in Figure 63.

In (5), the equation is unaltered if y is replaced by $-y$. Hence, the parabola is symmetric to the x-axis. It is called the *axis* of the parabola. The intersection of its axis and the parabola is called its *vertex*, which is the origin in Figure 63.

The preceding results for (5) are summarized in (6) on page 122. Other similar equations, which follow, can be obtained for a parabola in various positions whose vertex is the origin, and axis is a coordinate axis. The student may derive (7)–(9) later.

PARABOLAS FOR WHICH THE DISTANCE BETWEEN THE FOCUS AND THE DIRECTRIX IS $p > 0$.

Vertex (0,0); *axis* $y = 0$; *concave to right:* $y^2 = 2px.$ (6)

Vertex (0,0); *axis* $y = 0$; *concave to left:* $y^2 = -2px.$ (7)

Vertex (0,0); *axis* $x = 0$; *concave upward:* $x^2 = 2py.$ (8)

Vertex (0,0); *axis* $x = 0$; *concave downward:* $x^2 = -2py.$ (9)

Graphs of (7)–(9) are shown in Figures 64–66.

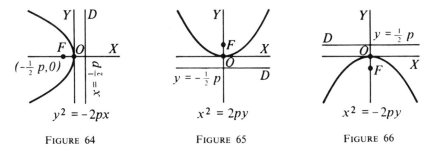

$y^2 = -2px$	$x^2 = 2py$	$x^2 = -2py$
FIGURE 64	FIGURE 65	FIGURE 66

In (6)–(9), we have verified the statements made without proof concerning the graphs of $y = Hx^2$ and $x = Ky^2$ in (4) and (5) on page 117.

By use of (8) and (9), we shall prove the following result, which justifies the Summary on page 106 for graphing a quadratic function of x.

If $a \neq 0$, the graph of

$$y = ax^2 + bx + c \qquad (10)$$

is a parabola whose axis is the line $x = -b/2a$. At the vertex of this parabola, $x = -b/2a$. Also, the parabola is concave upward if $a > 0$, and downward if $a < 0$.   (11)

Proof.   1. From (10),

$$y - c = a\left(x^2 + \frac{b}{a}x\right). \qquad (12)$$

To complete a square with the x-terms in (12), add $[\frac{1}{2}(b/a)]^2$ within the parentheses, or $b^2/4a$ to both sides:

$$y - \left(c - \frac{b^2}{4a}\right) = a\left(x^2 + \frac{b}{a}x + \frac{b^2}{4a^2}\right), \text{ or} \qquad (13)$$

$$y - k = a\left[x - \left(-\frac{b}{2a}\right)\right]^2, \qquad (14)$$

where $k = c - b^2/4a$.

2. Let $\qquad y' = y - k \quad and \quad x' = x - \left(-\dfrac{b}{2a}\right).$ (15)

Then (14) becomes $y' = ax'^2$, or

$$x'^2 = \frac{y'}{a}.$$ (16)

3. Suppose that $a > 0$ and let $2p = 1/a$. Then (16) becomes $x'^2 = 2py'$. Hence, from (8) and Figure 65, (16) and thus (10) represents a parabola concave *upward* whose vertex is $O':(x' = 0, y' = 0)$, with the line $x' = 0$ as the axis of symmetry. From (15), $x' = 0$ gives $x = -b/2a$. Thus, at the vertex for the parabola (10), we have $x = -b/2a$, and the equation of the axis also is $x = -b/2a$. Similarly, if $a < 0$ then (9) applies and the graph of (10) is concave *downward*. This completes the proof of (11).

Note 1. Parabolas arise frequently in applied mathematics. Thus, suppose that two celestial bodies are moving subject to their attraction according to the Newtonian law of gravitation. Then, under certain circumstances, the orbit of either body as observed from the second body will be a parabola with the second body at its focus. With idealized conditions (for instance with no air resistance), if a projectile is shot from an assumed plane surface as the earth, the path of the projectile will be a parabola.

Note 2. At any point P on a parabola with focus F and axis L, as in Figure 67, let SPT be the tangent line. Construct PR parallel to L, and draw PF. Then, in calculus it is proved that $\angle SPF = \angle TPR$. Now suppose that the parabola in Figure 67 is revolved about its axis to create a surface for a parabolic mirror. Then, if a source of light is at F, each ray $\overrightarrow{FP}$ of light to the mirror is reflected as a ray $\overrightarrow{PR}$ parallel to the axis of the mirror. This is true because angles SPF and TPR are the complements of the angles of incidence and reflection for the ray. Hence, the mirror sends out a beam of light parallel to the axis of the mirror.

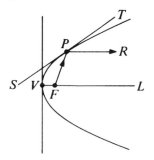

FIGURE 67

*EXERCISE 28

By direct use of Definition I as in Example 1 on page 121, obtain an equation for the parabola with the given focus F and directrix D.   Then draw the parabola.   Also specify its vertex and the equation of the axis of symmetry.   The constant p is positive.

1. $F:(4,0); D:\{x = -4\}$.
2. $F:(0,-3); D:\{y = 3\}$.
3. $F:(0,\frac{1}{2}p); D:\{y = -\frac{1}{2}p\}$.
4. $F:(-\frac{1}{2}p, 0); D:\{x = \frac{1}{2}p\}$.
5. $F:(5,2); D:\{x = 3\}$.
6. $F:(-2,5); D:\{y = 1\}$.

7. A parabola has the line $x - y = 2$ as the directrix, and $F:(3,3)$ as the focus in an xy-plane.   Construct the axis of the parabola and find its vertex.   Also, construct four other points on the parabola by use of lines parallel to the directrix and arcs of circles drawn with F as a center.   Finally, describe a construction for finding points on a parabola in any plane if the focus and directrix are known (the use of a co-ordinate system in the plane is not involved).

8. The chord of a parabola perpendicular to its axis at the focus is called the **focal chord** (or **latus rectum**) of the parabola.   Find the length of the focal chord of the parabola $y^2 = 2px$.

★45. ELLIPSE DEFINED BY MEANS OF FOCAL RADII

*DEFINITION II.   In a plane, an **ellipse** is the set of points $\{P\}$ for which the sum of the undirected distances from P to two fixed points F and F', called the **foci,** is a constant greater than the distance between F and F'.*

EXAMPLE 1.   Find an equation of the ellipse with foci $F:(4,0)$ and $F':(-4,0)$ if the sum of the focal radii for any point on the ellipse is 10.

Solution.   1. Let $P:(x,y)$ be any point in the plane, and let the lengths of the segments PF and PF', as in Figure 68, be h and h', respectively, where

$$h = \sqrt{(x - 4)^2 + y^2} \quad and \quad h' = \sqrt{(x + 4)^2 + y^2}. \qquad (1)$$

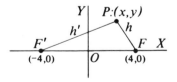

FIGURE 68

2. To state that $P:(x,y)$ is on the ellipse is equivalent to stating that $h + h' = 10$. Or, an equation for the ellipse is

$$\sqrt{(x - 4)^2 + y^2} + \sqrt{(x + 4)^2 + y^2} = 10, \, or \qquad (2)$$

$$\sqrt{(x - 4)^2 + y^2} = 10 - \sqrt{(x + 4)^2 + y^2}. \qquad (3)$$

In (3), square both sides and simplify:

$$(x - 4)^2 + y^2 = 100 - 20\sqrt{(x + 4)^2 + y^2} + (x + 4)^2 + y^2, \, or$$

$$25 + 4x = 5 \sqrt{(x + 4)^2 + y^2}. \qquad (4)$$

Square and simplify in (4):

$$(25 + 4x)^2 = 25(x + 4)^2 + 25y^2, \, or$$

$$9x^2 + 25y^2 = 225. \qquad (5)$$

3. If (x,y) is a solution of (2), then (x,y) is a solution of (4), and then of (5). *Conversely,* as discussed in the following comment, if (x,y) satisfies (5) then (x,y) satisfies (2). Thus, (2) and (5) are equivalent equations, so that (5) is an equation for the specified ellipse, as shown in Figure 70 on page 126 if scale units there are chosen so that $a = 5$ and $b = 3$.

Comment.   With (2) thought of as $h + h' = 10$, notice that any solution of $\pm h \pm h' = 10$ satisfies (5), because all variations due to choices of the ambiguous signs disappear in the *squaring* operations leading from (2) to (5). Also, if the steps from (2) to (5) should be reversed by extracting square roots, it would be seen that any solution (x,y) of (5) satisfies $\pm h \pm h' = 10$ for *some* choice or choices of the ambiguous signs. Then, we observe that $(+h + h') = 10$ is the only tenable possibility, because $h \geq 0$, $h' \geq 0$, and $\{h, h', 8\}$ are the lengths* of the sides of a triangle. Hereafter, such reasoning will be omitted in rationalizing equations similar to (2).

DERIVATION OF A STANDARD EQUATION FOR AN ELLIPSE

1. In Definition II, let the distance between the foci F' and F be $2c$ where $c \geq 0$. In the plane involved, take the line $F'F$ as the x-axis, with

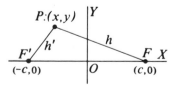

FIGURE 69

*Hence $|h - h'| \leq 8$ and thus $h - h' \neq 10$ and $h' - h \neq 10$.

the origin of coordinates at the midpoint of segment $F'F$, and with F having a positive x-coordinate. Then the foci are $F':(-c,0)$ and $F:(c,0)$. Let the sum of the focal radii of P, as mentioned in Definition II, be $2a$ where $a > c$.

2. If $P:(x,y)$ is any point in the plane, let h and h' be the lengths of PF and PF' as in Figure 69. Then we have

$$h = \sqrt{(x - c)^2 + y^2}; \qquad h' = \sqrt{(x + c)^2 + y^2}.$$

A point P is on the ellipse when and only when $h + h' = 2a$, or an equation for the ellipse is

$$\sqrt{(x - c)^2 + y^2} + \sqrt{(x + c)^2 + y^2} = 2a. \tag{6}$$

3. On rationalizing in (6) by transposing one radical and squaring both sides, simplifying, and then rationalizing again as in Example 1, we obtain

$$(a^2 - c^2)x^2 + a^2y^2 = a^2(a^2 - c^2). \tag{7}$$

Let a positive number $b \leq a$ be defined by

$$b^2 = a^2 - c^2. \tag{8}$$

Then (7) becomes

$$(\textit{foci on OX}) \qquad b^2x^2 + a^2y^2 = a^2b^2, \quad \textit{or} \quad \frac{x^2}{a^2} + \frac{y^2}{b^2} = 1, \tag{9}$$

where both sides of the equation at the left were divided by a^2b^2. A graph of (9) is in Figure 70.

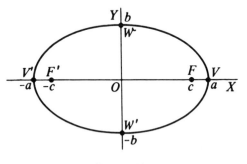

FIGURE 70

In terminology introduced on page 97, the major axis of the ellipse in Figure 70 is $V'V$, and the minor axis is $W'W$. The foci are on the major axis. The endpoints V and V' of the major axis are called the *vertices* of the ellipse.

Consider a repetition of the discussion leading to (9), with the line $F'F$ taken as the y-axis, so that the foci are $F:(0,c)$ and $F':(0,-c)$. Then, the equation of the ellipse is found by interchanging x and y in (9), which gives

$$\text{(foci on } OY) \qquad a^2x^2 + b^2y^2 = a^2b^2, \qquad \text{or} \qquad \frac{x^2}{b^2} + \frac{y^2}{a^2} = 1. \qquad (10)$$

In the details leading to (9), if $c = 0$ then $b = a$ and the two foci coincide, at the origin. Thus, with $b^2 = a^2$, (9) becomes $a^2x^2 + a^2y^2 = a^4$, or $x^2 + y^2 = a^2$. Hence the ellipse is a circle with its center at the origin. That is, according to Definition II, a circle is included as a special case of an ellipse.

In (9) on page 102, we saw that, when Case I of page 99 applies, the equation $Ax^2 + By^2 = C$ can be written in the equivalent form (9) or (10) of the present section, which was called the *intercept form* on page 102. Hence, our results in (9) and (10) justify the statements about an ellipse made in Case I on page 99.

Note 1.   It can be proved by use of calculus that the focal radii to any point P on an ellipse make equal angles with the tangent to the ellipse at P, as in Figure 71, where angles RPF' and TPF are equal. Now consider a reflecting surface (an *ellipsoid*) made by revolving the ellipse about its major axis. Suppose that random rays of light, or sound waves, issue from one focus F. Then, because of the property just mentioned for the tangents, the rays will be reflected to the other focus F'. Also, all rays will reach F' at the same time because the sum of the focal radii is a constant, $2a$. This property accounts for the phenomenon observed in certain "*whispering galleries,*" where a low sound made at one point (a focus) can easily be heard at another point (the other focus) in the room.

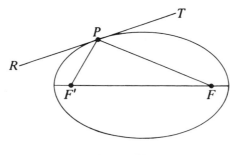

FIGURE 71

★EXERCISE 29

By use of Definition II *as in Example* 1 *on page* 124, *derive an equation for the ellipse satisfying the given conditions.*

1. Foci $(\pm 3,0)$; sum of the focal radii of any point is 10.
2. Foci $(0,\pm 8)$; sum of the focal radii of any point is 20.
3. Foci $(0,\pm c)$; sum of the focal radii of any point is $2a$ where $a > c$.
4. Foci $(2,3)$ and $(8,3)$; sum of the focal radii of any point is 10.
5. Foci $(3,2)$ and $(3,10)$; sum of the focal radii of any point is 10.
6. With particular values of a and c such that $a > 0$, $c > 0$, and $c < a$, prepare a loop of string with length $(2c + 2a)$. Pass the loop around tacks at two points F' and F on paper in a plane where $\overline{F'F} = 2c$. With a pencil stretching the loop tightly to a point P, where the pencil point is located, move P completely around the two tacks. Give remarks to demonstrate that P traces out an ellipse with F' and F as the foci.
7. The chord of an ellipse perpendicular to its axis at a focus is called a **focal chord** (or **latus rectum**) of the ellipse. Prove that the length of a focal chord of the ellipse (9) on page 126 is $2b^2/a$.

★46. HYPERBOLA DEFINED BY USE OF FOCAL RADII

DEFINITION III.   A **hyperbola** *is the set of points* $\{P\}$ *for which the* **absolute value** *of the difference of the undirected distances of P from two distinct fixed points* F' *and* F, *called the* **foci,** *is a constant, not zero, less than the distance between* F' *and* F.

EXAMPLE 1.   Find an equation for the hyperbola with foci $F':(-5,0)$ and $F:(5,0)$, if the absolute value of the difference of the focal radii for any point on the hyperbola is 6.

Solution.   1. In Figure 72, we have $h = |\overline{PF}|$ and $h' = |\overline{PF'}|$. Point $P:(x,y)$ is on the hyperbola when and only when $|h - h'| = 6$. This is true if and only if

$$h - h' = 6 \qquad or \qquad h' - h = 6. \qquad (1)$$

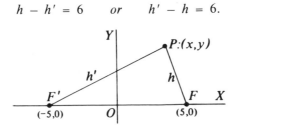

FIGURE 72

2. The equation $h - h' = 6$, or $h = 6 + h'$, becomes

$$\sqrt{(x - 5)^2 + y^2} = 6 + \sqrt{(x + 5)^2 + y^2}. \tag{2}$$

On rationalizing in (2), we obtain

$$16x^2 - 9y^2 = 144. \tag{3}$$

Use of the other equation $h' - h = 6$ from (1) leads to the same result, equation (3). Hence, (3) is the equation of the hyperbola. A graph of (3) is seen in Figure 74 on page 130 if the scale unit there is chosen so that $a = 3$.

DERIVATION OF A STANDARD EQUATION FOR A HYPERBOLA

1. Let the distance between the foci F' and F of Definition III be $2c$. Let the line through F' and F be the x-axis, with the origin at the midpoint of segment $F'F$, and with F in the positive direction from the origin on OX, as in Figure 73. Then the foci are $F':(-c,0)$ and $F:(c,0)$. Let the constant referred to in Definition III be $2a$ where $0 < a < c$.

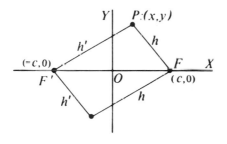

FIGURE 73

2. If $P:(x,y)$ is any point in the plane, let h and h' be the lengths of PF and PF', as in Figure 73, where $(h - h') > 0$ for one position of P and $(h - h') < 0$ for the other position of P. Then, P is on the hyperbola when and only when $|h - h'| = 2a$, that is, when and only when

$$h - h' = 2a \quad or \quad h' - h = 2a. \tag{4}$$

3. The equation $h - h' = 2a$, or $h = 2a + h'$, becomes

$$\sqrt{(x - c)^2 + y^2} = \sqrt{(x + c)^2 + y^2} + 2a. \tag{5}$$

On rationalizing in (5), we obtain

$$(c^2 - a^2)x^2 - a^2y^2 = a^2(c^2 - a^2). \tag{6}$$

If the equation $h' - h = 2a$ is operated on similarly, we again obtain (6). Hence, (6) is an equation for the hyperbola described in Definition III.

4. Define a number $b > 0$ by the equation

$$b^2 = c^2 - a^2. \tag{7}$$

Then, (6) becomes $b^2x^2 - a^2y^2 = a^2b^2$, *or*

(*foci on OX*) $$\frac{x^2}{a^2} - \frac{y^2}{b^2} = 1, \tag{8}$$

whose graph consists of the two branches in Figure 74. The x-intercepts are $\pm a$. The dimensions of the fundamental rectangle are $2a$ and $2b$. The transverse axis $V'V$ has length $2a$, and the endpoints V' and V are called the **vertices** of the hyperbola.

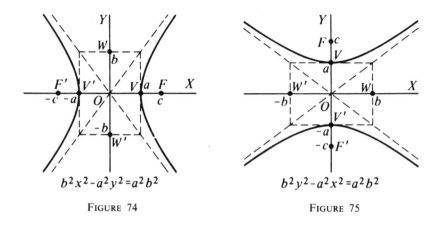

$$b^2x^2 - a^2y^2 = a^2b^2 \qquad\qquad b^2y^2 - a^2x^2 = a^2b^2$$

FIGURE 74   FIGURE 75

Without proof, as earlier in this chapter, we accept the fact that the hyperbola (8) has two asymptotes for which a single equation is

$$\frac{x^2}{a^2} - \frac{y^2}{b^2} = 0. \tag{9}$$

In the details leading to (8), if we let the line through F' and F be the y-axis, we obtain the hyperbola located as in Figure 75, with the equation $b^2y^2 - a^2x^2 = a^2b^2$, *or*

(*foci on OY*) $$\frac{y^2}{a^2} - \frac{x^2}{b^2} = 1. \tag{10}$$

Equations (8) and (10) are called *standard forms* for the equations of hyperbolas in the most simple positions in an xy-plane.

In (11) and (12) on pages 102 and 103, we saw that, when Case II of page 100 applies, the equation $Ax^2 + By^2 = C$ can be written in the form (8) or (10), which was called the *intercept form* on page 102. Hence, our results here in (8) and (10) justify the statements about a hyperbola made in Case II on page 100.

Note 1. A single definition called the *eccentricity-directrix* definition, as follows, can be given for *all* nondegenerate conics: *A conic is the set of all points {P} such that the undirected distance of P from a fixed point F is equal to a constant e > 0 times the undirected distance of P from a fixed line D, which does not go through F.* Then, we call *F* the *focus,* *e* the *eccentricity,* and *D* the *directrix* of the conic. Also, the conic is called a *parabola* when $e = 1$, an ellipse when $e < 1$, and a hyperbola when $e > 1$. In case $e = 1$, this definition duplicates Definition I on page 120 for a parabola. When $e < 1$ or $e > 1$, the new definition leads to the same standard equations as obtained previously for ellipses and hyperbolas, with the new "*focus*" being either one of the former *foci.* In the case of either an ellipse or a hyperbola, with the standard notations $\{a, b, c\}$ as used in Sections 45 and 46, it is found that the eccentricity e as just described is given by $e = c/a$. Thus, for an ellipse, with $b^2 = a^2 - c^2$, or $c^2 = a^2 - b^2$,

$$e = \sqrt{1 - \frac{b^2}{a^2}}. \tag{11}$$

For a hyperbola, with $b^2 = c^2 - a^2$, or $c^2 = a^2 + b^2$,

$$e = \sqrt{1 + \frac{b^2}{a^2}}. \tag{12}$$

If we consider $e = c/a$ apart from the new definition for a conic, we may refer to e as a "*shape constant.*" Then, for an ellipse, where $c^2 = a^2 - b^2$, with $e = c/a$ we make the following observations. Let a be fixed and suppose that $c \to a$ for a variable ellipse. Then the foci approach the vertices of the ellipse and $e \to 1$ as the ellipse becomes more slender in the direction perpendicular to the major axis. In the eccentricity-directrix definition for an ellipse, the value $e = 0$ is not permissible and hence, according to this definition, a circle (where $c = 0$) cannot be called a special case of an ellipse. However, with $e = c/a$ considered merely as a *shape constant,* we may say that $e = 0$ corresponds to the case of a *circle.* Thus, with a fixed value for a, if $b \to a$ the ellipse approaches a circle and $e \to 0$.

Note 2. Standard equations have been derived for the nondegenerate conics as described in Definitions I, II, and III, under the assumption that the scale units on the coordinate axes are equal. It can be proved that each of the standard equations represents a conic of the corresponding type if the equation is graphed in a coordinate plane where the scale units on the axes are *unequal.* In each case, the required proof can be accomplished by use of a transformation of coordinates of the form ($x = hx'$, $y = ky'$), which gives a new $x'y'$-system where the scale units are equal, but the type form of the equation remains unaltered. We shall not consider such proofs.* With unequal scale units, the structural formulas which can be derived for the locations of foci and any directrix no longer can be assumed to apply with the standard equations.

★EXERCISE 30

By use of Definition III *as in Example* 1 *on page* 128, *without using a standard form, derive the equation of the hyperbola to satisfy the data, and graph the equation.*

1. Foci $(0, \pm 5)$; absolute value of the difference of the focal radii is 6.
2. Foci $(\pm 5, 0)$; absolute value of the difference of the focal radii is 8.
3. Foci $(0, \pm c)$; absolute value of the difference of the focal radii is $2a$, where $0 < a < c$.
4. Foci $(2,3)$ and $(12,3)$; absolute value of the difference of the focal radii is 8.
5. Foci $(3,2)$ and $(3,10)$; absolute value of the difference of the focal radii is 6.
6. Suppose that the scale units on the coordinate axes in an xy-plane are equal. By use of Definition III, find the equation of the hyperbola with the foci (k,k) and $(-k,-k)$, where $k > 0$, if the absolute value of the difference of the focal radii of any point $P:(x,y)$ on the hyperbola is $2k$. Then show that, if $h > 0$, the graph of $xy = h$ is a hyperbola whose foci are the points $(\sqrt{2h}, \sqrt{2h})$ and $(-\sqrt{2h}, -\sqrt{2h})$. Prove that this hyperbola is equilateral, or that $a = b$ in the notation of (8) on page 130, and hence the asymptotes are the coordinate axes. (These results justify the statements in Case IV on page 101.)
7. Let a *focal chord* (or, *latus rectum*) for a hyperbola be defined as for an ellipse. Prove that the length of such a chord for the hyperbola (8) on page 130 is $2b^2/a$.

*For their discussion, see William L. Hart, *Algebra, Elementary Functions, and Probability* (Lexington, Mass.: D. C. Heath and Company, 1965) Theorem I, p. 167.

chapter 5

Introduction to Polynomial Functions

47. POLYNOMIALS AND EQUATIONS OF DEGREE *n*

Recall that, if f is a polynomial function of degree n in x, then

$$f(x) = a_0 + a_1 x + a_2 x^2 + \cdots + a_n x^n, \tag{1}$$

where $n \geq 0$ and $\{a_0, a_1, \cdots, a_n\}$ are constants with $a_n \neq 0$. A polynomial equation in x of degree n is of the form $P(x) = Q(x)$ where each of $P(x)$ and $Q(x)$ is 'a polynomial or is 0, and the equation $P(x) = Q(x)$ is equivalent to

$$a_0 + a_1 x + a_2 x^2 + \cdots + a_n x^n = 0, \tag{2}$$

with $n \geq 1$ and $a_n \neq 0$. We have referred to polynomial functions of degrees one, two, and three as linear, quadratic, and cubic functions, and to equations of these degrees as linear, quadratic, and cubic equations, respectively. There are corresponding names when the degree $n > 3$.

In this chapter, any functional symbol $f(x)$ will represent a polynomial in x unless otherwise specified. In any theorem and its proof, the numbers may be any complex numbers, unless exceptions are mentioned. Also, in any polynomial (1), as a rule we shall assume that $n > 0$.

48. CERTAIN BASIC THEOREMS

THEOREM I. **(The remainder theorem.)** *If r is a constant and if a polynomial $f(x)$ is divided by $(x - r)$ until a constant remainder is obtained, then this remainder is $f(r)$.*

Proof. After $f(x)$ is divided by $(x - r)$, let $q(x)$ be the partial quotient and let R be the constant remainder. At any stage in the division, the dividend, partial quotient and remainder are defined by

dividend = (divisor) · (quotient) + (remainder).

Hence, $$f(x) = (x - r)q(x) + R. \tag{1}$$

Since (1) is true for all values of x, we may use $x = r$ in (1), to obtain $f(r) = 0 \cdot q(r) + R$, or $R = f(r)$.

ILLUSTRATION 1.   With $f(x) = 3x^2 + 2x - 8$, the following long division of $f(x)$ by $(x - 3)$ illustrates Theorem I.

$$
\begin{array}{r}
3x + 11 = q(x) \\
x - 3 \overline{\smash{\big)}\; 3x^2 + 2x - 8} \\
\underline{3x^2 - 9x} \\
11x - 8 \\
\underline{11x - 33} \\
25 = R
\end{array}
$$

By substitution, $f(3) = 3(9) + 6 - 8 = 25 = R$.

THEOREM II.   (**The factor theorem.**) *If $f(r) = 0$ then $(x - r)$ is a factor of $f(x)$. That is, if r is a solution of $f(x) = 0$, then $(x - r)$ is a factor of $f(x)$.*

Proof.   In (1), $R = f(r)$ and hence $R = 0$. Therefore, the division of $f(x)$ by $(x - r)$ is exact or $f(x) = (x - r)q(x)$, which states that $(x - r)$ is a factor of $f(x)$.

THEOREM III.   (**Converse of the factor theorem.**) *If $(x - r)$ is a factor of $f(x)$, then $f(r) = 0$, or r is a solution of $f(x) = 0$.*

Proof.   If $f(x)$ is divided by $(x - r)$, the division is exact, and gives a quotient $q(x)$ which is a polynomial, so that $f(x) = (x - r)q(x)$. Hence, $f(r) = (r - r)q(r)$, or $f(r) = 0$, as stated in the theorem.

49. SYNTHETIC DIVISION

A telescopic method of division is available for abbreviating division of a polynomial $f(x)$ by a divisor $(x - r)$. The method is called *synthetic division,* and is introduced as follows.

ILLUSTRATION 1.   Consider the long division of $(3x^3 - 12x^2 + 8x - 2)$ by $(x - 2)$, given in the usual form in I below, and abbreviated in II and III as described later.

I.

$$
\begin{array}{r}
3x^2 - 6x - 4 = quotient \\
3x^3 - 12x^2 + 8x - 2 \,\big|\, x - 2 \\
\underline{\star 3x^3 - 6x^2} \\
- 6x^2 + 8x\star \\
\underline{\star - 6x^2 + 12x} \\
- 4x - 2\star \\
\underline{\star - 4x + 8} \\
Remainder = -10
\end{array}
$$

II.

$$3x^2 \quad - \quad 6x \quad - \quad 4 = quotient$$

$3x^3$	$- 12x^2$	$+ 8x$	$- 2$	$x - 2$
	$- 6x^2$	$+ 12x$	$+ 8$	
	$- 6x^2$	$- 4x$	$- 10$	

III.

$+3$	$- 12$	$+ 8$	$- 2$	1	$- 2$
	$- 6$	$+ 12$	$+ 8$		
$+3$	$- 6$	$- 4$	$- 10$		

In $(x - 2)$, the coefficient of x is 1. Therefore, at each stage of the division, the coefficient of the highest power of x in the remainder is the next coefficient of the quotient. Form II is obtained by omitting each term with a star $\star$ in form I, and then condensing form I into three lines. Form III is obtained from form II by writing only the coefficient in place of each term. The coefficient 3 is given in the third row of form III so that all coefficients of the quotient, and the remainder, are seen in the third row. Form III suggests the final form IV which illustrates synthetic division. In form IV, we use $+2$ instead of -2 at the right, to act as a multiplier, so that we may *add* instead of *subtract* in obtaining the fourth row.

IV.

$+3$	$- 12$	$+ 8$	$- 2$	$+2$
	$+ 6$	$- 12$	$- 8$	
$+3$	$- 6$	4	$- 10$	

Quotient $= 3x^2 - 6x - 4$.   *Remainder* $= -10$.

The following outline can be verified to be a description of the usual details involved in obtaining the coefficients in the quotient, and the constant remainder, when a polynomial $f(x)$ is divided by $(x - r)$.

Outline for the synthetic division of $f(x)$ by $(x - r)$.

1. *Write $f(x)$ in descending powers of x, with each missing power written with 0 as the coefficient. Then arrange the following calculations in three rows as specified.*
2. *In the first row, write the coefficients $\{a_n, a_{n-1}, \cdots, a_0\}$ of $f(x)$ in their listed order. In the third row, write a_n in the first place.*
3. *Multiply a_n by r, write ra_n in the second row, second place, and add ra_n to a_{n-1}. Write the sum in the third row.*
4. *Multiply the last mentioned sum by r, add the product to the next coefficient of the first row, and write the sum in the third row; etc. to the last coefficient (the constant term) of $f(x)$.*
5. *The final number at the right in the third row is the remainder, and the other numbers in the third row are the coefficients of the powers of x in the quotient, arranged in descending powers of x.*

EXAMPLE 1.   Divide: $(3x^4 + 11x^3 + 2x - 8) \div (x + 3)$.

Solution.   1. First we write $(x + 3) = [x - (-3)]$, so that $r = -3$ in the synthetic division given below. Let $f(x) = 3x^4 + 11x^3 + 2x - 8$.

$$\begin{array}{r|rrrrr|r} |+3 & +\,11 & 0 & +\,2 & -\,8 & \underline{-3} \\ & -\,9 & -\,6 & +\,18 & -\,60 \\ \hline |+3 & +\,2 & -\,6 & +\,20 & -\,68 & = f(-3) \end{array}$$

Quotient $= 3x^3 + 2x^2 - 6x + 20;$ *remainder* $= -68;$

$$\frac{3x^4 + 11x^3 + 2x - 8}{x + 3} = 3x^3 + 2x^2 - 6x + 20 - \frac{68}{x + 3}. \tag{1}$$

In Example 1, as a by-product of the division, we obtained $f(-3) = -68$. Thus, we could look upon the division as a means to solve the problem: "*Find $f(-3)$.*" This required division by $[x - (-3)]$. These remarks illustrate the following use of synthetic division.

> *To obtain the value of a polynomial $f(x)$ when $x = r$, divide $f(x)$ by $(x - r)$ by synthetic division to obtain the remainder $R = f(r)$.*   (2)

EXERCISE 31

Divide by long division and also by synthetic division.   Check the remainder by finding the value of the dividend for the value $x = r$ in the divisor when it is written in the form $(x - r)$.

1. $(4x^3 - 3x + 5) \div (x - 2)$.   **2.** $(2x^3 + 4x^2 - 7) \div (x + 3)$.

By synthetic division, find the quotient and the remainder.   Summarize the result as in (1) *on this page.*

3. $(5x^2 - 3x + 2) \div (x - 2)$.   **4.** $(3x^2 - 7) \div (x + 3)$.

5. $(4x^3 - 2x^2 + 5x - 6) \div (x + 2)$.

6. $(3x^4 - 2x^2 + 4x - 7) \div (x - 3)$.

7. $(2x^3 - 5x^2 + 4x - 5) \div (x - \frac{1}{2})$.

Solve by use of synthetic division.

8. If $f(x) = 3x^3 + 2x^2 - 7x - 5$, find $f(2); f(-1)$.

9. If $f(x) = 4x^4 - 3x^2 + x - 7$, find $f(3); f(-2)$.

10. If $f(x) = 2x^3 - 5x^2 - 10x + 8$, find $f(\frac{1}{2}); f(-\frac{1}{2})$.

11. If $g(y) = y^3 + 4y + 8$, find $g(.2); g(-.3)$.

12. If $h(z) = z^3 - 2z^2 + z - 5$, find $h(.3); h(-.2)$.

13. Obtain $(x^3 - x^2 - 8x + 12) \div (x - 2)^2$ by dividing by $(x - 2)$ twice, with synthetic division employed.

Answer the question by computing a value of the function involved and applying the factor theorem. If the answer is "yes," find another factor. Use synthetic division where convenient.

14. If $f(x) = x^3 - 5x^2 + 11x - 10$, is $(x - 2)$ a factor of $f(x)$?

15. If $g(y) = y^3 - 3y^2 - 5y + 10$, is $(y - 3)$ a factor of $g(y)$?

16. Is $(x - 3)$ a factor of $(x^3 - 27)$?

17. Is $(x + 2)$ a factor of $(x^5 + 32)$?

18. Is $(z + 2)$ a factor of $(z^6 - 64)$; of $(z^6 + 64)$?

19. If n is an *even* positive integer, prove that $(x - a)$ is a factor of $(x^n - a^n)$. Then find the other factor by synthetic division when $n = 6$.

20. Repeat Problem 19 when $(x + a)$ instead of $(x - a)$ is involved.

21. If n is a positive *odd* integer, prove that $(x + a)$ is a factor of $(x^n + a^n)$. Then find the other factor by synthetic division when $n = 7$.

Find the values of the constant h for which $(x - 3)$ is a factor of $f(x)$.

22. $f(x) = 2hx^2 - 5x + 3$.   **23.** $f(x) = h^2x^2 - 3hx - 10$.

50. THE ZEROS AND FACTORS OF A POLYNOMIAL

The following theorem was proved first in 1799 by the great German mathematician Johann Karl Friedrich Gauss (1777–1855). The proof is outside the scope of this book.

THEOREM IV. (**Fundamental theorem of algebra.**) *If the coefficients in a polynomial $f(x)$ of degree $n > 0$ are any complex numbers, then there exists at least one complex number which is a solution of the polynomial equation $f(x) = 0$. Or, every polynomial $f(x)$ of degree $n > 0$ with complex coefficients has at least one zero, real or imaginary.*

By use of Theorem IV, the following theorems will be proved.

THEOREM V. *If $f(x)$ is a polynomial of degree $n > 0$, there exist n factors, linear in x, whose product is $f(x)$.*

Proof.   1. Suppose that

$$f(x) = a_0 + a_1x + \cdots + a_nx^n, \tag{1}$$

where $a_n \neq 0$. By Theorem IV, $f(x)$ has at least one zero; let it be r_1. Then, by the factor theorem, $(x - r_1)$ is a factor of $f(x)$. Let $Q_1(x) = f(x) \div (x - r_1)$. Then $Q_1(x)$ is a polynomial of degree $(n - 1)$ whose term of highest degree is a_nx^{n-1}, and

$$f(x) = (x - r_1)Q_1(x). \tag{2}$$

2. By Theorem IV, $Q_1(x)$ has a zero r_2, and $Q_1(x) = (x - r_2)Q_2(x)$, where $Q_2(x)$ is a polynomial of degree $(n - 2)$ whose term of highest degree is $a_n x^{n-2}$. Then, by (2),

$$f(x) = (x - r_1)(x - r_2)Q_2(x).$$

3. By n steps of the preceding nature, we obtain n complex numbers $\{r_1, r_2, \cdots, r_n\}$, where there may be repetitions, and a polynomial $Q_n(x)$ whose term of highest degree is $a_n x^{n-n}$, or simply a_n, such that

$$f(x) = a_n(x - r_1)(x - r_2)\cdots(x - r_n). \tag{3}$$

In (3), we can absorb a_n into any chosen factor, for instance $a_n(x - r_1) = (a_n x - a_n r_1)$. Then, in (3), $f(x)$ is expressed as the product of n linear factors, which we desired to prove.

THEOREM VI.   *Any equation $f(x) = 0$ of degree $n > 0$ has at most n distinct solutions.*

Proof.   If $x = r_i$, where r_i is any one of the n numbers $\{r_1, r_2, \cdots, r_n\}$ in (3), then $f(r_i) = 0$, because at least one factor on the right in (3) is zero when $x = r_i$. There may be duplications among $\{r_1, r_2, \cdots, r_n\}$. Hence, these numbers constitute *at most n* distinct solutions of $f(x) = 0$.

2. If r is different from all of $\{r_1, r_2, \cdots, r_n\}$, then

$$f(r) = a_n(r - r_1)(r - r_2)\cdots(r - r_n) \neq 0,$$

because no factor on the right is zero. Hence, the only solutions of $f(x) = 0$ are the *distinct* numbers in the list $\{r_1, r_2, \cdots, r_n\}$, or there are *at most n distinct solutions* for the equation.

If a solution R occurs just once among $\{r_1, r_2, \cdots, r_n\}$ in (3), then R is called a *simple solution, or simple root,* of $f(x) = 0$. If R occurs exactly k times, that is, if $(x - R)^k$ is the highest power of $(x - R)$ which is a factor of $f(x)$, then R is called a solution, or root, of *multiplicity k* of $f(x) = 0$. Thus, Theorem VI may be restated as follows:

> *Every equation of degree $n > 0$ has exactly n solutions $\{r_1, r_2, \cdots, r_n\}$, where a solution of multiplicity k is counted as k solutions.*

Note 1.   Let $S = \{c_1, c_2, \cdots, c_k\}$ be the *solution set* of an equation $f(x) = 0$ of degree $n > 0$. In referring to S as the solution set, it is understood that any particular solution, say c_i, is listed *just once,* although c_i might occur more than once in the list $\{r_1, r_2, \cdots, r_n\}$ of numbers appearing in (3). Then, by Theorem IV, S is *not empty;* by Theorem VI, S consists of *at most n numbers.*

On account of Theorem V, any equation $f(x) = 0$ of degree n, with the solutions $\{r_1, r_2, \cdots, r_n\}$ from (3), is equivalent to an equation

$$a(x - r_1)(x - r_2) \cdots (x - r_n) = 0, \tag{4}$$

where $a \neq 0$ may be chosen arbitrarily. By use of (4), we can obtain a polynomial equation $f(x) = 0$ if its distinct solutions and their multiplicities are given.

EXAMPLE 1.   Obtain a polynomial equation in x with integral coefficients whose solutions are as follows: a simple solution $x = 2/3$, and a solution $x = 2$ of multiplicity two.

Solution.   1. By (4), the equation is of the form
$$a(x - \tfrac{2}{3})(x - 2)^2 = 0, or$$

$$a\left(\frac{3x - 2}{3}\right)(x - 2)^2 = 0. \tag{5}$$

In (5), choose $a = 3$ to eliminate fractions. Then the desired equation is

$$(3x - 2)(x^2 - 4x + 4) = 0, \qquad or \qquad 3x^3 - 14x^2 + 20x - 8 = 0.$$

Note 2.   If a and b are real, with $b \neq 0$, recall that the complex number $(a + bi)$ is called an *imaginary number*. Then, each of the imaginary numbers $(a + bi)$ and $(a - bi)$ is called the **conjugate** of the other number, and the two are referred to as **conjugate imaginary numbers**. Thus, the conjugate of $(3 - 4i)$ is $(3 + 4i)$.

In college algebra, the following theorem is proved, and we accept it for later use.

THEOREM VII.   *If an imaginary number $(a + bi)$, with a and b real and $b \neq 0$, is a solution of a polynomial equation $f(x) = 0$ with real coefficients, then the conjugate imaginary number $(a - bi)$ also is a solution of $f(x) = 0$.*

That is, by Theorem VII, imaginary solutions of $f(x) = 0$ occur in *conjugate pairs* if the coefficients in the polynomial $f(x)$ are real numbers.
Notice that, if a and b are real, then

$$[x - (a + bi)][x - (a - bi)] = \tag{6}$$

$$[(x - a) - bi][(x - a) + bi] = (x - a)^2 - (bi)^2$$
$$= x^2 - 2ax + a^2 - b^2i^2 = x^2 - 2ax + a^2 + b^2, \tag{7}$$

because $i^2 = -1$. Hence, in (4), if there are two conjugate imaginary solutions $(a + bi)$ and $(a - bi)$, the corresponding linear factors can be multiplied as in (6) to yield a quadratic factor (7) where all coefficients are real. Therefore, we have the following result:

> Every polynomial $f(x)$ with real coefficients can be written as a product of linear and quadratic factors having real numbers as coefficients.   (8)

EXAMPLE 1.   Obtain a polynomial equation of degree three with real coefficients where $x = 1$ and $x = 2 - 3i$ are solutions.

Solution.   Since imaginary solutions occur in conjugate pairs, another solution is $x = 2 + 3i$. Hence, the desired equation is

$$(x - 1)[x - (2 - 3i)][x - (2 + 3i)] = 0, \; or$$

$$(x - 1)[(x - 2) + 3i][(x - 2) - 3i] = 0, \; or \quad (9)$$

$$(x - 1)[(x - 2)^2 - 9i^2] = 0, \quad or \quad (x - 1)(x^2 - 4x + 13) = 0.$$

The equation is $x^3 - 5x^2 + 17x - 13 = 0$. Notice the convenience of the rearrangement of terms in (9) to exhibit a product of the sum and the difference of two numbers, producing the difference of their squares.

EXERCISE 32

Solve without multiplying given factors.   Use the quadratic formula where appropriate.
1. $(x - 2)(x + 3)(x - 1) = 0$.   2. $(2x^2 + 3x)(x^2 + 2x + 5) = 0$.
3. $(4x^2 - 25)(x^2 + 36) = 0$.   4. $(2x^2 + 4x - 3)(9x^2 + 1) = 0$.

If $f(x)$ has real coefficients, and the equation $f(x) = 0$ has the given solution, what other solution is possessed by the equation?
5. $(3 - 5i)$.   6. $(2 + 6i)$.   7. $(-4 - 3i)$.

Form an equation with integral coefficients having the given solutions.
8. $1; 1; -3$.   9. $3; 2; 2$.
10. $3; (1 + \sqrt{2})(1 - \sqrt{2})$.   11. $\frac{2}{3}; \pm i$.
12. $\pm 2; \pm 3i$.   13. $2: (3 + i); (3 - i)$.
14. $-2; (1 \pm i\sqrt{2})$.   15. $2; (-1 \pm \frac{1}{2}\sqrt{3})$.
16. $1; (2 \pm i\sqrt{3})$.   17. 3 as a double solution.
18. -2 as a solution of multiplicity three.

Form an equation $f(x) = 0$ with real coefficients to satisfy the condition.
19. A cubic equation with the solutions 3 and $(1 + i)$.
20. A quartic equation with the solutions $3i$ and $(2 - i)$.
21. Prove that a cubic equation with real coefficients has either three real solutions, or one real and two conjugate imaginary solutions.
22. State and prove theorems similar to the results in Problem 21 for quartic (degree four) equations, and quintic (degree five) equations.

51. RATIONAL SOLUTIONS OF EQUATIONS

Suppose that the coefficients in the equation

$$a_0 + a_1x + a_2x^2 + \cdots + a_nx^n = 0 \tag{1}$$

are integers and that $x = p/q$ is a rational solution of (1), where p/q is a fraction in lowest terms. On substituting $x = p/q$ in (1), where we shall specialize to the case $n = 3$ for clarity, we obtain

$$a_0 + a_1\frac{p}{q} + a_2\frac{p^2}{q^2} + a_3\frac{p^3}{q^3} = 0. \tag{2}$$

On multiplying both sides of (2) by q^3 we find that

$$a_0q^3 + a_1pq^2 + a_2p^2q + a_3p^3 = 0. \tag{3}$$

From (3), we obtain the following two equations:

$$-a_0q^3 = a_1pq^2 + a_2p^2q + a_3p^3; \tag{4}$$

$$-a_3p^3 = a_0q^3 + a_1pq^2 + a_2p^2q. \tag{5}$$

In (4) and (5), the value of each side is an integer, because the coefficients in (1), p, and q are integers. In (4), observe that p is a factor of the right-hand side. Hence, p must be a factor of the left-hand side, $-a_0q^3$. But, unless $p = \pm 1$, p is not a factor of q^3 because p/q is in lowest terms. Hence, p is a factor of a_0. Similarly, from (5), we find that q is a factor of a_3. In the preceding remarks, we have proved the following theorem, for the special case where $n = 3$.

THEOREM VIII.   Suppose that the coefficients in an equation of degree n are integers, and that p/q is a solution of the equation, where p and q are integers, q $\neq$ 0, and p/q is in lowest terms. Then p is a factor of the constant term in the equation and q is a factor of the coefficient of the term of highest degree.

If p/q is a negative rational solution of (1), we agree to take $q > 0$. By Theorem VIII, if $a_n = 1$ in (1), and p/q is a solution, then $q = 1$ because this is the only positive factor of a_n. Thus, with $a_n = 1$, *any rational solution of* (1) *is an integer p which is a factor of the constant term.*

In solving a polynomial equation by use of Theorem VIII, whenever a rational solution r is found, *depress the degree* of the equation $f(x) = 0$ by removing (dividing out) the linear factor $(x - r)$ which $f(x)$ possesses. Then, continue to search for other possible rational solutions by carrying on the work with the depressed equation.

To investigate an equation $f(x) = 0$ for rational solutions of the form p/q, first we list the possible values for p and for q, and all possible values of p/q, with q taken positive. Then, systematically, we test each possible rational solution p/q by using synthetic division to compute $f(p/q)$, except when it is easier not to use synthetic division.

EXAMPLE 1.   Solve: $x^4 - 3x^3 - 9x^2 + 9x + 14 = 0$.   (6)

Solution.   1. Any rational solution will be an integer which is a factor of 14. The possibilities are ± 1; ± 2; ± 7; ± 14.

2. Let $f(x)$ be the left-hand side of (6). Without synthetic division, $f(1) = 12 \neq 0$. Hence, $x = 1$ is *not* a solution. To test the value $x = -1$, we shall compute $f(-1)$ by dividing $f(x)$ by $[x - (-1)]$ by synthetic division:

$$
\begin{array}{r|rrrr|r}
+1 & -3 & -9 & +9 & +14 & \underline{-1} \\
   & -1 & +4 & +5 & -14 & \\
\hline
+1 & -4 & -5 & +14 & 0 = f(-1)
\end{array}
$$

Hence $x = -1$ is a solution of (6). Also, above, we found that

$$f(x) = (x + 1)(x^3 - 4x^2 - 5x + 14).$$ (7)

The remaining solutions of (6) are solutions of the depressed equation

$$q(x) = x^3 - 4x^2 - 5x + 14 = 0.$$ (8)

3. The possible rational solutions of (8) are $\{-1; \pm 2; \pm 7; \pm 14\}$. (We no longer list 1 because it was ruled out above.) Without synthetic division, $q(-1) = 14 \neq 0$. Hence, $x = -1$ is *not* a solution. Also, we find by synthetic division that $q(2) = -4$; hence, $x = 2$ is not a solution. To compute $q(-2)$ we shall divide $q(x)$ by $[x - (-2)]$ by synthetic division:

$$
\begin{array}{r|rrr|r}
+1 & -4 & -5 & +14 & \underline{-2} \\
   & -2 & +12 & -14 & \\
\hline
+1 & -6 & +7 & 0 = q(-2)
\end{array}
$$

Hence $x = -2$ is a solution of $q(x) = 0$ and thus is a solution of (6). Also, above, we found that

$$q(x) = (x^2 - 6x + 7)(x + 2).$$

Hence the other solutions of $q(x) = 0$ are solutions of the depressed equation

$$x^2 - 6x + 7 = 0.$$ (9)

4. To solve (9), use the quadratic formula:

$$x = \frac{6 \pm \sqrt{36 - 28}}{2}, \qquad or \qquad x = 3 \pm \sqrt{2}.$$

Hence the solutions of (6) are $\{-1; -2; (3 \pm \sqrt{2})\}$.

EXAMPLE 2.   Solve: $2x^3 - 3x^2 - 2x + 3 = 0.$ (10)

Solution.   1. Let $f(x)$ be the left-hand member of (10). If p/q is a solution of (10), the possible values for p are the factors of 3, and the possible values of q are the positive factors of 2. Thus, the possibilities for p are ± 1 and ± 3; for q are 1 and 2. Hence, the possibilities for p/q are $\{\pm 1; \pm 3; \pm\frac{1}{2}, \pm\frac{3}{2}\}$.

2. Without synthetic division, we find $f(1) = 0$. Hence $x = 1$ is a solution of (10). We divide $f(x)$ by $(x - 1)$ to obtain the other factor of $f(x)$:

$$
\begin{array}{c|cccc}
+2 & -3 & -2 & +3 & \underline{1} \\
 & +2 & -1 & -3 & \\
\hline
+2 & -1 & -3 & 0 = f(1) &
\end{array}
$$

Thus, $f(x) = (x - 1)(2x^2 - x - 3).$

The depressed equation is

$$2x^2 - x - 3 = 0.$$ (11)

3. On factoring in (11) we obtain $(2x - 3)(x + 1) = 0$. Hence, the solutions of (11) are $\frac{3}{2}$ and -1. The solutions of (10) are $\{1; -1; \frac{3}{2}\}$.

52. POSITIVE AND NEGATIVE SOLUTIONS OF *f(x) = 0*

Consider a polynomial equation of degree n,

$$f(x) = 0, \quad or \quad a_0 + a_1 x + a_2 x^2 + \cdots + a_n x^n = 0.$$ (1)

For convenience, we shall continue with the special case $n = 3$. If we place $x = -X$ in (1), we obtain

$$f(-X) = 0, \quad or \quad a_0 - a_1 X + a_2 X^2 - a_3 X^3 = 0.$$ (2)

If $x = r$ satisfies (1), then $X = -r$ will satisfy (2). Thus, the solutions of (2) are the *negatives* of the solutions of (1). We state our preceding conclusion as follows, with x used in place of X in (2).

> *To obtain an equation whose solutions are the negatives of the solutions of a given equation $f(x) = 0$, replace x by $-x$ in the equation, to obtain $f(-x) = 0$.*   (3)

Sometimes (3) is useful in the investigation of the negative solutions of an equation.

In this section, if we refer to a number as having a *plus* sign, we shall mean that the number is *positive.* To state that it has a *minus* sign, will mean that the number is *negative.*

Note 1.   Consider the equation

$$x^4 - 3x^2 + 2x - 7 = 0. \tag{4}$$

Its coefficients, with the terms arranged in descending powers of x, have the signs $(+, -, +, -)$. In reading from left to right, these signs show a change, or *variation,* from $+$ to $-$, then a *variation* from $-$ to $+$, and then from $+$ to $-$. Hence, we say that (4) shows *three variations in sign.* In counting such variations in an equation, any missing power with zero as its coefficient is disregarded.

The following result is proved in advanced texts on the theory of equations. We accept the result without discussion. In any polynomial equation $f(x) = 0$, before counting variations in signs, it is understood that the left-hand member is arranged in descending powers of x.

THEOREM IX.   (**Descartes' rule of signs.**)   *If $f(x)$ is a polynomial with real coefficients, the number of positive solutions of the equation $f(x) = 0$ cannot exceed the number of variations of sign in $f(x)$, and in any case differs from this number by an even integer.*

With the aid of (3), the following consequence of Theorem IX is obtained.

COROLLARY 1.   *The number of negative solutions of $f(x) = 0$ cannot exceed the number of variations of sign in $f(-x)$ and, in any case, differs from this number by an even integer.*

Without solving a given polynomial equation, we may be able to obtain useful information about its solutions by use of Descartes' rule of signs and other theorems which we have discussed.

EXAMPLE 1.   Without solving, investigate facts about solutions of

$$3x^4 + 2x^2 - 3x - 5 = 0. \tag{5}$$

Solution.   1. Let $f(x)$ represent the left-hand member. We notice that $f(x)$ exhibits just one variation of sign. Theorem IX states that (5) has *at most one* positive solution. Also, however, Theorem IX states that the actual number of positive solutions differs from 1 at most by an *even integer.* Hence, it cannot be true that there is *no* positive solution because 0 differs from 1 by an odd integer. Therefore, (5) has *exactly one* positive solution.

2. We obtain $f(-x) = 3x^4 + 2x^2 + 3x - 5$.   The equation $f(-x) = 0$ shows one variation of sign. Hence, $f(-x) = 0$ has *exactly*

one positive solution. It is the *negative* of a solution of $f(x) = 0$. Therefore (5) has *exactly one negative solution.*

3. Since (5) is of degree 4, there are four solutions, of which one is positive and one is negative. Hence there are two imaginary solutions, which are conjugate imaginary numbers. *Conclusion:* (5) has one positive solution, one negative solution, and two conjugate imaginary solutions.

EXERCISE 33

Find all rational solutions. If their determination leads to a depressed equation which is a quadratic equation, then find all solutions. If there are no rational solutions, demonstrate this fact.

1. $x^3 + x^2 - 8x - 12 = 0$.
2. $x^3 - 3x^2 - 16x - 12 = 0$.
3. $x^3 - 2x^2 - 9x + 4 = 0$.
4. $x^4 - 15x^2 - 10x + 24 = 0$.
5. $x^3 - x^2 - 6x + 2 = 0$.
6. $x^4 - 6x^2 - 15x - 4 = 0$.
7. $2x^3 + 3x^2 - 7x + 6 - 0$.
8. $3x^3 + 4x^2 - 12x - 5 = 0$.
9. $3x^3 - 7x^2 - 3x + 2 = 0$.
10. $2x^3 - 5x^2 - 8x + 5 = 0$.

Without solving the equation, investigate the nature of its solutions by use of available theorems.

11. $3x^2 - 5x - 7 = 0$.
12. $4x^3 - 2x^2 + 3x = 7$.
13. $2x^4 - 5x = 2$.
14. $3x^4 + 2x^2 + 3 = 2x^3$.
15. $5x^4 + 7x^2 = 3$.
16. $3x^6 + 8 = 0$.
17. $2x^3 + 5 = 0$.
18. $2x^7 + 8 = 0$.
19. $2x^4 + 3x^3 + 4x^2 = 8$.
20. $5x^3 + 3x^2 - 2 = 0$.
21. $2x^5 + 3x^2 = 7 + 4x$.
22. $4x^5 + 5x^2 - 2x^3 = 1$.

★ *Let x be any nth root of H, where n is a positive integer and H is a real number, not zero. In the following problems, recall that* $x^n = H$. *Use preceding theorems.*

23. If $K > 0$ and n is even, prove that all nth roots of $-K$ are imaginary.
24. If $K > 0$ and n is even, prove that one nth root of K is positive, one is negative, and any other nth root of K is imaginary.
25. If $K > 0$ and n is odd, prove that one nth root of K is positive and any other nth root of K is imaginary.
26. If $K > 0$ and n is odd, prove facts similar to those in Problem 25 about the nth roots of $-K$.

53. POLYNOMIAL GRAPHS AND APPLICATIONS

If $P(x)$ is a polynomial of degree n where $n > 2$, no special name is given to the graph of $y = P(x)$. In calculus, powerful methods are developed for graphing functions of any variety. Hence, our present discussion of polynomial graphs will be incomplete.

We recall the remarks about applications of the graph of $y = f(x)$ on page 108. Those statements are useful when $f(x)$ is a polynomial of any degree, particularly when $f(x)$ is a product of powers of real linear factors.

EXAMPLE 1.   Graph the polynomial $(x^3 - 27x + 6)$. Then, by means of the graph, obtain approximations to the solutions of the following equation, and the solution set of each inequality.

$$x^3 - 27x + 6 = 0; \quad x^3 - 27x + 6 < 0; \quad x^3 - 27x + 6 \geqq 0. \quad (1)$$

Solution.   1. With $P(x) = x^3 - 27x + 6$, the preceding statements become $P(x) = 0$; $P(x) < 0$; $P(x) \geqq 0$.

2. Since it appears that $P(x)$ cannot be factored conveniently, our only device for obtaining a graph of $y = P(x)$ is to compute solutions (x,y) by substituting values for x in $P(x)$. The following table of solutions is the basis for the graph of $y = P(x)$ in Figure 76.

$y =$	-48	50	60	52	32	6	-20	-40	-48	-38	-4	60
$x =$	-6	-4	-3	-2	-1	0	1	2	3	4	5	6

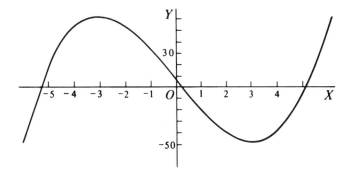

FIGURE 76

3. In Figure 76, the *x-intercepts* of the graph, T, or the *zeros* of $P(x)$, or the *solutions* of $P(x) = 0$, are approximately $\{-5.2, .2, 5.1\}$.

4. T is *below* OX, or $P(x) < 0$, when $x < -5.2$, and when $.2 < x < 5.1$. Hence, the solution set of $P(x) < 0$ is the *union* of two intervals, or $\{x < -5.2\} \cup \{.2 < x < 5.1\}$. T is *at or above* the x-axis, or $P(x) \geqq 0$, when $-5.2 \leqq x \leqq .2$ and when $5.1 \leqq x$. Hence, the solution set of $P(x) \geqq 0$ is $\{-5.2 \leqq x \leqq .2\} \cup \{5.1 \leqq x\}$.

Suppose that r is a constant and h is a positive integer such that

$(x - r)^h$ is the highest power of $(x - r)$ which is a factor of a polynomial $P(x)$. In calculus, the following facts concerning the graph of $y = P(x)$ are demonstrated. We shall use these results hereafter.

    I. *If $h = 1$, the graph cuts the x-axis sharply at $x = r$, and thus is not tangent to the x-axis at $x = r$.*

    II. *If $h > 1$, the graph is tangent to the x-axis at $x = r$.*

    III. *If $h > 1$ and h is even, the graph is entirely on one side of the x-axis near $x = r$.*

    IV. *If $h > 1$ and h is odd, the graph crosses the x-axis at $x = r$ and is tangent to the axis from below and above.*

Under the circumstances for a graph T near $x = r$ as described in IV, it is said that T has an **inflection point** at $x = r$.

If a polynomial $P(x)$ is a product of real linear factors, the x-intercepts of the graph, T, of $y = P(x)$ can be obtained by inspection of $P(x)$. Then T can be drawn quickly with the aid of a few of its points intermediate between the intercept points.

    EXAMPLE 2.   If $P(x) = (x + 2)(x - 1)(x - 3)$, obtain the graph of $y = P(x)$, and the solution sets of $P(x) < 0$ and $P(x) > 0$.

    Solution.   1. Let T be the graph of $y = P(x)$. The x-intercepts of T are $\{-2, 1, 3\}$. The following table of solutions of $y = P(x)$ were obtained by substituting values for x in $P(x)$. The table is the basis for the graph T in Figure 77.

$y =$	-24	0	8	6	0	-4	0	18
$x =$	-3	-2	-1	0	1	2	3	4

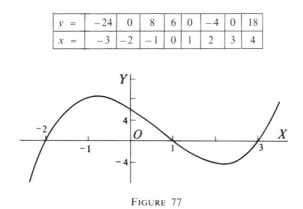

FIGURE 77

    2. From Figure 77, the solution set of $P(x) < 0$ is the following *union* of intervals: $\{x < -2\} \cup \{1 < x < 3\}$. The solution set of $P(x) > 0$ is $\{-2 < x < 1\} \cup \{3 < x\}$.

Suppose that $P(x)$ is a polynomial whose linear factors are known and whose zeros are real* numbers. Then, the inequalities $P(x) < 0$ and $P(x) > 0$ can be solved conveniently without a graph of $y = P(x)$, if desired, by the method employed in Example 3 on page 109 for the case where P is a quadratic function.

EXAMPLE 3.   With $P(x) = (x + 2)(x - 1)(x - 3)$ as in Example 2, obtain the solution set of $P(x) < 0$ without using a graph of $y = P(x)$.

Solution.   1. To solve $P(x) < 0$, or

$$[x - (-2)](x - 1)(x - 3) < 0, \tag{2}$$

first observe that the zeros $x = -2$, $x = 1$, and $x = 3$ of $P(x)$ divide the number scale into the following open intervals, as seen in Figure 78:

$$\{x < -2\}; \quad \{-2 < x < 1\}; \quad \{1 < x < 3\}; \quad \{3 < x\}. \tag{3}$$

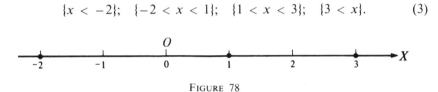

FIGURE 78

2. By use of (9) and (10) on page 109, learn the nature, positive or negative, of each factor of the left-hand side of (2) on each interval of (3):

On $\{x < -2\}$: $[x - (-2)] < 0;$ $(x - 1) < 0;$ $(x - 3) < 0.$ (4)

On $\{-2 < x < 1\}$: $[x - (-2)] > 0;$ $(x - 1) < 0;$ $(x - 3) < 0.$ (5)

On $\{1 < x < 3\}$: $x - (-2) > 0;$ $(x - 1) > 0;$ $(x - 3) < 0.$ (6)

On $\{3 < x\}$: $x - (-2) > 0;$ $(x - 1) > 0;$ $(x - 3) > 0.$ (7)

Hence, from (2), $P(x) < 0$ on $\{x < -2\}$ and on $\{1 < x < 3\}$. Or, the solution set of $P(x) < 0$ is $\{x < -2\} \cup \{1 < x < 3\}$.

EXAMPLE 4.   Graph: $y = (x + 1)^2(x - 2)^3.$

Solution.   1. The x-intercepts are $x = -1$ and $x = 2$.

2. We recall statements II–IV concerning the effect of a factor $(x - r)^h$. Also, we compute a few values for y with $x < -1$, with $-1 < x < 2$, and with $2 < x$, as in the following table. In Figure 79, the graph was made tangent to the x-axis and below it at $x = -1$, because of III. At $x = 2$, the graph was drawn tangent to the x-axis from above when $x > 2$, and from below when $x < 2$, because of IV. Thus the tangent line *crosses the graph* where $x = 2$, with the curve concave downward to the left and concave upward to the right. The graph has an *inflection point* where $x = 2$. The graph as thus obtained may be referred

---

*This condition is not essential but will apply here.

to as being only *qualitatively accurate,* because it was based on knowledge
of its general shape, and just a few accurate points.

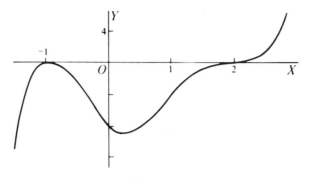

FIGURE 79

$y =$	-64	0	-8	-4	0	16
$x =$	-2	-1	0	1	2	3

EXERCISE 34

*In some problems it may be desirable to use a large scale unit on the
x-axis, and a very small scale unit on the y-axis in the xy-plane.*

In Problems 1 and 2, obtain meticulously accurate graphs.

1. If $P(x) = -x^3 + 12x - 3$, obtain a graph of the equation $y = P(x)$,
   with $x = 2$ and $x = -2$ among the values for x in the table of solutions (x, y) which is used. Then solve graphically $P(x) = 0$; $P(x) < 0$;
   $P(x) \geq 0$.
2. If $P(x) = 2x^3 + 3x^2 - 12x - 2$, graph the function $P(x)$, with
   $x = -2$ and $x = 1$ among the values used in the table of coordinates for points on the graph. Then solve graphically $P(x) = 0$;
   $P(x) \leq 0$; $P(x) > 0$.

*On the same coordinate system, draw merely a qualitatively accurate
graph of each equation. Factor first if desirable.*

3. $y = (x + 1)(x - 2)(x - 4)$; $(x + 1)(x - 2)(x - 4) = 0$.
4. $y = x^3 + x^2 - 6x$; $x(x - 2)(x + 3) = 0$.
5. Solve graphically: $(x + 2)(x + 1)(x - 4) > 0$.
6. First solve the equation $x^3 - 2x^2 - 3x = 0$ by use of factoring.
   Then solve $x^3 - 2x^2 - 3x \leq 0$ graphically, or without a graph.

Graph the function with qualitative accuracy.

7. x^3.   8. x^4.   9. $(x - 1)^2$.   10. $(x - 2)^3$.

Obtain a qualitatively accurate graph of the equation.

11. $y = (x + 1)^2(x - 3)$.   **12.** $y = (x + 2)^2(x - 2)^2$.

Find the solution set of each inequality without drawing a graph.

13. $(x + 3)(x - 2)(x - 4) < 0$; $(x + 3)(x - 2)(x - 4) > 0$.

14. $(2x + 5)(x - 1)(3 - x) < 0$; $(2x + 5)(x - 1)(3 - x) > 0$.

Note 1. To state that a function $f(x)$ is an **even function** of x means that $f(-x) = f(x)$ for each number x in the domain of f. To state that a function $f(x)$ is an **odd function** of x means that $f(-x) = -f(x)$ for each number x in the domain of f.

Prove that f is an odd function or an even function. Then test the equation $y = f(x)$ for the various types of symmetry of its graph, and draw it.

15. $f(x) = x^3$.   **16.** $f(x) = x^4$.   **17.** $f(x) = x^3 - 4x$.

18. $f(x) = 9x^2 - x^4$.

Comment. The results in Problems 15–18 indicate that the names *even function* and *odd function* are appropriate when polynomial functions are involved. However, an odd or an even function need not be a polynomial function.

19. Prove that the graph of the equation $y = f(x)$ is symmetric with respect to the y-axis if f is an even function, and to the origin if f is an odd function.

Find the solution set of the inequality by any method.   Perhaps first factor, after solving a corresponding equation.

20. $x^3 - 7x + 6 < 0$.   **21.** $x^3 - x^2 - 8x + 12 \geq 0$.

chapter 6

Trigonometric Functions of Angles

54. NUMERICAL AND ANALYTIC TRIGONOMETRY

The subject called *trigonometry* involves the study of certain functions, which originated as *functions of angles*. However, as we shall observe later, the geometric concept of an *angle* is not essential in the introduction and use of these functions in calculus. The Greek mathematician Hipparchus (most active scientific work about 146–127 B.C.) is credited with being the originator of trigonometry. The oldest existing book including it is the *Almagest*, by the Greek-Egyptian mathematician and astronomer Ptolomey, who lived in the second century A.D. Many outstanding developments in trigonometry occurred in the work of various Arabian and Persian mathematicians and astronomers in the cleven centuries after the time of Ptolomey. Trigonometry was first treated as a subject apart from astronomy by the Persian mathematician Nasir addin at-Tusi in the thirteenth century. The first treatment of trigonometry as a separate field in the Western world was given by the German mathematician Regiomontanus (1436–76), whose work was done independently without knowledge of the achievements of the Persian mathematicians just mentioned.

The early development of trigonometry was pointed at the use of the subject in the solution of plane and spherical triangles, and their applications, particularly in surveying and astronomy. This side of the subject now is referred to as *numerical trigonometry* and remains of great importance. In the development of calculus, it was found that the trigonometric functions, which form the skeleton of trigonometry, are indispensable in connection with analytic features of calculus which have no contact with numerical trigonometry. Thus, it is impossible to develop calculus thoroughly for the *algebraic* functions without introduction of the *trigonometric* functions. The resulting necessary study of the trigonometric functions apart from their use in numerical trigonometry may be referred to as *analytic trigonometry*. That phase of the subject is our primary concern in this text.

55. DIRECTED ANGLES

Let *P* be any point on a given line *L*. Then, the set of points consisting of *P* and all points on *L* in either one of the two directions* on *L* is referred to as a **ray,** or **half-line,** with *P* as the *initial point.* Thus, in Figure 80, the thick part of the line represents a ray.

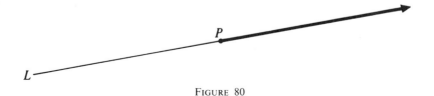

FIGURE 80

In a plane, suppose that a ray with the initial point *O* revolves about *O* in the plane in either a clockwise or a counterclockwise sense from an *initial position OI* to a *terminal position OT.* Then, this rotation is said to *generate an angle IOT* with *vertex O, initial side OI,* and *terminal side OT,* as shown in Figure 81, where the curved arrow indicates the nature of the rotation. We write "∠*IOT*" for "*angle IOT.*" To measure rotation, one degree (1°) is defined as† 1/360 of a complete revolution of *OI* about *O*; one minute (1') as 1/60 of 1°; one second (1") as 1/60 of 1'. We define ∠*IOT* as the *amount of the rotation* shown by the geometric configuration consisting of the sides *OI* and *OT,* and the curved arrow indicating the rotation involved. Regardless of the unit employed, an angle is assigned a *positive value,* and then is called a *positive angle,* if it is generated by *counterclockwise rotation.* An angle is assigned a *negative value,* and then is called a *negative angle,* if it is generated by *clockwise rotation.* An angle *IOT* is said to have the value *zero* if the initial side *OI* and terminal side *OT* coincide‡ and *no* rotation from *OI* to *OT* is in-

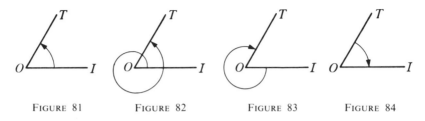

| FIGURE 81 | FIGURE 82 | FIGURE 83 | FIGURE 84 |

*We accept "*direction*" on *L* as an undefined term.

†Measurement of rotation in degrees, minutes, and seconds is called the **sexagesimal system,** because of the divisor 60 in the subdivision of units.

‡*OI* and *OT* may coincide when the angle does not have the value zero, for instance when the angle is 360°.

volved. Thus, any angle is a *real number* (of angular units), which may be positive, negative, or zero, for which we may construct a geometric representation as in Figure 81 when desired. To add angles, we may combine their amounts of rotation geometrically, or perform the equivalent operation on the values of the angles to obtain their sum. If an angle $w°$ is represented geometrically as in Figure 81, we write $\angle IOT = w°$. In any reference to the initial and terminal sides of an angle $w°$, or in any geometric situation where $w°$ is met, without special comment it is to be inferred that a geometric representation of $w°$, as in Figure 81, is involved.

ILLUSTRATION 1.   In each of Figures 81–84, the relative locations of the initial and terminal sides of $\angle IOT$ are the same. In Figure 81, the measure of $\angle IOT$ is 60°, or $\angle IOT = 60°$. In Figure 82, $\angle IOT$ is formed by rotation of OI through 360°, and then through 60°, so that

$$\angle IOT = 360° + 60° = 420°.$$

In Figure 83, the measure of $\angle IOT$ is negative, and $\angle IOT = -300°$. In Figure 84, $\angle TOI = -60°$.

Note 1.   In elementary geometry, an angle, say $\angle IOT$, is a *static* entity, described by a configuration consisting of rays OI and OT with a common initial point O, but with no rotation involved. In this section, we have introduced "*signed*" or "*directed*" angles.

A positive angle is referred to as an *acute angle* if its value lies between 0° and 90°, and an *obtuse angle* if its value lies between 90° and 180°. If any angle is represented by merely drawing two rays as its sides radiating from the vertex V of the angle, it will be understood that the measure of the angle is positive, at most equal to 180°. Thus, in any triangle, each angle will be said to have positive measure.

56. RADIAN MEASURE FOR ANGLES

The sexagesimal system of angular measurement, where the unit is 1° (with its subunits), is very convenient in many applications of trigonometry. However, in the study of calculus, and in related applications of trigonometry, it is found that the most convenient* unit for angular measurement is a *radian,* described as follows.

DEFINITION I.   One **radian** *is the measure of a positive angle such that, if its vertex is at the center of a circle, rotation of the initial side to the terminal side will sweep out on the circumference an arc whose length is the radius of the circle.*

*A comment on the nature of the convenience will occur in a later section.

Thus, in a circle, a central angle of one radian intercepts on the circumference an arc whose length is the circle's radius. Figure 85 shows an angle of 1 radian, with the length of arc AB equal to the circle's radius.

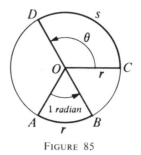

FIGURE 85

Let H radians be the measure in radians of the angle generated by revolving radius OA of Figure 85 about O through a complete revolution counterclockwise. From elementary geometry, the measure of a central angle in a circle is proportional to the length of the arc intercepted by the angle on the circle's circumference. The angle H radians intercepts the whole circumference, with length "$2\pi r$," and 1 radian intercepts an arc of length r. Hence

$$\frac{H}{1} = \frac{2\pi r}{r}, \quad or \quad H = 2\pi. \tag{1}$$

Thus the measure of the angle formed by a complete counterclockwise revolution of a ray about its initial point is 2π *radians.* This angle also has the measure 360°. Hence

$$\mathbf{360° = (2\pi\ radians)},\ or \tag{2}$$

$$\mathbf{180° = (\pi\ radians)}. \tag{3}$$

An "*equation*" such as (2), or (3), is not an equation in the usual sense, expressing the equality of two numbers. Thus, (3) is merely a convenient abbreviation of the statement that an angle of 180° is the same as an angle of 2π radians. If we indicated "2π radians" as "$2\pi^{(r)}$," with the superscript "r" meaning *radians,* then (2) would become $360° = 2\pi^{(r)}$. From (3), on dividing both sides first by 180, and second by π, or $3.14159\cdots$, taken as 3.1416 approximately, we obtain the following statements:

$$1° = \frac{\pi}{180}\ radian = .017453\ radian,\ approximately; \tag{4}$$

$$(1\ radian) = \frac{180°}{\pi} = 57.296°,\ approximately. \tag{5}$$

Equalities (4) and (5) give the following rules for transformation from degrees to radians, and from radians to degrees.

> *To change from degrees to radians, multiply the number of degrees by $\pi/180$.* } (6)

> *To change from radians to degrees, multiply the number of radians by $180/\pi$.* } (7)

Instead of using (6) or (7), we may recall (3), which justifies the statement that *any multiple of* 180° *is the same multiple of* π *radians.*

ILLUSTRATION 1. $45° = \frac{1}{4}(180°) = \frac{1}{4}\pi$ rad.:

$60° = \frac{1}{3}(180°) = (\frac{1}{3}\pi$ rad.); $\qquad (\frac{1}{2}\pi$ rad.) $= \frac{1}{2}(180°) = 90°$:

$(\frac{3}{2}\pi$ rad.) $= \frac{3}{2}(180°) = 270°$; $\qquad (\frac{5}{4}\pi$ rad.) $= \frac{5}{4}(180°) = 225°$.

ILLUSTRATION 2. From (6) and (4).

$$11° = 11 \cdot \frac{\pi}{180} rad. = 11(.017453) rad. = .1920 \, rad., approximately.$$

From (7) and (5),

$$5 \text{ rad.} = 5 \cdot \frac{180°}{\pi} - 5(57.296°) = 286.48°.$$

Instead of using (6) or (7), Table VI and the associated auxiliary tables are available for converting degrees to radians, or radians to degrees.

ILLUSTRATION 3. To convert 17°25′ to radians, we may use the conversion tables for degrees, and for minutes, following Table VI:

$$17° = .29671 \, rad.; \qquad 25' = .00727 \, rad.$$

Hence, $\qquad 17°25' = (.29671 + .00727) \, rad. = .30398 \, rad.$

In a circle, as in Figure 85, a central angle of 1 radian intercepts on the circumference an arc of length r. Hence, a central angle of* θ radians intercepts on the circumference an arc of length s where

$$\frac{\theta}{1} = \frac{s}{r}, \qquad or \qquad s = r\theta; \ or \tag{8}$$

$$\textbf{arc} = (\textbf{radius} \times \textbf{angle}, in \, radians). \tag{9}$$

ILLUSTRATION 4. In a circle of radius 5, a central angle of 2 radians intercepts on the circumference an arc whose length is 5×2 or 10 units.

*The Greek letter "θ" is read "*theta.*"

Hereafter in this text, if a number, say θ, is specified as an *angle,* it will be understood that the measure of the angle is θ *radians.* If it is desired to describe an angle in degree measure, the usual notation indicating a number of degrees will be used. Thus, if we wish to specify that an angle θ is the same as $w°$, where w is a number, we shall write $\theta = w°$, meaning that the degree measure of θ radians is $w°$.

ILLUSTRATION 5.   If θ is an angle, and $\theta = .83$, the angle is .83 radians. By inspection of the first two columns of Table VI, we find that the degree measure of θ is $47°33.3'$, or $\theta = 47°33.3'$.

In this text we shall employ both degree measure and radian measure for angles, with the major emphasis placed on use of radian measure, because of its importance in calculus.

ILLUSTRATION 6.   If θ is an acute angle, then $0 < \theta < \frac{1}{2}\pi$. If θ is an obtuse angle then $\frac{1}{2}\pi < \theta < \pi$.

EXERCISE 35

Change the radian measure of the angle to measure in degrees. Possibly use Table VI.

1. $\frac{1}{6}\pi$.   2. $\frac{1}{3}\pi$.   3. $\frac{2}{3}\pi$.   4. $\frac{3}{4}\pi$.   5. $\frac{5}{6}\pi$.

6. $\frac{11}{6}\pi$.   7. $\frac{5}{4}\pi$.   8. $\frac{1}{4}\pi$.   9. $-\frac{2}{3}\pi$.   10. $-\pi$.

11. 3π.   12. 4.   13. -3.   14. 1.5.   15. 2.7.

Find the radian measure of the angle. Possibly use the conversion tables given with Table VI.

16. $60°$.   17. $30°$.   18. $45°$.   19. $135°$.   20. $-90°$.
21. $-180°$.   22. $210°$.   23. $150°$.   24. $225°$.   25. $270°$.
26. $315°$.   27. $330°$.   28. $300°$.   29. $-150°$.   30. $-240°$.
31. $2°20'$.   32. $40°17'$.   33. $82°38'$.   34. $62°11'$.

35. By use of a protractor and preliminary conversion of radian measure to degree measure, construct an angle of 1 rad.; 1.5 rad.; 2 rad.
36. If $\theta = 1.7$, is θ acute or obtuse?
37. If $\theta = 1.42$, is θ acute or obtuse?

57. DEFINITION OF THE TRIGONOMETRIC FUNCTIONS

An angle θ is said to be in its **standard position** on an xy-system* of coordinates in case the vertex of θ is at the origin and the initial side of θ

---

*In all of this chapter, except where otherwise mentioned, the scale units will be equal on the axes of any coordinate system involved, and the same unit will be used for distance in an arbitrary direction.

lies on that part of the *x*-axis where $x > 0$. Certain angles are shown in standard positions in Figure 86. An angle θ is said to lie *in a certain quadrant* when the terminal side of θ falls in that quadrant if θ is in its standard position in the *xy*-plane. An angle θ is called a **quadrantal angle** if its terminal side falls on a coordinate axis when θ is in its standard position in the *xy*-plane.

ILLUSTRATION 1.   The angles $0°$, $90°$, $180°$, $270°$, $360°$, $-90°$, $-180°$, etc. are quadrantal angles. In radian measure, the following are quadrantal angles: $0, \frac{1}{2}\pi, \pi, \frac{3}{2}\pi, 2\pi, -\frac{1}{2}\pi, -\pi$, etc. Any quadrantal angle is of the form $n \cdot (\frac{1}{2}\pi)$ radians, or $n \cdot (90°)$, where *n* is an integer.

ILLUSTRATION 2.   Figure 86 shows an acute angle, $40°$; an obtuse angle, $140°$; an angle $320°$ in quadrant IV. Any acute angle is in quadrant I. Any obtuse angle is in quadrant II.

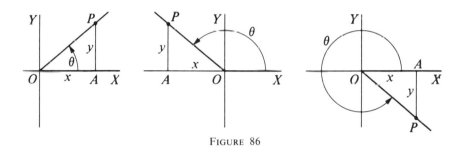

FIGURE 86

Let θ be a variable whose domain is all real numbers. Then, for each angle θ, six numbers (with exceptions noted later) sine θ, cosine θ, tangent θ, etc., are defined as follows.

DEFINITION II.   *Place angle θ in its standard position in an xy-plane, and let $P:(x,y)$ be any point, not the origin, on the terminal side of θ, with*

$$| \overline{OP} | = r = \sqrt{x^2 + y^2},$$

*where r will be called the **radius vector** of P. Then the direct* trigonometric functions of θ are as follows:†*

---

*"*Direct*" will be omitted usually hereafter. This name, *direct,* serves to distinguish the corresponding functions from "*inverse*" trigonometric functions as defined later.

†Except when the denominator involved is zero, for the tangent, cotangent, secant, and cosecant.

$$\text{sine } \theta = \frac{\text{ordinate of } P}{\text{radius vector of } P}, \quad \text{or} \quad \sin \theta = \frac{y}{r};$$

$$\text{cosine } \theta = \frac{\text{abscissa of } P}{\text{radius vector of } P}, \quad \text{or} \quad \cos \theta = \frac{x}{r};$$

$$\text{tangent } \theta = \frac{\text{ordinate of } P}{\text{abscissa of } P}, \quad \text{or} \quad \tan \theta = \frac{y}{x};$$

$$\text{cotangent } \theta = \frac{\text{abscissa of } P}{\text{ordinate of } P}, \quad \text{or} \quad \cot \theta = \frac{x}{y}; \qquad (1)$$

$$\text{secant } \theta = \frac{\text{radius vector of } P}{\text{abscissa of } P}, \quad \text{or} \quad \sec \theta = \frac{r}{x};$$

$$\text{cosecant } \theta = \frac{\text{radius vector of } P}{\text{ordinate of } P}, \quad \text{or} \quad \csc \theta = \frac{r}{y}.$$

In (1), $\tan \theta$ and $\sec \theta$ are not defined when θ is a quadrantal angle having its terminal side on the y-axis, because then $x = 0$ so that y/x and r/x are meaningless. Similarly, $\cot \theta$ and $\csc \theta$ are not defined when θ has its terminal side on the x-axis. For instance, $\sec \frac{1}{2}\pi$, $\cot 0$, $\tan 90°$ and $\csc 0°$ are not defined.

For any angle θ associated with Definition II, as in Figure 86, we shall refer to $\triangle OAP$ as the **reference triangle** for θ.

EXAMPLE 1.   If the terminal side of an angle θ, in standard position, passes through $P:(-4,3)$, find the trigonometric functions of θ.

Solution.   In Figure 87,

$$r = \sqrt{x^2 + y^2} = \sqrt{16 + 9} = 5. \qquad (2)$$

Hence, with $r = 5$, $x = -4$, and $y = 3$ in (1), we obtain

$$\sin \theta = \tfrac{3}{5}; \quad \cos \theta = -\tfrac{4}{5}; \quad \tan \theta = -\tfrac{3}{4};$$

$$\csc \theta = \tfrac{5}{3}; \quad \sec \theta = -\tfrac{5}{4}; \quad \cot \theta = -\tfrac{4}{3}.$$

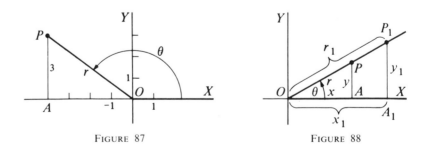

FIGURE 87   FIGURE 88

The ratios in (1) are properly referred to as *functions* of θ because the values of the fractions depend only on the terminal side of θ and not on the particular point P used on that side. Thus, suppose that we select any two distinct points $P:(x,y)$ and $P_1:(x_1,y_1)$, not at the origin, on the terminal side of an angle θ, as in Figure 88, with $r = |\overline{OP}|$ and $r_1 = |\overline{OP_1}|$. By (1), from $\triangle$'s OAP and OA_1P_1, respectively,

$$\sin \theta = \frac{y}{r}; \qquad \sin \theta = \frac{y_1}{r_1}. \tag{3}$$

Since $\triangle$'s OAP and OA_1P_1 are similar, we have $(y_1/r_1) = (y/r)$. Thus, the choice of P in Definition I is immaterial, and the ratios depend only on the value of θ, that is, only on its terminal side.

EXAMPLE 2.   If $\cos \theta = -\frac{4}{5}$ and θ is in quadrant II, obtain all trigonometric functions of θ.

Solution.   Construct a particular angle θ in quadrant II roughly to scale in Figure 87 with $\cos \theta = -\frac{4}{5}$, by drawing a reference $\triangle AOP$ with $\overline{OA} = -4$ and $\overline{OP} = 5$. Then

$$\overline{AP}^2 + 16 = 25; \qquad \overline{AP}^2 = 9.$$

Hence $\overline{AP} = 3$. With $x = -4$, $y = 3$, and $r = 5$, the functions of θ are obtained from (1): $\tan \theta = -\frac{3}{4}$; $\sin \theta = \frac{3}{5}$; etc., as in Example 1.

By use of (1), we obtain the following fundamental relations between the trigonometric functions. We refer to these equations as *trigonometric identities,* because each one is true if θ is any angle for which all function values involved in the equation exist. Equations (4)–(6) are called the *reciprocal relations.* Equations (7)–(9) are called the *relations involving squares.*

$$\csc \theta = \frac{1}{\sin \theta}, \qquad or \qquad \sin \theta = \frac{1}{\csc \theta}; \tag{4}$$

$$\sec \theta = \frac{1}{\cos \theta}, \qquad or \qquad \cos \theta = \frac{1}{\sec \theta}; \tag{5}$$

$$\cot \theta = \frac{1}{\tan \theta}, \qquad or \qquad \tan \theta = \frac{1}{\cot \theta}; \tag{6}$$

$$\sin^2 \theta + \cos^2 \theta = 1; \tag{7}$$

$$\tan^2 \theta + 1 = \sec^2 \theta; \tag{8}$$

$$1 + \cot^2 \theta = \csc^2 \theta; \tag{9}$$

$$\tan \theta = \frac{\sin \theta}{\cos \theta}; \qquad \cot \theta = \frac{\cos \theta}{\sin \theta}. \tag{10}$$

ILLUSTRATION 3.   In (1), csc $\theta = r/y$, which is the reciprocal of y/r. Hence (4) is true. Similarly, (5) and (6) are true. From (1),

$$\frac{\sin \theta}{\cos \theta} = \frac{y}{r} \div \frac{x}{r} = \frac{y}{r} \cdot \frac{r}{x} = \frac{y}{x} = \tan \theta,$$

which agrees with (10). Similarly, we may prove the other identity in (10).

ILLUSTRATION 4.   In (1), $\qquad\qquad\qquad x^2 + y^2 = r^2.$ (11)

On dividing both sides of (11) by r^2, we obtain

$$\frac{x^2}{r^2} + \frac{y^2}{r^2} = 1, \qquad or \qquad \sin^2 \theta + \cos^2 \theta = 1,$$

which proves (7). Similarly, on dividing both sides of (11) by x^2 and using (1), we obtain (8). On dividing by y^2 in (11), we obtain (9).

In Figure 86 on page 157, for any angle θ in standard position in an xy-plane, with $\triangle OAP$ as the reference triangle for θ, it is seen that $|\overline{OA}|$, or $|\overline{AP}|$, is not greater than $|\overline{OP}|$. Thus, in

$$\sin \theta = \frac{y}{r} \qquad and \qquad \cos \theta = \frac{x}{r},$$

we have $|x| \leq r$ and $|y| \leq r$, so that $|\sin \theta| \leq 1$ and $|\cos \theta| \leq 1$. Similarly, from

$$\csc \theta = \frac{r}{y} \qquad and \qquad \sec \theta = \frac{r}{x},$$

we have $|\csc \theta| \geq 1$ and $|\sec \theta| \geq 1$.

Two angles θ and ϕ (read *phi*) are said to be **coterminal** if θ and ϕ have the same terminal side when both angles are in their standard positions on a coordinate system.

ILLUSTRATION 5.   In Figure 89, it is seen that 120°, 480°, and $-240°$ are coterminal.

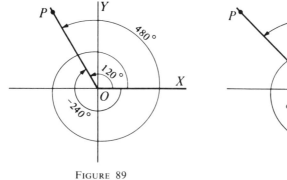

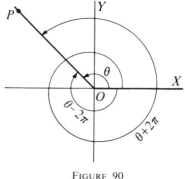

FIGURE 89   FIGURE 90

If two angles are coterminal, their trigonometric functions are identical, because Definition I involves only the terminal side of the angle θ.

ILLUSTRATION 6.   From Figure 89,

$$\sin(-240°) = \sin 120° = \sin 480°.$$

For any angle θ, the angles θ, $(\theta + 2\pi)$, and $(\theta - 2\pi)$ are coterminal with θ, as seen in Figure 90. This fact proves the following identities:

$$\textbf{(any trig. funct. of } \theta) = [\textbf{same funct. of } (\theta + 2\pi)]; \tag{12}$$

$$\textbf{(any trig. funct. of } \theta) = [\textbf{same funct. of } (\theta - 2\pi)]. \tag{13}$$

For instance,

$$\sin(\theta + 2\pi) = \sin \theta; \qquad \tan(\theta - 2\pi) = \tan \theta. \tag{14}$$

ILLUSTRATION 7.   In degree measure, (12) and (13) become

$$(\textit{any trig. funct. of } w°) = [\textit{same funct. of } (w° \pm 360°)]. \tag{15}$$

For instance, $\sin 440° = \sin(440° - 360°) = \sin 80°.$

Because of (12) and (13), a trigonometric function of an angle θ of any size, positive or negative, is equal to the same named function of some angle ϕ where $0 \leq \phi \leq 2\pi$.

In (12) and (13), the angles θ and $(\theta + 2\pi)$ differ by 2π radians; θ and $(\theta - 2\pi)$ differ by 2π radians. Thus, either (12) or (13) states that, if two angles differ by 2π radians, the trigonometric functions of the specified two angles are identical. Hence, it is said that each of the trigonometric functions is *periodic,* with 2π radians, or 360°, as the *period.* Later, we shall see that the tangent and cotangent functions also have the *smaller period* π radians, or 180°. When graphs of the trigonometric functions are considered later, we shall discuss the graphical consequences of the periodicity properties just mentioned.

The signs of the trigonometric functions of angles in the various quadrants are summarized in Figure 91. For any angle θ, the numbers

FIGURE 91

sin θ and csc θ have the same sign because sin θ = 1/csc θ; cos θ and sec θ have the same sign; tan θ and cot θ have the same sign.

ILLUSTRATION 8.   For an angle θ in quadrant I, all trigonometric functions of θ are positive, because $x > 0$, $y > 0$, and $r > 0$ in Definition II. With attention merely to the signs of $\{x, y, r\}$, if θ is in quadrant IV, we have

$$\sin \theta = \frac{-}{+} = (-); \qquad \cos \theta = \frac{+}{+} = (+); \qquad \tan \theta = \frac{-}{+} = (-); \text{ etc.}$$

★*Note 1.*   Suppose that f is a function of the variable x, whose domain consists of one or more intervals of real numbers. In calculus, one of the two most basic concepts is referred to as the *derivative function* of f, and is represented by f'. If $f(x) = \sin x$, meaning the sine of x radians, it can be proved that

$$f'(x) = \cos x. \tag{16}$$

Now suppose that $g(x) = \sin x°$. Then it is found that

$$g'(x) = \frac{\pi}{180} \cos x°. \tag{17}$$

Similarly, in the case of each trigonometric function, an inconvenient constant factor, as in (17), arises in the corresponding derivative formula if degree measure is involved. Hence, great convenience results from the use of *radian measure* for angles whenever their trigonometric functions are involved in calculus.

EXERCISE 36

Find the trigonometric functions of the angle θ if its terminal side, in standard position on the xy-plane, passes through the specified point.

1. $(3, 4)$.
2. $(-15, 8)$.
3. $(6, -8)$.
4. $(-1, -\sqrt{3})$.
5. $(12, 5)$.
6. $(3, 3)$.
7. $(-7, 24)$.
8. $(-1, -1)$.
9. $(2, -5)$.
10. $(-\sqrt{3}, -1)$.
11. $(-2, 3)$.
12. $(\sqrt{3}, 1)$.

13. Verify the signs of the trigonometric functions as indicated for quadrants II and III in Figure 91.

Specify the quadrants in which θ may lie if the given condition is satisfied.

14. $\cos \theta < 0$.
15. $\tan \theta < 0$.
16. $\sin \theta < 0$.
17. $\sec \theta < 0$.

Specify the quadrant in which θ must lie if the conditions are satisfied.

18. $\cos \theta > 0$ and $\tan \theta < 0$.
19. $\sin \theta < 0$ and $\cot \theta > 0$.
20. $\sec \theta < 0$ and $\cot \theta > 0$.
21. $\tan \theta > 0$ and $\csc \theta < 0$.

Find another function of the angle θ by use of a reciprocal identity.

22. $\sin \theta = \frac{2}{5}$.   **23.** $\tan \theta = -\frac{5}{7}$.   **24.** $\cos \theta = -\frac{3}{7}$.   **25.** $\cot \theta = -\frac{5}{2}$.

26. $\sec \theta = \frac{5}{4}$ **27.** $\csc \theta = -3$.   **28.** $\tan \theta = 5$.   **29.** $\cot \theta = 6$.

With the given $\sin \theta$ *find* $\cos \theta$, *or with the given* $\cos \theta$ *find* $\sin \theta$, *by use of* $\sin^2 \theta + \cos^2 \theta = 1$. *Then obtain all of the other trigonometric functions of* θ *by use of the fundamental identities.*

30. $\sin \theta = \frac{3}{5}$ and θ is in quadrant I; in quadrant II.

31. $\cos \theta = -\frac{5}{13}$ and θ is in quadrant II; in quadrant III.

Find the other trigonometric functions of θ *by use of fundamental identities without constructing* θ.

32. $\tan \theta = \frac{8}{15}$ and θ is in quadrant III.

33. $\cot \theta = -\frac{7}{24}$ and θ is in quadrant IV.

58. CERTAIN CONVENIENT RIGHT TRIANGLES

In an isosceles right $\triangle ABC$ as in Figure 92, let $\overline{AC} = \overline{BC} = 1$. Then

$$\overline{AB}^2 = \overline{AC}^2 + \overline{BC}^2 = 1 + 1 = 2,$$

or $\overline{AB} = \sqrt{2}$. Frequently we shall use $\triangle ABC$ with dimensions as in Figure 92 as a convenient isosceles right triangle.

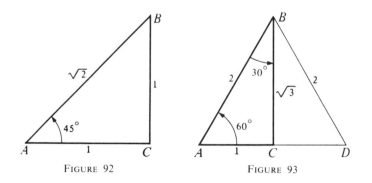

FIGURE 92   FIGURE 93

In the equilateral $\triangle ABD$ in Figure 93, let each side be 2 units long. The perpendicular BC from B to AD divides $\triangle ABD$ into two congruent right triangles having the acute angles 30° and 60°, with $\overline{AC} = \overline{CD} = 1$. Then, from $\triangle ABC$,

$$\overline{AC}^2 + \overline{BC}^2 = \overline{AB}^2, \qquad or \qquad \overline{BC}^2 = \overline{AB}^2 - \overline{AC}^2 = 4 - 1 = 3.$$

Hence $\overline{BC} = \sqrt{3}$. We shall use $\triangle ABC$ with sides as in Figure 93 when we desire a convenient right triangle with the acute angles 30° and 60°. In radian measure, we recall that $\{30°, 45°, 60°\}$ are $\{\frac{1}{6}\pi, \frac{1}{4}\pi, \frac{1}{3}\pi\}$. Then, we repeat the triangles ABC of Figures 92 and 93 with the acute angles given in radian measure, in Figures 94 and 95.

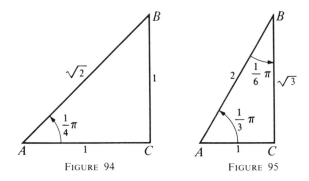

FIGURE 94   FIGURE 95

59. TRIGONOMETRIC FUNCTIONS OF SPECIAL ANGLES

EXAMPLE 1.   Find the trigonometric functions (*a*) of $\frac{1}{2}\pi$; (*b*) of π.

Solution.   (*a*) Figure 96 shows the angle $\frac{1}{2}\pi$ in standard position. For use of Definition II, select $P:(0,1)$ on the terminal side of $\frac{1}{2}\pi$. Then, in (1) on page 158, we have $x = 0, y = 1$, and $r = 1$. Hence,

$$\sin\frac{1}{2}\pi = \frac{1}{1} = 1; \cos\frac{1}{2}\pi = \frac{0}{1} = 0; \csc\frac{1}{2}\pi = \frac{1}{1} = 1; \cot\frac{1}{2}\pi = \frac{0}{1} = 0.$$

In $\tan\theta = y/x$ from Definition II, we have $x = 0$; hence $\tan\frac{1}{2}\pi$ is not defined, or does not exist. Similarly, $\sec\frac{1}{2}\pi$ does not exist. Later, we shall study $\tan\theta$ and $\sec\theta$ when θ is near $\theta = \frac{1}{2}\pi$.

(*b*) To apply Definition II in case $\theta = \pi$, we choose $R:(-1,0)$ as in Figure 96, on the terminal side of π. Then we have $x = -1, y = 0$, and $r = 1$ in Definition II. The student may complete the solution.

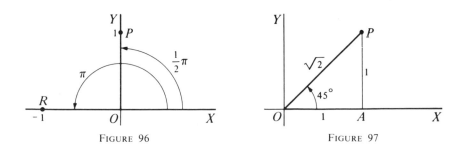

FIGURE 96   FIGURE 97

Similarly as in Example 1, it is found that, on the interval of angles $\{0 \leqq \theta \leqq 2\pi\}$, the undefined values of the trigonometric functions of θ are as follows:

$$\begin{array}{lllll} \left.\begin{array}{c} \cot 0 \\ \csc 0 \end{array}\right\}; & \left.\begin{array}{c} \tan \frac{1}{2}\pi \\ \sec \frac{1}{2}\pi \end{array}\right\}; & \left.\begin{array}{c} \cot \pi \\ \csc \pi \end{array}\right\}; & \left.\begin{array}{c} \tan \frac{3}{2}\pi \\ \sec \frac{3}{2}\pi \end{array}\right\}; & \left.\begin{array}{c} \cot 2\pi \\ \csc 2\pi \end{array}\right\}. \end{array}$$

The student will verify the preceding facts later.

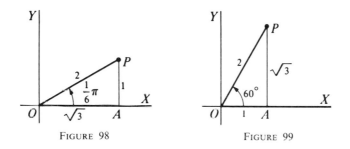

FIGURE 98   FIGURE 99

ILLUSTRATION 1.   The angle $\frac{1}{4}\pi$, or 45°, is shown in its standard position in Figure 97. The angles $\frac{1}{6}\pi$ and $\frac{1}{3}\pi$, or 30° and 60°, are in their standard positions in Figures 98 and 99. In Figure 97, P is selected with $x = 1$ so that the reference triangle AOP has the convenient dimensions seen in $\triangle ABC$ of Figure 94. We select P with coordinates $(\sqrt{3}, 1)$ in Figure 98, and coordinates $(1, \sqrt{3})$ in Figure 99. Then, the reference triangles AOP have the convenient dimensions seen in Figure 95. By use of Definition II, the student should verify all entries for trigonometric functions in the following table. Thus, from Figure 97,

$$\tan \tfrac{1}{4}\pi = \tan 45° = \frac{1}{1} = 1; etc.$$

ANGLE	SIN	COS	TAN	COT	SEC	CSC	ANGLE (*rad.*)
0°	0	1	0	*none*	1	*none*	0
30°	$\frac{1}{2}$	$\frac{\sqrt{3}}{2}$	$\frac{1}{\sqrt{3}}$	$\sqrt{3}$	$\frac{2}{\sqrt{3}}$	2	$\frac{1}{6}\pi$
45°	$\frac{1}{\sqrt{2}}$	$\frac{1}{\sqrt{2}}$	1	1	$\sqrt{2}$	$\sqrt{2}$	$\frac{1}{4}\pi$
60°	$\frac{\sqrt{3}}{2}$	$\frac{1}{2}$	$\sqrt{3}$	$\frac{1}{\sqrt{3}}$	2	$\frac{2}{\sqrt{3}}$	$\frac{1}{3}\pi$
90°	1	0	*none*	0	*none*	1	$\frac{1}{2}\pi$
180°	0	-1	0	*none*	-1	*none*	π
270°	-1	0	*none*	0	*none*	-1	$\frac{3}{2}\pi$

60. THE ACUTE REFERENCE ANGLE

It will be seen later that the trigonometric functions of an angle in any quadrant can be found if we have the means (for instance, in a table) for finding the trigonometric functions of *acute* angles. This result is a consequence of use of an auxiliary acute angle described as follows for any angle.

*DEFINITION III.   Let an angle θ in any quadrant be located in its standard position in an xy-plane. Then, the **reference angle** for θ is the acute angle ϕ (read phi) between the terminal side of θ and the x-axis.*

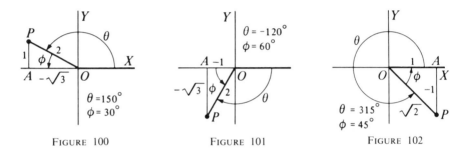

| FIGURE 100 | FIGURE 101 | FIGURE 102 |

ILLUSTRATION 1.   The reference angle for 150° in Figure 100 is 30°; for −120° in Figure 101 is 60°; for 315° in Figure 102 is 45°.

Suppose that the reference angle for a given angle $w°$ is 30°, 45°, or 60°, and $w°$ is in its standard position in an xy-plane. Then, the reference $\triangle OAP$ for $w°$ can be assigned the dimensions used for 30°, 45°, or 60°, respectively. Hence, the trigonometric functions of $w°$ are found to be the same, except possibly for signs, as the same named functions of the corresponding reference angle.

ILLUSTRATION 2.   From Figures 100–102, for 150°, −120°, and 315°:

$$\sin 150° = \frac{1}{2} = \sin 30°; \qquad \cos 150° = \frac{-\sqrt{3}}{2} = -\cos 30°.$$

$$\tan(-120°) = \frac{-\sqrt{3}}{-1} = \sqrt{3} = \tan 60°; \quad \tan 315° = \frac{-1}{+1} = -1 = -\tan 45°.$$

In Illustration 2, special cases of the following result were met. The theorem is true because, in applying Definition II to any angle $θ$ in standard position, we may use a reference $\triangle OAP$ which is congruent to a reference triangle which *could* be used for the acute reference angle $ϕ$ for $θ$.

THEOREM I. *Any trigonometric function of an angle θ in any quadrant is the same, except perhaps for sign, as the same named function of the acute reference angle, φ, for θ.*

EXERCISE 37

Place the given angle, θ, in its standard position in an xy-plane. Choose a convenient reference triangle for θ, and apply Definition II to obtain all trigonometric functions of θ which exist. If θ is not quadrantal, verify the truth of Theorem I by comparison with the table on page 165.

1. $0°$.   2. $\frac{3}{2}\pi$.   3. 2π.   4. $120°$.   5. $135°$.

6. $\frac{5}{4}\pi$.   7. $210°$.   8. $\frac{4}{3}\pi$.   9. $\frac{7}{4}\pi$.   10. $330°$.

11. $60°$.   12. $-45°$.   13. $405°$.   14. $510°$.   15. $\frac{5}{2}\pi$.

16. -3π.   17. $-\frac{2}{3}\pi$.   18. $-\frac{5}{4}\pi$.   19. $-270°$.   20. $\frac{9}{4}\pi$.

21. $-\frac{1}{4}\pi$.   22. $-\frac{1}{6}\pi$.   23. $-\frac{1}{3}\pi$.   24. $-\frac{3}{4}\pi$.

25. Specify all values of the angle $w°$ between $-360°$ and $360°$ for which the trigonometric functions are the same, except possibly for signs, as the same named functions of $30°$.
26. Specify all values of $θ$ between -2π and 2π for which the trigonometric functions of $θ$ have the same absolute values as the same named functions of $\frac{1}{4}\pi$; $\frac{1}{3}\pi$; $\frac{1}{6}\pi$.
27. Specify all angles $θ$ in the interval $-2\pi \le θ \le 2\pi$ for which $\tan θ$ is undefined; $\cot θ$ is undefined.

61. FUNCTIONS OF ACUTE ANGLES

In any right triangle ABC as in Figure 103 on page 168, let* $\{α,β,γ\}$ be the angles at the vertices $\{A,B,C\}$, respectively, where $γ = 90°$. Let $\{a,b,c\}$ be the sides, or the measures of the sides, opposite the angles $\{α,β,γ\}$ respectively. By the Pythagorean theorem,

$$c^2 = a^2 + b^2. \tag{1}$$

Any acute angle $α$ in a right triangle ABC, in Figure 103, can be placed on a coordinate system with $α$ in its standard position and with

*Read $\{alpha, beta, gamma\}$.

$\triangle ABC$ as the reference triangle for α, as in Figure 104. With B as the chosen point "P" of Definition II on the terminal side of α, the side of $\triangle ABC$ *opposite* α is the y-coordinate of B; the side *adjacent* to α is the x-coordinate of B. Then, from (1) on page 158 with α replacing θ, $y = a$, $x = b$, and $r = c$, we obtain the following results.

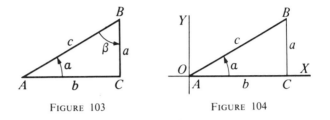

FIGURE 103   FIGURE 104

For an acute angle α located as an angle in a right triangle:

$$\left. \begin{array}{ll} \sin \alpha = \dfrac{\text{opposite side}}{\text{hypotenuse}} = \dfrac{a}{c}; & \cos \alpha = \dfrac{\text{adjacent side}}{\text{hypotenuse}} = \dfrac{b}{c}; \\[2ex] \tan \alpha = \dfrac{\text{opposite side}}{\text{adjacent side}} = \dfrac{a}{b}; & \cot \alpha = \dfrac{\text{adjacent side}}{\text{opposite side}} = \dfrac{b}{a}; \\[2ex] \sec \alpha = \dfrac{\text{hypotenuse}}{\text{adjacent side}} = \dfrac{c}{b}; & \csc \alpha = \dfrac{\text{hypotenuse}}{\text{opposite side}} = \dfrac{c}{a}. \end{array} \right\} \quad (2)$$

We emphasize that (2) applies ONLY WHEN α is an acute angle situated as an angle of a right triangle. Equations (2) must be completely discarded in any general consideration of the trigonometric functions. With appropriate changes in letters, the trigonometric functions of β in Figure 103 can be obtained from the verbal forms in (2). Thus

$$\sin \beta = \frac{side\ opposite}{hypotenuse} = \frac{b}{c} = \cos \alpha. \tag{3}$$

EXAMPLE 1.   In a right $\triangle ABC$ with $\gamma = 90°$ as in Figure 105, if $a = 3$ and $b = 5$ find $\sin \alpha$ and $\tan \beta$.

Solution.   In Figure 105, $c^2 = 9 + 25 = 34$; $c = \sqrt{34}$. From (2),

$$\sin \alpha = \frac{3}{\sqrt{34}} = \frac{3}{\sqrt{34}} \cdot \frac{\sqrt{34}}{\sqrt{34}} = \frac{3}{34}\sqrt{34}; \qquad \tan \beta = \frac{5}{3}.$$

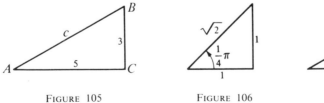
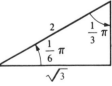

FIGURE 105   FIGURE 106   FIGURE 107

It now becomes essential to memorize the convenient right triangles in Figures 106 and 107, which have been used before. By visualization of these triangles, the student should be able to recall any desired trigonometric function of one of the angles $\{\frac{1}{6}\pi, \frac{1}{4}\pi, \frac{1}{3}\pi\}$ or, in degree measure, $\{30°, 45°, 60°\}$. Immediate practice with these triangles is desirable.

Suppose that α and β are any acute angles such that $\alpha + \beta = \frac{1}{2}\pi$. Then, α and β are referred to as **complementary angles,** with each called the **complement** of the other angle. In this case, α and β may be located as the acute angles of a right $\triangle ABC$ as in Figure 103. Then, with $\beta = \frac{1}{2}\pi - \alpha$, equation (3) becomes $\cos \alpha = \sin \left(\frac{1}{2}\pi - \alpha\right)$. In similar fashion, by use of (2), all of the following equations may be proved for all values of α such that $0 < \alpha < \frac{1}{2}\pi$. Later, these equations will be proved to be identities for *all* values of α.

$$\begin{aligned} \sin \alpha &= \cos \left(\tfrac{1}{2}\pi - \alpha\right); & \cos \alpha &= \sin \left(\tfrac{1}{2}\pi - \alpha\right); \\ \tan \alpha &= \cot \left(\tfrac{1}{2}\pi - \alpha\right); & \cot \alpha &= \tan \left(\tfrac{1}{2}\pi - \alpha\right); \\ \sec \alpha &= \csc \left(\tfrac{1}{2}\pi - \alpha\right); & \csc \alpha &= \sec \left(\tfrac{1}{2}\pi - \alpha\right). \end{aligned} \qquad (4)$$

Frequently, the trigonometric functions are grouped in pairs as follows: *sine* and **co***sine; tangent* and **co***tangent; secant* and **co***secant.* In each pair, either function is called the **cofunction** of the other function. Thus, (4) states that any trigonometric function of an acute angle α is equal to the *cofunction* of the complement of α. Equations (4) are referred to as the *cofunction identities.*

ILLUSTRATION 1.   From (4), $\tan 32° = \cot (90° - 32°) = \cot 58°$.

Suppose that the acute reference angle for an angle θ is one of the angles $\{\frac{1}{6}\pi, \frac{1}{4}\pi, \frac{1}{3}\pi\}$ or, in degree measure, $\{30°, 45°, 60°\}$. Then, by visualization of Figure 106 or Figure 107, the student should become able to obtain any trigonometric function of θ quickly, by use of Theorem I on page 167.

ILLUSTRATION 2.   *To find* $\tan 300°$: visualize $300°$ in standard position in an xy-plane; $300°$ is in quadrant IV where the tangent is negative; since the reference angle is $60°$, $\tan 300° = -\tan 60°, = -\sqrt{3}$.

ILLUSTRATION 3.   *To obtain* $\sin \frac{5}{6}\pi$: visualize the angle $\frac{5}{6}\pi$ in standard position; the angle is in quadrant II where the sine is positive, and the reference angle is $\frac{1}{6}\pi$. Hence $\sin \frac{5}{6}\pi = +\sin \frac{1}{6}\pi = \frac{1}{2}$.

It is found that, for many purposes, the sine, cosine, and tangent are more convenient in applications than the cotangent, secant, and cosecant. Also, the sine, cosine, and tangent occur more frequently in discussions than the other functions. Hence, hereafter, we shall give more emphasis to consideration of the sine, cosine, and tangent than to their reciprocals, the secant, cosecant and cotangent.

As a rule, any value of a trigonometric function is an endless decimal which is an irrational number. The trigonometric functions of any angle can be computed by advanced methods. The values of these functions, with some specified degree of accuracy, for angles spaced at a regular interval, then may be assembled in a table. On account of Theorem I, it is necessary to tabulate the function values *only for acute angles,* because the trigonometric functions of any angle θ differ at most in signs from the same named functions of the acute reference angle for θ.

Table IV is referred to as a three-place table of the trigonometric functions of angles from $0°$ to $90°$, inclusive, at intervals of $1°$. Each entry in Table IV was obtained by rounding off to three decimal places a number specified to several more decimal places. Each entry is a function of some acute angle α and, likewise, is the cofunction of the complement of α, because of the cofunction identities (4). Thus, each entry in Table IV serves a double purpose.

ILLUSTRATION 4.   From Table IV, $\sin 39° = .629 = \cos 51°$, where titles of columns are read at the bottom in the case of angles greater than $45°$.

ILLUSTRATION 5.   To obtain $\sin 317°$: the angle is in quadrant IV where the sine is negative; the reference angle is $43°$. Hence, from Table IV,

$$\sin 317° = -\sin 43° = -.682$$

ILLUSTRATION 6.   Suppose that θ is an acute angle and that we are given $\tan \theta = 2.747$. On searching for 2.747 in the columns labeled Tangent at the top or bottom in Table IV, we find that $\theta = 70°$.

EXAMPLE 2.   If $\tan \theta = -.424$ and θ is between $0°$ and $180°$, obtain θ from Table IV.

Solution.   1. *To find the acute reference angle,* ϕ, *for* θ: we have $\tan \phi = .424$. From Table IV, $\phi = 23°$.

2. *To obtain* θ: a rough sketch (or, visualization without a figure) shows that $\theta = 180° - 23°$ or $\theta = 157°$.

EXERCISE 38

Construct a right triangle roughly to scale with an angle α having the given trigonometric function. Then find all other trigonometric functions of α. Also obtain the trigonometric functions of β, the complement of α.

1. $\cot \alpha = \frac{4}{3}$.   **2.** $\cos \alpha = \frac{8}{17}$.   **3.** $\tan \alpha = \frac{24}{7}$.

4. $\sin \alpha = \frac{12}{13}$.   **5.** $\cos \alpha = \frac{7}{25}$.   **6.** $\tan \alpha = \frac{8}{15}$.

By use of a reference angle, and knowledge of the trigonometric functions of $\{\frac{1}{6}\pi, \frac{1}{4}\pi, \frac{1}{3}\pi\}$, *or* $\{30°, 45°, 60°\}$ *in degree measure, obtain the specified trigonometric function value.*

7. cos 120°.   **8.** sin 210°.   **9.** tan 135°.   **10.** cot 300°.

11. sin $\frac{5}{6}\pi$.   **12.** cos $\frac{4}{3}\pi$.   **13.** cot 120°.   **14.** tan 330°.

15. tan 210°.   **16.** cot 150°.   **17.** cos $\frac{3}{4}\pi$.   **18.** sin $\frac{5}{4}\pi$.

19. tan (−135°).   **20.** cos (−$\frac{5}{4}\pi$).   **21.** tan (−$\frac{1}{3}\pi$).   **22.** cot (−$\frac{3}{4}\pi$).

23. sin 405°.   **24.** cos 480°.   **25.** cot 495°.   **26.** tan 570°.

Obtain θ to satisfy the given condition. (First obtain the reference angle.)

27. $\pi < \theta < \frac{3}{2}\pi$; sin $\theta = -\frac{1}{2}$.   **28.** $\frac{1}{2}\pi < \theta < \pi$; cos $\theta = -\frac{1}{2}\sqrt{3}$.

29. $0 < \theta < 2\pi$; tan $\theta = 1$.   **30.** $0 < \theta < 2\pi$; cot $\theta = -1$.

31. $0 < \theta < 2\pi$; cos $\theta = \frac{1}{2}\sqrt{3}$.   **32.** $0 < \theta < 2\pi$; sin $\theta = \frac{1}{2}\sqrt{2}$.

Obtain the function value from Table IV.

33. cos 18°.   **34.** sin 49°.   **35.** tan 53°.   **36.** cot 71°.
37. sec 41°.   **38.** csc 82°.   **39.** cos 69°.   **40.** sin 26°.
41. tan 106°.   **42.** cot 188°.   **43.** cos 249°.   **44.** sin 316°.
45. sin 140°.   **46.** cos 98°.   **47.** sec 91°.   **48.** tan 103°.
49. cos 157°.   **50.** tan 200°.   **51.** cot 298°.   **52.** csc 112°.

Find the function value from Table VI. *Use* $\pi = 3.14$ *if a reference angle is to be found.*

53. sin .46.   **54.** cos 1.07.   **55.** tan 1.48.   **56.** sin 1.37.
57. cos 1.74.   **58.** sin 3.57.   **59.** tan 1.98.   **60.** cos 4.34.

Hint for Problem 57. The angle 1.74 radians is in quadrant II. The reference angle is (3.14 − 1.74) or 1.40 radians. Hence cos 1.74 = −cos 1.40.

Obtain the unknown angle w° by use of Table IV.
61. sin $w°$ = .545 and (a) $w°$ is acute; (b) 90 < w < 180.
62. cos $w°$ = .875 and (a) $w°$ is acute; (b) 270 < w < 360.
63. tan $w°$ = .781 and (a) $w°$ is acute; (b) 180 < w < 270.
64. sin $w°$ = −.829 and (a) 180 < w < 270; (b) 270 < w < 360.
65. cos $w°$ = −.438 and (a) 90 < w < 180; (b) 180 < w < 270.
66. tan $w°$ = −2.145 and (a) 90 < w < 180; (b) 270 < w < 360.

Note 1. Suppose that $0 < \theta < \frac{1}{2}\pi$. Construct θ in its standard position. Also, construct −θ in its standard position. What is the reference angle for −θ? Then, since the sine and tangent of an angle in quadrant IV are negative, and the cosine is positive, observe that

$$\sin(-\theta) = -\sin\theta; \quad \cos(-\theta) = \cos\theta; \quad \tan(-\theta) = -\tan\theta. \quad (1)$$

Later, we shall prove that the preceding equations are identities for all admissable values of θ. At present, we shall assume this fact on the basis of the preceding remarks.

By use of (1), *obtain the function value.*

67. $\sin(-\tfrac{1}{3}\pi)$.   **68.** $\cos(-\tfrac{1}{4}\pi)$.   **69.** $\tan(-\tfrac{1}{6}\pi)$.   **70.** $\sin(-\tfrac{1}{4}\pi)$.

71. $\sin(-\tfrac{3}{4}\pi)$.   **72.** $\tan(-\tfrac{1}{4}\pi)$.   **73.** $\cos(-\tfrac{2}{3}\pi)$.   **74.** $\cos(-\tfrac{1}{6}\pi)$.

62. RANGES OF THE TRIGONOMETRIC FUNCTIONS

In Figure 108 the circle has the radius $r = 1$, and the angle θ is in its standard position. Let $P:(x,y)$ be the intersection of the circle and the terminal side of θ. Then

$$\sin\theta = \frac{y}{r} = \frac{y}{1} = y; \quad \cos\theta = \frac{x}{1} = x. \quad (1)$$

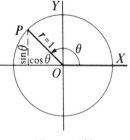

FIGURE 108

Suppose that θ increases from 0 to 2π. Then P moves completely around the circle starting from $(1,0)$. Meanwhile, the ordinate of P, or $\sin\theta$, increases continuously from 0 to 1, when $\theta = \tfrac{1}{2}\pi$; decreases from 1 to 0, when $\theta = \pi$; decreases from 0 to -1, when $\theta = \tfrac{3}{2}\pi$; increases from -1 to 0, when $\theta = 2\pi$. Thus, the range of values of $\sin\theta$ consists of all numbers from -1 to $+1$, inclusive. The same range is found for $\cos\theta$.

We recall that $\tan\tfrac{1}{2}\pi$ and $\sec\tfrac{1}{2}\pi$ are not defined. Suppose, now, that $0 \leq \theta < \tfrac{1}{2}\pi$, and that θ is in its standard position in an xy-plane. On the terminal side of θ, let $P:(x,y)$ be selected with $x = 1$, as in Figure 109. We have $r = \overline{OP}$ and

$$\tan\theta = \frac{y}{x} = \frac{y}{1} = y; \quad \sec\theta = \frac{r}{1} = \overline{OP}. \quad (2)$$

If θ increases continuously from 0, and $\theta \to \frac{1}{2}\pi$ (where "$\to$" is read *approaches*), the corresponding point $P:(x,y)$ of Figure 109 moves from $P:(1,0)$ *upward* beyond all bounds.   Then y, or tan θ, starts with the value 0 when $\theta = 0$, and increases without bound through all positive numbers.   Hence, we say that "tan θ *approaches plus infinity as* $\theta \to \frac{1}{2}\pi$ *with* $\theta < \frac{1}{2}\pi$." We abbreviate this by writing as follows, with "∞" meaning "*infinity*," and "$\theta \to \frac{1}{2}\pi(-)$" meaning "$\theta \to \frac{1}{2}\pi$ with $\theta < \frac{1}{2}\pi$":

$$\tan \theta \to +\infty \qquad as \qquad \theta \to \tfrac{1}{2}\pi(-). \qquad (3)$$

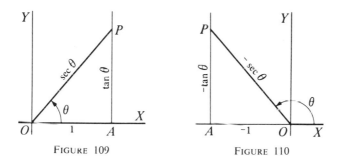

FIGURE 109   FIGURE 110

Thus, with $0 \leq \theta < \frac{1}{2}\pi$, the range of tan θ is the set of all positive numbers.   Similarly, sec θ starts with the value 1 when $\theta = 0$ and increases without bound through all numbers greater than 1 if $\theta \to \frac{1}{2}\pi(-)$.   Hence, with $0 \leq \theta < \frac{1}{2}\pi$, the range of sec θ is the set of all numbers greater than or equal to 1, and

$$\sec \theta \to +\infty \qquad as \qquad \theta \to \tfrac{1}{2}\pi(-). \qquad (4)$$

In place of (3) and (4), we may write

$$\lim_{\theta \to \frac{1}{2}\pi(-)} \tan \theta = +\infty; \qquad \lim_{\theta \to \frac{1}{2}\pi(-)} \sec \theta = +\infty. \qquad (5)$$

We read as follows in (5): "*the limit of* tan θ *is* $+\infty$ *as* θ *approaches* $\frac{1}{2}\pi$ *with* $\theta < \frac{1}{2}\pi$," and similarly for the other statement.

When $\frac{1}{2}\pi < \theta \leq \pi$, as in Figure 110, where the line $x = -1$ replaces $x = 1$ of Figure 109, similar reasoning leads to the following conclusions. On this domain for θ, the range of tan θ is the set of all negative numbers; the range of sec θ is the set of all numbers less than or equal to -1.   Let us write "$\theta \to \frac{1}{2}\pi(+)$" for "$\theta$ *approaches* 90° *with* $\frac{1}{2}\pi < \theta$."   Then, it is found that

$$\lim_{\theta \to \frac{1}{2}\pi(+)} \tan \theta = -\infty; \qquad \lim_{\theta \to \frac{1}{2}\pi(+)} \sec \theta = -\infty. \qquad (6)$$

If θ is a variable with the domain $\{0 \leq \theta \leq \pi\}$, previous results justify the following statements, where (7) is equivalent to (5) and (6) for $\tan \theta$, and (8) is equivalent to (5) and (6) for $\sec \theta$.

$$\left. \begin{aligned} &\tan \tfrac{1}{2}\pi \text{ is undefined: the range for } \tan \theta \text{ consists of all} \\ &\text{real numbers, and}^* \\ &\qquad\qquad \lim_{\theta \to \frac{1}{2}\pi} |\tan \theta| = +\infty. \end{aligned} \right\} \qquad (7)$$

$$\left. \begin{aligned} &\sec \tfrac{1}{2}\pi \text{ is undefined; the range for } \sec \theta \text{ consists of all} \\ &\text{real numbers with absolute value at least 1, and} \\ &\qquad\qquad \lim_{\theta \to \frac{1}{2}\pi} |\sec \theta| = +\infty. \end{aligned} \right\} \qquad (8)$$

Sometimes, in view of the facts summarized above, it is said that

$$\tan \tfrac{1}{2}\pi \text{ is infinite}, \qquad (9)$$

and, similarly, $\sec \tfrac{1}{2}\pi$ is infinite. Statement (9) abbreviates the facts that $\tan \tfrac{1}{2}\pi$ does not exist and also (7) is true.

The behavior of $\tan \theta$ and $\sec \theta$ when $\{\pi \leq \theta \leq 2\pi\}$ is similar to that observed when $\{0 \leq \theta \leq \pi\}$, with $\tan \tfrac{3}{2}\pi$ and $\sec \tfrac{3}{2}\pi$ undefined, and

$$\lim_{\theta \to \frac{3}{2}\pi} |\tan \theta| = +\infty; \qquad \lim_{\theta \to \frac{3}{2}\pi} |\sec \theta| = +\infty.$$

Thus, on the whole domain $\{0 \leq \theta \leq 2\pi\}$, the ranges of $\tan \theta$ and $\sec \theta$ are as specified for the domain $\{0 \leq \theta \leq \pi\}$.

63. THE CONCEPT OF A PERIODIC FUNCTION

Consider a function $f(x)$ whose domain consists of all real numbers, with the exception, possibly, of certain isolated numbers. Suppose that $p > 0$ and that

$$f(x) = f(x + p) \text{ at all admissable values of } x. \qquad (1)$$

Then it is said that f is **periodic** with p as a **period**. This means that the values of $f(x)$ *repeat* at intervals of length p in the values of x. By use of (1),

$$f(x) = f(x + p) = f((x + p) + p) = f(x + 2p). \qquad (2)$$

---

*The plus sign with "$+\infty$" is redundant, but emphasizes the fact that $|\tan \theta|$ is positive or 0. In (7), we may read "*the limit of* $|\tan \theta|$ *is plus infinity as* $\theta \to \tfrac{1}{2}\pi$," or "$|\tan \theta|$ *grows large without bound as* $\theta \to \tfrac{1}{2}\pi$."

Thus, $2p$ also is a period of f. Similarly, f has the periods $3p, 4p, \cdots, np$, where n is any positive integer. If f is periodic, its *smallest* period is called THE period of f. Sinc $(x - p) + p = x$, from (1) with $(x - p)$ replacing x we obtain

$$f(x - p) = f(x); \quad then \quad f(x) = f(x - p) = f(x - 2p) = etc. \quad (3)$$

If a function $f(x)$ has p as a period, let C be the graph of $y = f(x)$ over any interval of values of x having the length p, for instance over the interval $D = \{0 \leq x \leq p\}$. Then the complete graph of $y = f(x)$ consists of endless repetitions of C to the left and the right in the xy-plane.

ILLUSTRATION 1.   Suppose that the graph of a function $f(x)$, or of $y = f(x)$, consists of the ridge ABC in Figure 111 over $\{-3 \leq x \leq 3\}$, and endless repetitions of this ridge to the right and the left. Then, f is periodic with the period 6. For instance,

$$4 = f(0) = f(6) = f(12) = f(18) = \cdots;$$
$$= f(-6) = f(-12) - f(-18) = \cdots.$$

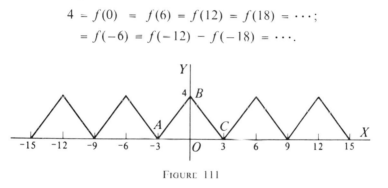

FIGURE 111

In (12) and (13) on page 161, it was emphasized that each trigonometric function, T, is periodic with 2π as a period. Hence, the values of $T(x)$ *repeat* at intervals of 2π in the values of x. Also, if we obtain the graph, C, of $y = T(x)$ for x on $\{0 \leq x \leq 2\pi\}$, or on any other interval of length 2π, then the complete graph of $y = T(x)$ will consist of endless repetitions of C to the left and the right in the xy-plane. We shall make use of this fact in the next section.

ILLUSTRATION 1.   We recall $\sin \frac{1}{6}\pi = \frac{1}{2}$. Since $\left(\frac{1}{6}\pi - 2\pi\right) = -\frac{11}{6}\pi$ and $\left(\frac{1}{6}\pi + 2\pi\right) = \frac{13}{6}\pi$, by use of (2) and (3) we obtain

$$\sin\left(-\frac{11}{6}\pi\right) = \sin\frac{13}{6}\pi = \frac{1}{2}.$$

64. GRAPHS OF THE TRIGONOMETRIC FUNCTIONS

EXAMPLE 1.   Obtain a meticulously accurate graph of the function $\sin x$ on the domain $\{0 \leq x \leq \pi\}$.

Solution.   1. By agreement, "sin x" means "*sine of x radians.*"

2. To obtain solutions (x,y) of* $y = \sin x$:   Let $w° = (x$ radians). For any value of x, we have $y = \sin w°$.   In the following table, values of w on $\{0 \leqq w \leqq 90\}$ were chosen.   Table IV gives $y = \sin w°$ and, if desired, also gives x such that $(x$ *radians*$) = w°$.   Corresponding to $w° = 75°$, the following table lists $75°$, and $105°$ whose reference angle is $75°$.   From Table IV,

$$\sin 75° = .97 = \sin 105° \qquad and \qquad 75° = (1.31 \; radians)$$

We obtain the entries for $30°$, $45°$, $60°$, and the corresponding angles in quadrant II from our knowledge of these convenient angles.   The values of $\sin w°$ for $0°$, $90°$, and $180°$ were recalled.   We use "$\doteq$" for "*approximately equals.*"

$w° =$	$0°$	$30°$	$45°$	$60°$	$75°$	$90°$	$105°$	$120°$	$135°$	$150°$	$180°$
$x =$	0	$\frac{1}{6}\pi$	$\frac{1}{4}\pi$	$\frac{1}{3}\pi$	$\frac{5}{12}\pi$	$\frac{1}{2}\pi$	$\frac{7}{12}\pi$	$\frac{2}{3}\pi$	$\frac{3}{4}\pi$	$\frac{5}{6}\pi$	π
$x \doteq$	0	$.52$	$.79$	1.05	1.31	1.57	1.83	2.09	2.36	2.62	2.14
$y = \sin x$	0	$.50$	$.71$	$.87$	$.97$	1.00	$.97$	$.87$	$.71$	$.50$	0

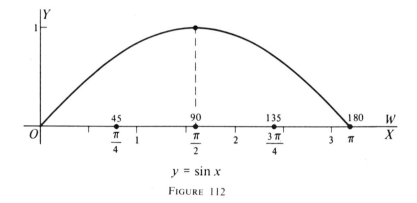

$y = \sin x$

Figure 112

3. *To draw the graph.*   In Figure 112, act as if the horizontal axis has *two sides.*   On it, *above*, choose a w-unit, with $0 \leqq w \leqq 180$.   *Below*, corresponding to each value of w, list x such that $(x$ *radians*$) = w°$. Indirectly, we thus select, as the x-unit, the distance from the origin to the point where $w = 57.296$, because one radian is $57.296°$.   This value may be thought of as approximately $60°$.   We take the scale unit on OY equal to the preceding x-unit, to harmonize with later agreements.   To plot

*Hereafter, x and y no longer have the restricted significance associated with Definition II on page 157.

points (x,y), we use the equivalent coordinates (w,y). Thus, the complicated decimal values of x in the table are *not* involved in the graphing. These values are listed as an introduction to later methods, and to emphasize that we have obtained the graph of $y = \sin x$, with x meaning x *radians*. Degree measure was introduced for preliminary convenience. Such use of degree measure is useful whenever a meticulously accurate graph of a trigonometric function is desired.

In Figure 112 we simultaneously have graphs of $y = \sin w°$, which is the usual sine graph as met in elementary trigonometry, and the graph of $y = \sin x$. In calculus, there rarely is need for the graph of $y = \sin w°$. Thus, when the graph of the *sine function* is referred to in calculus, this will mean the graph of $y = \sin x$, where* "x" is interpreted as x *radians."*

ILLUSTRATION 1.   A qualitatively accurate graph of $y = \sin x$ is in Figure 113, as obtained from the following table. For such a graph, it is unnecessary and inadvisable to include an auxiliary horizontal scale based on degree measure, as in Figure 112. The scale units are equal on the coordinate axes. This will be the case with all trigonometric graphs in this chapter, and is the rule in their applications in calculus. It is impossible to take account of highly accurate decimal values of multiples of π in graphing. The table indicates two convenient stages of approximations to π which might be used without measurably distorting the graph. The part of the graph on the domain $\{0 \leq x \leq 2\pi\}$ may be referred to as the *basic sine wave.* Since the function $\sin x$ is periodic with the period 2π, the complete graph of the function consists of endless repetitions of the basic wave to the left and the right. Without proof (which

$y = \sin x$	0	1	0	-1	0	1	0	-1	0
$x =$	-2π	$-\dfrac{3}{2}\pi$	$-\pi$	$-\dfrac{1}{2}\pi$	0	$\dfrac{1}{2}\pi$	π	$\dfrac{3}{2}\pi$	2π
$x \doteq$	-6	-4.5	-3	-1.5	0	1.5	3	4.5	6
$x \doteq$	-6.4	-4.8	-3.2	-1.6	0	1.6	3.2	4.8	6.4

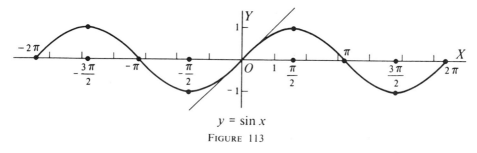

$y = \sin x$

FIGURE 113

*In the next chapter an equivalent of this interpretation is phrased differently.

requires methods from calculus), we notice that, at each point where a wave intersects the *x*-axis, the curve has a *tangent* with slope 1 or −1 cutting the curve, as shown in Figure 113 at the origin. We have called such a point on a curve an *inflection point*.

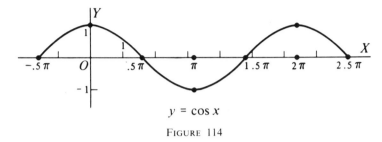

$$y = \cos x$$

FIGURE 114

ILLUSTRATION 2.   Later, the student will verify the graph of $y = \cos x$ in Figure 114. Notice that the basic wave in Figure 114 is a duplicate of the sine wave of Figure 113, but is shifted $\frac{1}{2}\pi$ horizontal units to the left. Thus it appears geometrically that $\cos x = \sin (x + \frac{1}{2}\pi)$. This identity will be proved in the next chapter. Since the cosine function has the period 2π, the graph of $y = \cos x$ consists of endless repetitions of its basic wave, for $\{-\frac{1}{2}\pi \leqq x \leqq \frac{3}{2}\pi\}$, to the right and the left.

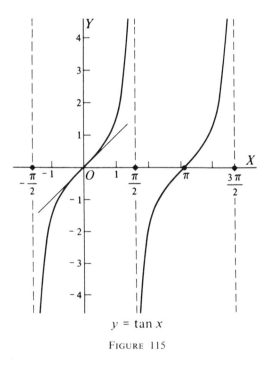

$$y = \tan x$$

FIGURE 115

ILLUSTRATION 3.   The graph of the function tan x in Figure 115 can be checked by use of the following brief table of coordinates. Any entry "∞" in the table should be read "*the tangent is infinite*." We recall that tan $\frac{1}{2} \pi$ does not exist and

$$\lim_{x \to \frac{1}{2}\pi(-)} \tan x = +\infty; \qquad \lim_{x \to \frac{1}{2}\pi(+)} \tan x = -\infty. \qquad (1)$$

As a consequence of (1), the vertical broken line $x = \frac{1}{2}\pi$ in Figure 115 is an *asymptote* of the graph. It recedes *upward* beyond all bounds $(y \to +\infty)$ as $x \to \frac{1}{2}\pi(-)$ (from the *left*), and *downward* beyond all bounds $(y \to -\infty)$ *as* $x \to \frac{1}{2}\pi(+)$(from the *right*). Observe the tangent line with slope 1 at the origin, which is an inflection point of the curve. The existence of the tangent at every x-intercept point is proved in calculus. In the table, notice that the values of tan x for $\{\frac{1}{2}\pi < x < \frac{3}{2}\pi\}$ repeat those for $\{-\frac{1}{2}\pi < x < \frac{1}{2}\pi\}$. This is a consequence of the fact that the tangent function is periodic with the *period* π, which we shall prove in

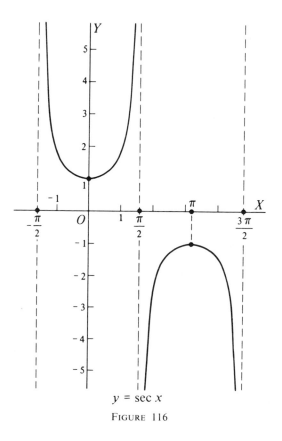

$y = \sec x$

FIGURE 116

the next chapter. The complete graph of $y = \tan x$ consists of endless repetitions to the right and the left of the branch obtained with $\{-\frac{1}{2}\pi < x > \frac{1}{2}\pi\}$. The table illustrates a possible basis for a rapid qualitative graph of $y = \tan x$. The values $x = \pm\frac{1}{4}\pi$, $\pm\frac{3}{4}\pi$, $\pm\frac{5}{4}\pi$, etc., are particularly convenient for rapid graphing. The graph of $y = \tan x$ has a vertical asymptote corresponding to each value of x for which $\tan x$ is infinite.

$y = \tan x$	∞	-1	0	1	∞	-1	0	1	∞
$x =$	$-\frac{1}{2}\pi$	$-\frac{1}{4}\pi$	0	$\frac{1}{4}\pi$	$\frac{1}{2}\pi$	$\frac{3}{4}\pi$	π	$\frac{5}{4}\pi$	$\frac{3}{2}\pi$
$x \doteq$	-1.6	$-.8$	0	.8	1.6	2.4	3.2	4.0	4.8

The student will obtain graphs of the cotangent and cosecant functions in the next exercise, and also will verify the graph of the secant as given in Figure 116. The graph has a vertical asymptote corresponding to each value of x for which $\sec x$ is infinite.

EXERCISE 39

In graphing any equation $y = f(x)$ in the following problems, use equal scale units on the coordinate axes.

1. Obtain a meticulously accurate graph of the function $\cos x$ on the domain $\{-\frac{1}{2}\pi \le x \le \frac{1}{2}\pi\}$ with a large scale unit used on the coordinate axes. With $w° = (x \text{ radians})$, use values of $w°$ spaced at intervals of $15°$.

Obtain a qualitatively accurate graph based on very few points.

2. Graph the function $\sin x$ on $\{-3\pi \le x \le \pi\}$.
3. Graph the function $\cos x$ on $\{-\frac{5}{2}\pi \le x \le \frac{3}{2}\pi\}$.
4. Graph the function $2 \sin x$ on $\{0 \le x \le 2\pi\}$.
5. Graph the function $-2 \cos x$ on $\{-\frac{1}{2}\pi \le x \le \frac{3}{2}\pi\}$.

6. Obtain a meticulously accurate graph of the function $\tan x$ on the interval $\{-\frac{1}{2}\pi < x < \frac{1}{2}\pi\}$. Show the asymptotes. If $w° = (x \text{ radians})$, use points corresponding to values of $w°$ spaced at intervals of $15°$, and also $85°$ and $-85°$.

Obtain a qualitatively accurate graph of the equation. On any branch of the graph, locate at least three accurate points. Draw all asymptotes.

7. $y = \tan x$ on $\{-\frac{3}{2}\pi \le x \le \frac{3}{2}\pi\}$.
8. $y = \cot x$ on $\{0 \le x \le 2\pi\}$.
9. $y = \sec x$ on $\{-\frac{1}{2}\pi \le x \le \frac{5}{2}\pi\}$; on the same plane, $y = \cos x$.
10. $y = \csc x$ on $\{0 \le x \le 2\pi\}$; on the same plane, $y = \sin x$.

11. $y = 2 - \sin x$ on $\{0 \leqq x \leqq 2\pi\}$.

Hint. Think of first graphing the function $-\sin x$, or $y_1 = -\sin x$. How may the requested graph be obtained?

By recalling Definition II *on page* 157, *or graphs, or convenient reference angles, find all values of x on* $\{-2\pi \leqq x \leqq 2\pi\}$ *for which the given equation is true.*

12. $\sin x = 0$. **13.** $\cos x = 1$. **14.** $\sin x = -1$.

15. $\tan x = 0$. **16.** $\tan x = -1$. **17.** $\sin x = \frac{1}{2}$.

18. $\cos x = -\frac{1}{2}$. **19.** $\sin x = \frac{1}{2}\sqrt{3}$. **20.** $\cos x = \frac{1}{2}\sqrt{2}$.

21. $\cos x = -\frac{1}{2}\sqrt{2}$. **22.** $\tan x = \sqrt{3}$. **23.** $\tan x = -\sqrt{3}$.

24. $\sin x = -\frac{1}{2}\sqrt{3}$. **25.** $\cos x = \frac{1}{2}\sqrt{3}$. **26.** $\tan x = \frac{1}{3}\sqrt{3}$.

Obtain a qualitatively accurate graph of the equation, with y on the given domain, and with the x-axis horizontal.

27. $x = \sin y$; $\{0 \leqq y \leqq 4\pi\}$. **28.** $x = \cos y$; $\{-\frac{1}{2}\pi \leqq y \leqq \frac{5}{2}\pi\}$.
29. $x = \tan y$; $\{-\frac{1}{2}\pi \leqq y \leqq \frac{3}{2}\pi\}$. **30.** $x = \sec y$; $\{-\frac{1}{2}\pi \leqq y \leqq \frac{5}{2}\pi\}$.

chapter 7

The Standard Trigonometric Functions

65. A NEW VIEWPOINT

In the preceding chapter the trigonometric functions were described as functions of angles, referred to then as *numbers* (of angular units). In calculus, trigonometric functions will arise without any necessity for a background involving angles. Hence it is desirable to show that trigonometric functions of real numbers can be defined without mentioning angles.

Note 1. A sound definition of length of arc on a general curve, and thus in particular on a circle, cannot be given conveniently without use of calculus. However, the concept of length of arc was used in the measurement of angles when the trigonometric functions of angles were introduced in the preceding chapter. This concept also will be used in our future discussion in the present chapter. Hence, the same logical objection can be raised to the methods of both chapters. In calculus, a procedure becomes available for defining the trigonometric functions without presupposing acquaintance with the concept of length of arc.

In Figure 117 a single unit of length will apply on the coordinate axes and in measuring all distances, including arc lengths on the circle, K, whose radius is one unit. The figure shows a vertical number scale labeled as an s-axis, with the origin at $P_0:(1,0)$, and the positive direction upward. Conceive of winding the half-line of the s-scale where $s \geq 0$ around the circumference of K, counterclockwise. Then, to each number $s \geq 0$, there corresponds just one point P_s on the circle such that the directed arc length $P_0 P_s$ measured *counterclockwise,* and considered *positive* in this direction, is s units. We write $(arc\ P_0 P_s) = s$. Similarly, conceive of winding the half-line of the s-axis where $s \leq 0$ about the circumference of K, clockwise. Then, to each number $s < 0$ there corresponds just one point P_s such that the directed arc length $P_0 P_s$ measured *clockwise,* and considered *negative* in this direction, is s units. Again we write $(arc\ P_0 P_s) = s$. Thus, to each real number $s \neq 0$ there corresponds just one point P_s on K such that $(arc\ P_0 P_s) = s$. If $s = 0$, then $P_s = P:(0,1)$.

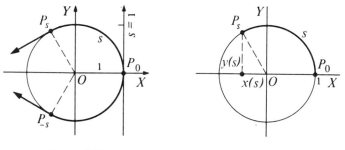

FIGURE 117   FIGURE 118

ILLUSTRATION 1.   The circumference of the circle K in Figure 117 is 2π, because the radius is 1. Hence, the coordinates of $P_{\frac{1}{2}\pi}$ are $(0,1)$; of P_π are $(-1,0)$; of $P_{\frac{3}{2}\pi}$ are $(0,-1)$.

For any number s, let the coordinates of P_s be $(x(s), y(s))$, as in Figure 118. Thus, two functions $x(s)$ and $y(s)$ are introduced, where the domain of s is *all real numbers*. Then, as follows, six functions are defined:* {s̄ine, c̄osine, t̄angent, c̄otangent, s̄ecant, c̄osecant}, where we read "*sine bar,*" etc., and later abbreviate the names as usual in writing.

$$\bar{\sin} s = y(s); \qquad \bar{\cos} s = x(s); \qquad \bar{\tan} s = \frac{y(s)}{x(s)};$$

$$\bar{\csc} s = \frac{1}{y(s)}; \qquad \bar{\sec} s = \frac{1}{x(s)}; \qquad \bar{\cot} s = \frac{x(s)}{y(s)}. \tag{1}$$

The functions in (1) will be called the *standard trigonometric functions of real numbers*. They are the trigonometric functions used in calculus. Sometimes they are called the **circular functions** because of their relation to a circle. At present, we have no right to identify these functions with any which have been met previously in this text. The desired identification is a consequence of the following result.

THEOREM I.   The value of any one of the standard trigonometric functions with the argument s is the same as the value of the similarly named trigonometric function of s radians, or†

$$\bar{\sin} s = \sin s^{(r)}; \qquad \bar{\cos} s = \cos s^{(r)}; \qquad \bar{\tan} s = \tan s^{(r)}; etc. \tag{2}$$

Proof.   1. In Figure 118, let w radians be the measure of the angle formed by revolving the line OP about O from the position where P is

---

*The bar over the first letter is a temporary notation.

†For emphasis and contrast here, we use $s^{(r)}$ for the angle s *radians*. Later, we shall omit the superscript "(r)."

P_0:(1,0), through all points on the directed arc $P_0 P_s$ of the circle. From page 155, recall that

$$(length\ of\ arc) = (radius) \cdot (measure\ of\ angle\ in\ radians).$$

For the angle w radians, since $(arc P_0 P_s) = s$ and the radius is 1,

$$s = 1 \cdot w = w. \tag{3}$$

Hence, the angle with the terminal side OP_s has the measure s radians.

2. From Definition II on page 157, with P_s:$(x(s), y(s))$ of Figure 118 chosen on the terminal side OP_s of s radians, where $|\overline{OP_s}| = 1$, we have

$$\sin s^{(r)} = \frac{y(s)}{1} = \bar{\sin}\, s; \qquad \cos s^{(r)} = \frac{x(s)}{1} = \bar{\cos}\, s;$$

$$\tan s^{(r)} = \frac{y(s)}{x(s)} = \bar{\tan}\, s; \qquad \sec s^{(r)} = \frac{1}{x(s)} = \bar{\sec}\, s;\ etc.$$

Thus Theorem I has been proved.

Hereafter we shall omit the bar over the first letter in $\bar{\sin}\, s$, $\bar{\cos}\, s$, $\bar{\tan}\, s$, etc., and write simply $\sin s$, $\cos s$, $\tan s$, etc. Also, as a rule, we shall prefer to use x instead of s as the independent variable. Thus, we have the new symbols $\{\sin x, \cos x, \tan x, \cot x, \sec x, \csc x\}$. They have the same values, respectively, as $\{\sin x^{(r)}, \cos x^{(r)}, \cdots\}$. Then, we discard the preceding notations by making the following agreement, specialized for the case of the sine.

> *"sin x" will mean either the standard sine function of numbers defined in* (1), *or the sine of x radians. In either case, the value of sin x is the sine of x radians.* (4)

An agreement like (4) applies for each symbol representing a standard trigonometric function. Whenever such a symbol is used, the context will indicate which meaning is involved. Regardless of the choice, there is only one value for the symbol, because of the equalities in (2).

As a consequence of (4), if a number x without a degree mark is met as the argument of any trigonometric function of angles, the symbol represents a function of x radians, as was the case in Chapter 6. Thus, "$\tan 3$" means the tangent of 3 radians.

By (4), the values of any trigonometric function of numbers can be found by use of a trigonometric table, or perhaps by recollection of the functions of various convenient angles, with the aid of reference angles.

ILLUSTRATION 2.   We have the following results, obtained by interpreting each function's argument as the radian measure of an angle.

$$\cos 0 = 1; \qquad \sin \tfrac{1}{2}\pi = 1; \qquad \cos \tfrac{1}{2}\pi = 0;$$

$$\tan \tfrac{3}{4}\pi = -1; \qquad \tan \tfrac{1}{2}\pi\ is\ infinite.$$

Hereafter, unless otherwise mentioned, if a trigonometric function is referred to, or is specified in notation implied by (4), it will be understood that a standard trigonometric function is involved. The word *standard* no longer will be used, except possibly for emphasis. When desirable, the value of any trigonometric function of x will be obtained, without comment, by interpreting the result as a function of the angle x radians. The number x will be referred to as being on a certain *quadrant interval* in case the angle x radians is in that quadrant. A number z, such that $0 < z < \frac{1}{2}\pi$, will be called the *reference number* for x in case z radians is the reference angle for x radians. The graphs obtained for the trigonometric functions $\sin x$, $\cos x$, etc. on pages 176–179 now may be referred to as graphs of the standard trigonometric functions.

ILLUSTRATION 3.   To obtain $\sin\left(-\frac{2}{3}\pi\right)$, visualize the angle $-\frac{2}{3}\pi$ radians in its standard position on a coordinate system. The reference angle is $\frac{1}{3}\pi$, whose sine is $\sqrt{3}/2$, and the sine is negative in quadrant III. Hence $\sin\left(-\frac{2}{3}\pi\right) = -\sqrt{3}/2$. Although we referred to angles in preceding sentences, our result is a value of the standard sine function.

It is instructive to recall important facts about the trigonometric functions in their new setting, where angles are not involved. Thus, let T be any trigonometric function. Then, we observe that T is periodic with the period 2π because the corresponding trigonometric function of angles in radian measure has the period 2π radians. Hence

$$T(x + 2\pi) = T(x), \text{ and} \tag{5}$$

$$T(x - 2\pi) = T(x). \tag{6}$$

Identities (5) and (6) are equivalent. Each one states that, if two numbers differ by 2π, the trigonometric functions of these numbers are equal. Each of (5) and (6) abbreviates six identities, one corresponding to each of the six trigonometric functions. The functions $\tan x$ and $\sec x$ are undefined if x is any quadrantal number which is an odd multiple of $\pi/2$, and we say that the tangent or secant of such a number is *infinite*. In particular, $|\tan x| \to +\infty$ and $|\sec x| \to +\infty$ when $x \to \frac{1}{2}\pi$. The range of $\sec x$ is all numbers with absolute value at least one. Etc. A rehearsal of such facts about the trigonometric functions is best associated with their graphs.

ILLUSTRATION 4.   A graph of $y = \sin x$ with x on the domain $\{-2\pi \leqq x \leqq 2\pi\}$ is given in Figure 119, which is a duplicate of Figure 113 from page 177. *By inspection of* Figure 119, we reach the following conclusions:

Range of the sine: If x increases from $x = 0$ to $x = \frac{1}{2}\pi$, then $\sin x$ increases from 0 to 1, as a maximum value; then decreases to 0 at

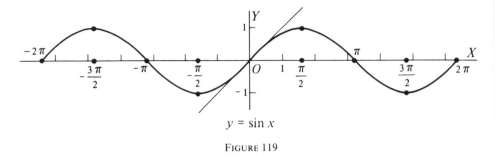

$$y = \sin x$$

FIGURE 119

$x = \pi$; to -1 at $x = \frac{3}{2}\pi$; and then increases from -1 to 0 at $x = 2\pi$.

On the domain $\{-2\pi \leqq x \leqq 2\pi\}$, the indicated solutions are obtained for the following equations:

$\sin x = 0$: *Solutions* $\{-2\pi, -\pi, 0, \pi, 2\pi\}$, x-intercepts of the graph.

$\sin x = 1$: *Solutions* $\{-\frac{3}{2}\pi, \frac{1}{2}\pi\}$, at crests of waves.

$\sin x = \frac{1}{2}$: Place a ruler edge horizontally on the graph through $y = \frac{1}{2}$ on the y-axis. One solution recalled quickly is $x = \frac{1}{6}\pi$, at the intersection of the ruler edge nearest the y-axis. The other intersections give the solutions $\{\frac{5}{6}\pi, -\frac{11}{6}\pi, -\frac{7}{6}\pi\}$.

ILLUSTRATION 5. From Table VI, $\sin .63 = .58914$. If the argument, x, of $\sin x$ is negative or is greater than (approximately) $\frac{1}{2}\pi$ or 1.57, we interpret the value of $\sin x$ as the sine of x radians and first obtain an acute reference angle. In such a case, with Table VI involved, use $\pi = 3.14$. Thus, to obtain $\sin 3.32$: since $\pi = 3.14$, an angle of 3.32 radians is in quadrant III, with the reference angle as $(3.32 - 3.14)$ or .18 radians, and the sine is negative in quadrant III. Hence, by Table VI, $\sin 3.32 = -\sin .18 = .17903$. (Draw a figure.)

EXERCISE 40

If it is necessary to approximate π in any problem, use $\pi = 3.14$.
Draw a circle of radius 1 in an xy-plane. Then, as on page 183, locate P_s on the circle for the given value of s. Also find $\sin s$, $\cos s$, and $\tan s$.

1. $s = \frac{1}{3}\pi$. 2. $s = \frac{1}{4}\pi$. 3. $s = -\frac{1}{4}\pi$. 4. $s = -\frac{1}{2}\pi$.

5. $s = \frac{5}{6}\pi$. 6. $s = -\frac{2}{3}\pi$. 7. $s = -\pi$. 8. $s = \frac{3}{4}\pi$.

9. $s = \frac{2}{3}\pi$. 10. $s = 5\pi$. 11. $s = -\frac{1}{6}\pi$. 12. $s = -\frac{5}{6}\pi$.

13. Specify all values of s on the interval $\{-2\pi \leqq s \leqq 2\pi\}$ such that $\tan s$ is undefined; $\sec s$ is undefined; $\csc s$ is undefined.

By use of only a few accurate points, obtain a qualitatively accurate graph of the function cos x for x on the interval $\{-2\pi \leq x \leq 2\pi\}$. *Then, by use of the graph and other information, solve the following problems.*

14. Describe how the values of cos x vary as x varies from 0 to 2π. What is the range of the cosine function?

15. Find all solutions on the interval $\{-2\pi \leq x \leq 2\pi\}$ of each of the following equations and observe the location of the corresponding points of the graph (perhaps draw a horizontal line with a ruler): cos $x = 1$; cos $x = -1$; cos $x = 0$; cos $x = \frac{1}{2}$; cos $x = -\frac{1}{2}\sqrt{3}$; cos $x = \frac{1}{2}\sqrt{2}$; cos $x = -\frac{1}{2}\sqrt{2}$.

By use of only a few accurate points, obtain qualitatively accurate graphs of the functions tan x *and* sec x *on the interval* $\{-\frac{3}{2}\pi \leq x \leq \frac{3}{2}\pi\}$, *and draw any asymptote possessed by either graph. Then, solve the following problems.*

16. Describe how the values of tan x vary as x varies from 0 to $\frac{1}{2}\pi$; from 0 to $-\frac{1}{2}\pi$. What is the range of the tangent function? Repeat the preceding details for the case of the secant function.

17. Find all solutions on the interval $\{-\frac{3}{2}\pi \leq x \leq \frac{3}{2}\pi\}$ of each of the following equations or statements: tan $x = 0$; tan $x = 1$; sec $x = -1$; sec $x = 1$; sec $x = \sqrt{2}$; tan $x = \frac{1}{3}\sqrt{3}$; the function tan x is infinite; the secant function is infinite; tan $x = -1$; tan $x = -\sqrt{3}$; tan $x = \frac{1}{3}\sqrt{3}$; sec $x = -\sqrt{2}$; sec $x = 2$.

Obtain qualitatively accurate graphs of the functions cot x *and* csc x *on the interval* $\{-\pi \leq x \leq 3\pi\}$, *and draw any asymptote possessed by either graph. Then, solve the following problems.*

18. Describe how the values of cot x vary as x varies from 0 to 2π. What is the range of the function cot x? Repeat the problem for csc x.

19. Solve each of the following equations or statements: cot $x = 1$; cot $x = -1$; cot $x = \sqrt{3}$; cot $x = -\frac{1}{3}\sqrt{3}$; csc $x = 1$; csc $x = \sqrt{2}$; the function cot x is infinite; the function csc x is infinite.

By use of Table VI, *find the value of the function, or the unknown argument x on* $\{0 \leq x \leq 2\pi\}$. *Use* $\pi = 3.14$ *where needed.*

20. sin 1.38.   **21.** tan .84.   **22.** cos 1.43.

23. sin 3.70.   **24.** tan 2.29.   **25.** cos 5.22.

26. sin $x = .21823$.   **27.** cos $x = .62941$.   **28.** tan $x = 2.5722$.

29. sin $x = -.84683$. **30.** cos $x = -.72484$. **31.** tan $x = -.23414$.

66. BASIC TRIGONOMETRIC IDENTITIES

A remarkable number of identities can be proved for the trigonometric functions. The most common identities which are used either primarily in trigonometry itself, or in its applications in calculus, are given below. Each of these identities will be referred to hereafter in this text by its Roman numeral. Identities (I)–(X) arise so frequently that they deserve memorization. Each identity is true if each variable involved has a value such that all function values in the identity exist. Identities (I)–(III) were proved in Chapter 6. The other numbered identities below will be proved either in the text discussion or in later exercises.

Summary of Trigonometric Identities

$$\sec x = \frac{1}{\cos x}; \qquad \csc x = \frac{1}{\sin x}; \qquad \cot x = \frac{1}{\tan x}. \tag{I}$$

$$\tan x = \frac{\sin x}{\cos x}; \qquad \cot x = \frac{\cos x}{\sin x}. \tag{II}$$

$$\sin^2 x + \cos^2 x = 1; \quad \tan^2 x + 1 = \sec^2 x; \quad 1 + \cot^2 x = \csc^2 x. \tag{III}$$

$$\sin(-x) = -\sin x; \quad \cos(-x) = \cos x; \quad \tan(-x) = -\tan x. \tag{IV}$$

Addition formulas:

("+" *with* "+") $\sin(x \pm y) = \sin x \cos y \pm \cos x \sin y.$ (V)

("+" *with* "−") $\cos(x \pm y) = \cos x \cos y \mp \sin x \sin y.$ (VI)

Each trigonometric function has 2π as a period:
[any trig. funct. of $(x \pm 2\pi)$] = (same function of x).   (VII)

The tangent and cotangent functions have π as a period:
$$\tan(x \pm \pi) = \tan x; \qquad \cot(x \pm \pi) = \cot x. \tag{VIII}$$

$$\sin 2x = 2 \sin x \cos x. \tag{IX}$$

$$\cos 2x = \cos^2 x - \sin^2 x = 2\cos^2 x - 1 = 1 - 2\sin^2 x. \tag{X}$$

Cofunction identities:

$$[any \ trig. \ function \ of (\tfrac{1}{2}\pi - x)] = (cofunction \ of \ x). \tag{XI}$$

$$\tan(x + y) = \frac{\tan x + \tan y}{1 - \tan x \tan y}; \qquad \tan(x - y) = \frac{\tan x - \tan y}{1 + \tan x \tan y}. \tag{XII}$$

$$\tan 2x = \frac{2 \tan x}{1 - \tan^2 x}. \tag{XIII}$$

$$2\cos^2 \tfrac{1}{2}x = 1 + \cos x; \qquad 2\sin^2 \tfrac{1}{2}x = 1 - \cos x. \tag{XIV}$$

$$\tan^2 \tfrac{1}{2}x = \frac{1 - \cos x}{1 + \cos x}. \tag{XV}$$

In order to prove (IV), it is convenient to return to the notation of page 183 and rewrite (IV) as follows.

$$\sin(-s) = -\sin s; \quad \cos(-s) = \cos s; \quad \tan(-s) = -\tan s. \tag{1}$$

Proof of (I).   1. A circle K of radius 1 is shown in Figure 120. For any number s, as on page 183, we have on K the trigonometric points

$$P_s:(x(s), y(s)) \quad and \quad P_{-s}:(x(-s), y(-s)). \tag{2}$$

By the definitions in (1) on page 183, (2) becomes

$$P_s:(\cos s, \sin s) \quad and \quad P_{-s}:(\cos (-s), \sin (-s)). \tag{3}$$

2. In Figure 120, P_s and P_{-s} are *symmetrical to the x-axis.* Hence the x-coordinates of the two points are the *same;* the y-coordinate of $P(-s)$ is the *negative* of the y-coordinate of $P(s)$. Thus, from (3), $\cos s = \cos (-s)$ and $\sin (-s) = -\sin s$. Then, by identity (II),

$$\tan (-s) = \frac{\sin (-s)}{\cos (-s)} = -\frac{\sin s}{\cos s} = -\tan s.$$

This completes the proof of (1), or (IV).

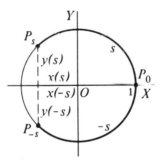

FIGURE 120

ILLUSTRATION 1.   Since $\sin (-x) = -\sin x$, the function $\sin x$ is an *odd* function, as defined on page 150. In the equation $y = \sin x$, suppose that we replace x by $-x$ and y by $-y$. By use of (IV), we obtain $-y = \sin (-x) = -\sin x$, or $-y = -\sin x$, which is equivalent to $y = \sin x$. Hence, if $P:(x,y)$ is on the graph of $y = \sin x$, the point $Q:(-x,-y)$ also is on the graph, and thus it is *symmetric to the origin.* This conclusion checks with the graph in Figure 119 on page 186.

67. ADDITION FORMULAS AND RELATED IDENTITIES

We shall arrive at (V) and (VI) by proving a sequence of results.

THEOREM II.   *Let P_1 and P_2 be points on a circle of radius 1, where the length of any arc will be considered positive if traced counterclockwise, and negative if traced clockwise. Suppose that a path on the circle* *from P_1 to P_2 has measure s. Then*

$$\overline{P_1P_2}^2 = 2 - 2\cos s. \qquad (1)$$

Proof.   1. In the plane of the circle with center O in Figure 121, create an xy-system of coordinates, oriented as usual, with line OP_1 as the x-axis, and P_1 having the coordinates $(+1,0)$. By the definitions of the trigonometric functions on page 183, with $(arc\ P_1P_2) = s$, P_2 has the coordinates $(\cos s, \sin s)$.

2. By use of the distance formula of page 57 applied to $P_1:(1,0)$ and $P_2:(\cos s, \sin s)$,

$$\overline{P_1P_2}^2 = (1 - \cos s)^2 + \sin^2 s$$
$$= (\sin^2 s + \cos^2 s) + 1 - 2\cos s = 2 - 2\cos s,$$

because $\sin^2 s + \cos^2 s = 1$.

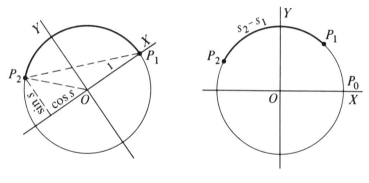

FIGURE 121   FIGURE 122

THEOREM III.   *For any two numbers s_1 and s_2,*

$$\cos(s_2 - s_1) = \cos s_1 \cos s_2 + \sin s_1 \sin s_2. \qquad (2)$$

Proof.   1. Figure 122 shows a circle of radius 1 as used on page 183. Let the trigonometric points on the circle corresponding to the numbers s_1 and s_2 be $P_1:(\cos s_1,\ \sin s_1)$ and $P_2:(\cos s_2,\ \sin s_2)$. Then we have

$$(arc\ P_0P_1) = s_1; \quad (arc\ P_0P_2) = s_2; \quad (arc\ P_1P_2) = s_2 - s_1. \qquad (3)$$

In (3),* arc P_1P_2 is directed counterclockwise if $s_2 > s_1$, and clockwise if $s_2 < s_1$.

*It is not necessary to have $|s_1| < 2\pi$ and $|s_2| < 2\pi$.

2. By Theorem II, with s of Theorem II replaced by $(s_2 - s_1)$,

$$\overline{P_1 P_2}^2 = 2 - 2 \cos (s_2 - s_1). \tag{4}$$

By the distance formula of page 57,

$$\overline{P_1 P_2}^2 = (\cos s_2 - \cos s_1)^2 + (\sin s_2 - \sin s_1)^2$$
$$= (\cos^2 s_2 + \sin^2 s_2) + (\cos^2 s_1 + \sin^2 s_1)$$
$$-2(\cos s_1 \cos s_2 + \sin s_1 \sin s_2), \textit{ or}$$
$$\overline{P_1 P_2}^2 = 2 - 2(\cos s_1 \cos s_2 + \sin s_1 \sin s_2). \tag{5}$$

From (4) and (5) we obtain (2), which proves (VI) for $\cos (x - y)$.

Proof of (VI) for $\cos (x + y)$. From (2) with $s_2 = x$ and $s_1 = -y$,

$$\cos (x + y) = \cos x \cos (-y) + \sin x \sin (-y). \tag{6}$$

From (IV), $\cos (-y) = \cos y$ and $\sin (-y) = -\sin y$. Then (6) gives (VI):

$$\cos (x + y) = \cos x \cos y - \sin x \sin y.$$

By use of (2), we shall prove the following identities:

$$\cos \left(\tfrac{1}{2}\pi - x\right) = \sin x. \tag{7}$$

$$\sin \left(\tfrac{1}{2}\pi - x\right) = \cos x. \tag{8}$$

Proof of (7) and (8). 1. From (2) with $s_2 = \tfrac{1}{2}\pi$ and $s_1 = x$,

$$\cos \left(\tfrac{1}{2}\pi - x\right) = \cos \tfrac{1}{2}\pi \cos x + \sin \tfrac{1}{2}\pi \sin x$$
$$= 0 \cdot \cos x + 1 \cdot \sin x = \sin x,$$

because $\sin \tfrac{1}{2}\pi = 1$ and $\cos \tfrac{1}{2}\pi = 0$. Hence, (7) is true.

2. On replacing x by $\left(\tfrac{1}{2}\pi - x\right)$ on both sides of (7), we obtain (8) as follows:

$$\cos \left[\tfrac{1}{2}\pi - \left(\tfrac{1}{2}\pi - x\right)\right] = \sin \left(\tfrac{1}{2}\pi - x\right), \textit{ or}$$
$$\cos x = \sin \left(\tfrac{1}{2}\pi - x\right).$$

Proof of (V) for $\sin (x + y)$. On replacing x by $(x + y)$ in (7), we find
$$\sin (x + y) = \cos \left(\tfrac{1}{2}\pi - x - y\right) = \cos \left[\left(\tfrac{1}{2}\pi - x\right) - y\right]. \tag{9}$$

In (9) on the right, use (2) with $s_2 = \left(\tfrac{1}{2}\pi - x\right)$ and $s_1 = y$; then use (7) and (8), to obtain (V) for $\sin (x + y)$:

$$\sin (x + y) = \cos \left(\tfrac{1}{2}\pi - x\right) \cos y + \sin \left(\tfrac{1}{2}\pi - x\right) \sin y, \textit{ or}$$

$$\sin (x + y) = \sin x \cos y + \cos x \sin y. \tag{10}$$

If y is replaced by $-y$ in (10) and (IV) is used, the student may verify that (V) is obtained for $\sin (x - y)$.

Identities (V) and (VI) are known as the *addition and subtraction formulas*. Identity (VII) was proved in Chapter 6. Identities (VIII)–(XV) will be proved by use of (V) and (VI).

THEOREM IV.   *The tangent and cotangent functions have π as a period. That is,* (VIII) *is true.*

Proof.   1. By use of (II), and then (V) and (VI) for sin $(x + \pi)$ and cos $(x + \pi)$, with cos $\pi = -1$ and sin $\pi = 0$,

$$\tan (x + \pi) = \frac{\sin (x + \pi)}{\cos (x + \pi)} = \frac{\sin x \cos \pi + \cos x \sin \pi}{\cos x \cos \pi - \sin x \sin \pi}, \text{ or}$$

$$\tan (x + \pi) = \frac{\sin x}{\cos x} = \tan x, \tag{11}$$

which proves (VIII) for tan $(x + \pi)$.   A similar proof applies for tan $(x - \pi)$.

2. By use of (I) and (11),

$$\cot (x + \pi) = \frac{1}{\tan (x + \pi)} = \frac{1}{\tan x} = \cot x,$$

which proves (VIII) for cot $(x + \pi)$.   Similarly, (VIII) is true for cot $(x - \pi)$.

Proof of (IX).   Place $y = x$ in (10).

Proof of (X).   With y replaced by x in (VI), we obtain

$$\cos 2x = \cos^2 x - \sin^2 x. \tag{12}$$

On using $\cos^2 x = 1 - \sin^2 x$ from (III) in (12), we find that

$$\cos 2x = 1 - \sin^2 x - \sin^2 x = 1 - 2 \sin^2 x.$$

With $\sin^2 x = 1 - \cos^2 x$ used in (12), we obtain $\cos 2x = 2 \cos^2 x - 1$. Hence (X) is true.

The student will prove (XII)–(XV) in the exercises. Identity (XI) will be met in the next section.

EXAMPLE 1.   Expand $\tan \left(\frac{1}{4} \pi + x\right)$ and insert known function values.

Solution.   By use of (XII) and $\tan \frac{1}{4} \pi = 1$,

$$\tan \left(\tfrac{1}{4} \pi + x\right) = \frac{\tan \frac{1}{4} \pi + \tan x}{1 - \tan \frac{1}{4} \pi \tan x} = \frac{1 + \tan x}{1 - \tan x}.$$

EXAMPLE 2.   By use of the known values for trigonometric functions of 30° and 135°, find sin 165°.

Solution. We interpret (V) as an identity involving trigonometric functions of the angles x and y, in either radian or degree measure. From (V), since $30° + 135° = 165°$,

$$\sin 165° = \sin 30° \cos 135° + \cos 30° \sin 135°$$

$$= \frac{1}{2} \cdot \left(-\frac{1}{2}\sqrt{2}\right) + \frac{1}{2}\sqrt{3}\left(\frac{1}{2}\sqrt{2}\right) = -\frac{\sqrt{2}}{4} + \frac{\sqrt{6}}{4}.$$

Note 1. Identities (IX), (X), and (XIII) are called the *double argument* identities or, when interpreted as identities for functions of angles, are called the *double angle* formulas. Similarly, (XIV) and (XV) are called *half-argument* identities or, for angles, the *half-angle* formulas.

EXERCISE 41

Expand by use of addition formulas and insert known function values.

1. $\cos\left(\frac{1}{2}\pi + x\right).$ **2.** $\sin\left(x - \frac{1}{2}\pi\right).$ **3.** $\sin(x + \pi).$

4. $\cos\left(x - \frac{3}{2}\pi\right).$ **5.** $\tan\left(x + \frac{1}{4}\pi\right).$ **6.** $\tan\left(\frac{3}{4}\pi + x\right).$

7. $\tan\left(\frac{3}{4}\pi - \frac{1}{3}\pi\right).$ **8.** $\sin\left(\frac{1}{4}\pi + \frac{1}{3}\pi\right).$ **9.** $\sin\left(\frac{1}{4}\pi - x\right).$

10. $\tan\left(x - \frac{1}{4}\pi\right).$ **11.** $\cos(\pi - x).$ **12.** $\cos\left(\frac{1}{3}\pi + x\right).$

13. $\tan(2\pi - x).$ **14.** $\tan\left(\frac{5}{4}\pi - x\right).$ **15.** $\sin\left(x - \frac{1}{6}\pi\right).$

16. $\cos\left(\frac{5}{6}\pi + x\right).$ **17.** $\tan\left(\frac{1}{3}\pi - x\right).$ **18.** $\tan\left(x - \frac{1}{6}\pi\right).$

19. $\cos\left(x - \frac{1}{6}\pi\right).$ **20.** $\sin\left(x - \frac{5}{6}\pi\right).$ **21.** $\tan\left(\frac{1}{3}\pi - \frac{1}{4}\pi\right).$

By use of an addition formula, find the specified results by use of functions with the indicated arguments.

22. Obtain $\sin\frac{3}{4}\pi$ and $\cos\frac{3}{4}\pi$ by use of functions of $\frac{1}{2}\pi$ and $\frac{1}{4}\pi$.

23. Obtain $\cos\frac{3}{4}\pi$ and $\tan\frac{3}{4}\pi$ by use of functions of π and $\frac{1}{4}\pi$.

24. Obtain $\sin\frac{2}{3}\pi$ and $\tan\frac{2}{3}\pi$ by use of functions of $\frac{1}{3}\pi$.

25. Obtain $\sin\frac{7}{12}\pi$ and $\cos\frac{7}{12}\pi$ by use of functions of $\frac{1}{3}\pi$ and $\frac{1}{4}\pi$.

26. Obtain $\sin\frac{5}{12}\pi$ and $\cos\frac{5}{12}\pi$ by use of functions of $\frac{3}{4}\pi$ and $\frac{1}{3}\pi$.

Find the sine, cosine, and tangent of the first angle by use of trigonometric functions of the other angles and addition formulas.

27. Of 75°, by use of 30° and 45°. **28.** Of 15°, by use of 30° and 45°.

29. Of 165°, by use of 225° and 60°. **30.** Of 195°, by use of 135° and 60°.

Express the sine, cosine, and tangent of the first number in terms of functions of the second number by use of double-argument or half-argument identities, where possible.

31. $\frac{2}{3}\pi$ by use of $\frac{1}{3}\pi$. **32.** $\frac{1}{2}\pi$ by use of $\frac{1}{4}\pi$.

33. $\frac{4}{3}\pi$ by use of $\frac{2}{3}\pi$. **34.** $\frac{1}{6}\pi$ by use of $\frac{1}{3}\pi$.

35. $\frac{1}{4}\pi$ by use of $\frac{1}{2}\pi$.   **36.** $\frac{1}{3}\pi$ by use of $\frac{2}{3}\pi$.

37. $3x$ by use of $\frac{3}{2}x$.   **38.** $4x$ by use of $2x$.

39. $2x$ by use of $4x$.   **40.** $\frac{3}{4}x$ by use of $\frac{3}{2}x$.

41. Prove identity (XII) for $\tan(x + y)$, and then for $\tan(x - y)$ by substituting $-y$ for y.

Hint. $\tan(x + y) = \sin(x + y)/\cos(x + y)$. Use (V) and (VI); then divide both numerator and denominator by $\cos x \cos y$.

42. Prove identity (XIII) by use of an earlier identity of the list.

43. Prove (XIV), and then (XV).

Hint.   In (X), replace x by $\frac{1}{2}x$.

44. By use of (IV), investigate the nature of the symmetry exhibited by the graph of $y = \tan x$; of $y = \cos x$. Use tests for symmetry.

68. REDUCTION FORMULAS

If a number w is an integral multiple of $\frac{1}{2}\pi$, then w will be called a *quadrantal number,* because w radians is a *quadrantal angle.* It is found that, if w is any quadrantal number, each trigonometric function of $(\pm x + w)$ is equal to either *plus or minus a function of x.* The resulting identities frequently are called *reduction formulas.* For instance, the periodicity identities (VII) and (VIII), and the cofunction identities (XI) are reduction formulas. Any reduction formula can be proved by first applying (V) and (VI) for the sine and cosine, and then using (I) or (II) when needed.

Example 1.   Obtain the reduction formulas for the trigonometric functions* of $\left(x - \frac{3}{2}\pi\right)$.

Solution.   With $\cos \frac{3}{2}\pi = 0$ and $\sin \frac{3}{2}\pi = -1$, by use of (V), (VI), and then (II) we obtain

$$\sin\left(x - \tfrac{3}{2}\pi\right) = \sin x \cos \tfrac{3}{2}\pi - \cos x \sin \tfrac{3}{2}\pi = \cos x; \tag{1}$$

$$\cos\left(x - \tfrac{3}{2}\pi\right) = \cos x \cos \tfrac{3}{2}\pi + \sin x \sin \tfrac{3}{2}\pi = -\sin x; \tag{2}$$

$$\tan\left(x - \tfrac{3}{2}\pi\right) = \frac{\sin\left(x - \tfrac{3}{2}\pi\right)}{\cos\left(x - \tfrac{3}{2}\pi\right)} = \frac{\cos x}{-\sin x} = -\cot x. \tag{3}$$

Comment.   We could not obtain $\tan\left(x - \frac{3}{2}\pi\right) = -\cot x$ by use of (XII) because $\tan \frac{3}{2}\pi$ is undefined.

---

*Unless otherwise stated, this will mean only the sine, cosine, and tangent. Then, by use of (I), the corresponding formulas for the secant, cosecant, and cotangent could be obtained.

EXAMPLE 2.   Obtain the reduction formulas for the trigonometric functions of $(-x + 3\pi)$.

Solution.   By use of (V), (VI), $\cos 3\pi = -1$, $\sin 3\pi = 0$, and then (II),

$$\sin (3\pi - x) = \sin 3\pi \cos x - \cos 3\pi \sin x = \sin x; \tag{4}$$

$$\cos (3\pi - x) = \cos 3\pi \cos x + \sin 3\pi \sin x = -\cos x; \tag{5}$$

$$\tan (3\pi - x) = \frac{\sin (3\pi - x)}{\cos (3\pi - x)} = \frac{\sin x}{-\cos x} = -\tan x.$$

In Example 1, an *odd* multiple of $\frac{1}{2}\pi$ is involved; each result obtained is plus, or minus, the *cofunction* of x. Example 2 involves 3π, or $6 \cdot \frac{1}{2}\pi$, an *even* multiple of $\frac{1}{2}\pi$; each result is plus, or minus, the *same function* of x. Also, the signs, $+$ or $-$, on the right in each formula obtained *did not depend on the value of x.* The results in Examples 1 and 2 are seen to be consequences of the facts that $\sin (n \cdot \frac{1}{2}\pi)$ is 0 if n is an *even* integer, and is ± 1 if n is *odd*; $\cos (n \cdot \frac{1}{2}\pi)$ is ± 1 if n is *even* and is 0 if n is *odd*. Hence, without added discussion, we state the following result, where n is any integer.

$$\begin{bmatrix} Any \ trig. \ func. \\ of \left(\pm x + n \cdot \frac{1}{2}\pi\right) \end{bmatrix} = \begin{bmatrix} \pm \ \textbf{(same func. } of \ x, \ n \textbf{ even);} \\ \pm \ \textbf{(cofunc. } of \ x, \ n \textbf{ odd)} \end{bmatrix}. \tag{6}$$

In (6), for $+x$ or for $-x$ on the left, and a specified value of n, just *one sign* applies on the right, with the same sign involved for *all values of x*. Suppose that we desire a particular case of (6) without a detailed solution as in Examples 1 and 2. Then, first, we write the special case of (6) with an ambiguous sign $\pm$ on the right. Second, we interpret the variable x as the measure of an *acute angle,* and check signs in the equality in order to choose the correct sign on the right.

EXAMPLE 3.   Express $\sin \left(\frac{3}{2}\pi - x\right)$ in terms of a trigonometric function of x by use of (6).

Solution.   1. From (6), since $\frac{3}{2}\pi$ is an odd multiple of $\frac{1}{2}\pi$,

$$\sin \left(\tfrac{3}{2}\pi - x\right) = \pm\cos x. \tag{7}$$

2. Interpret $\frac{3}{2}\pi$ and x as the radian measures of angles. Assume that x is an *acute angle;* then $\left(\frac{3}{2}\pi - x\right)$ is an angle in quadrant III where the sine is *negative.* In this case, the left-hand side is negative in (7) and $\cos x > 0$. Hence, to make (7) an equality, the *minus sign* must be used on the right. Or, finally,

$$\sin \left(\tfrac{3}{2}\pi - x\right) = -\cos x. \tag{8}$$

ILLUSTRATION 1.   Without applying (6), although it could be used, from the periodicity identity (VIII), and then (IV),

$$\tan(\pi - x) = \tan(-x + \pi) = \tan(-x) = -\tan x.$$

If a function $f(x)$ has a period p, then on page 175 it was seen that $f(x)$ also has the periods $2p, 3p, \cdots, np$, where n is any positive integer. Thus,

$$f(x) = f(x + p) = f[(x + p) + p] = f(x + 2p),$$

so that $2p$ is a period. This feature of periodicity should be remembered in using (VII) and (VIII).

ILLUSTRATION 2.   Since the secant has the period 2π, then $\sec(x + 4\pi) = \sec x$.   Since the cotangent has the period π, then $\cot(x + 3\pi) = \cot x$. Also, $\cot(x - 3\pi) = \cot x$.

Without reduction formulas, the student is expected to learn the trigonometric function values of quadrantal numbers. By use of reduction formulas, if x is a number which is *not* quadrantal, any trigonometric function of x can be expressed as a function of some related number w where $0 < w < \frac{1}{2}\pi$. With x interpreted as the angle x radians, our previous use of a reference angle in finding any trigonometric function of x has amounted essentially to a graphical equivalent of the use of a reduction formula. With w as above, the reference angle would be w radians or its complement. The student may prefer to continue to use reference angles instead of reduction formulas when trigonometric function values are to be obtained from a table or otherwise.

ILLUSTRATION 3.   To obtain $\tan 243°$, notice that $243° = 180° + 63°$. Since the tangent has the period $180°$, from Table IV we obtain $\tan 243° = \tan 63° = 1.963$.

ILLUSTRATION 4.   To obtain $\sin 653°$:

$$\sin 653° = \sin(720° - 67°) = \sin(-67°),$$

because the function $\sin w°$ has the period $2(360°)$ or $720°$. Hence, by (IV) and Table IV, $\sin 653° = -\sin 67° = -.921$.

EXERCISE 42

Apply (V), (VI), *or* (XII), *to obtain the reduction formula for the specified function value. Do not use* (6) *on page* 195.

1. $\cos(x - \pi)$.   **2.** $\tan(\pi - x)$.   **3.** $\sin(x + \frac{3}{2}\pi)$.
4. $\cos(\frac{1}{2}\pi + x)$.   **15.** $\sin(x - \frac{5}{2}\pi)$.   **6.** $\cos(\frac{3}{2}\pi - x)$.

7. $\sin (3\pi - x)$.   **8.** $\cos \left(x - \frac{7}{2}\pi\right)$.   **9.** $\tan (x - 3\pi)$.

10. $\sin (\pi - x)$.   **11.** $\cos (\pi - x)$.   **12.** $\cot (\pi - x)$.

For the given number, prove the reduction formulas for all of its trigonometric functions, including the cotangent, secant, and cosecant. Start by using (V) *and* (VI), *and do not use* (6) *on page* 195.

13. $\left(\frac{1}{2}\pi - x\right)$, the cofunction identities.

14. $\left(x - \frac{1}{2}\pi\right)$.   **15.** $\left(-x + \frac{3}{2}\pi\right)$.   **16.** $(x + \pi)$.

By use of (I), (IV), *periodicity of the functions, the general reduction formula* (6) *on page* 195, *or any other method, express the result as plus or minus a trigonometric function value with the argument w, x, or y.*

17. $\cot (-x)$.   **18.** $\sec (-x)$.   **19.** $\csc (-x)$.

20. $\sin (2\pi - x)$.   **21.** $\tan (x + 2\pi)$.   **22.** $\cos (x - 2\pi)$.

23. $\tan \left(\frac{1}{2}\pi + x\right)$.   **24.** $\sin \left(\frac{1}{2}\pi + x\right)$.*   **25.** $\sec (2\pi - y)$.

26. $\sin (5\pi - x)$.   **27.** $\cos (-x + 2\pi)$.   **28.** $\tan (-x + 3\pi)$.

29. $\tan \left(x + \frac{5}{2}\pi\right)$.   **30.** $\sin \left(\frac{3}{2}\pi + y\right)$.   **31.** $\cot \left(w + \frac{1}{2}\pi\right)$.

32. Obtain an expression for $\sin 3x$ and $\cos 3x$ in terms of $\sin x$ and $\cos x$ by first obtaining $\sin 2x$ and $\cos 2x$, and then using the fact that $3x = 2x + x$.

33. Obtain $\sin 4x$ and $\cos 4x$ in terms of $\sin x$ and $\cos x$.

69. PROOFS OF TRIGONOMETRIC IDENTITIES

By use of identities (I) and II on page 188, the values of all trigonometric functions of x can be expressed in terms of $\sin x$ and $\cos x$ without use of radicals. This type of transformation of a trigonometric expression frequently is useful in calculus. Also, such a transformation sometimes can be employed in proving trigonometric identities.

ILLUSTRATION 1.   To change the expression at the left below to a form involving no trigonometric functions except $\sin x$ and $\cos x$, first we use (XII), and then (II):

$$\tan \left(x - \tfrac{1}{4}\pi\right) = \frac{\tan x - \tan \frac{1}{4}\pi}{1 + \tan x \tan \frac{1}{4}\pi} = \frac{\tan x - 1}{1 + \tan x} =$$

*We find $\sin \left(x + \frac{1}{2}\pi\right) = \cos x$. This identity was mentioned on page 178 as proving that the graph of $y = \cos x$ is obtained if the graph of $y = \sin x$ is shifted $\frac{1}{2}\pi$ units to the left.

$$\left(\frac{\sin x}{\cos x} - 1\right) \div \left(1 + \frac{\sin x}{\cos x}\right) = \frac{\sin x - \cos x}{\cos x} \div \frac{\sin x + \cos x}{\cos x}$$

$$= \frac{\sin x - \cos x}{\cos x} \cdot \frac{\cos x}{\sin x + \cos x} = \frac{\sin x - \cos x}{\sin x + \cos x}.$$

Note 1.   We have agreed to emphasize $\{\sin x, \cos x, \tan x\}$. Hence, if sec x, csc x, or cot x occurs, as a rule we shall use the reciprocal relations in (I) to change to use of $\{\sin x, \cos x, \tan x\}$.

In applications of analytic trigonometry, it may become necessary to prove that one expression in terms of trigonometric functions has the same value as another expression at all values of the variables involved. With the equality expressed as an equation, the following methods are recommended for proof of the identity, with method A usually being more desirable than method B on account of the nature of the typical application.

Methods for Proving an Identity

A.   *With a first side left unaltered, use basic identities to change the form of the other side until it takes on the same form as the first side.*

B.   *Change both sides until they assume identical forms.*

EXAMPLE 1.   Prove the identity:

$$\tan x + \frac{2}{\tan x} = \frac{2 - \sin^2 x}{\sin x \cos x}. \tag{1}$$

Solution.   *By Method* A: Leave the right-hand side of (1) unaltered. By use of (II) from page 188, on the left in (1) we obtain

$$\frac{\sin x}{\cos x} + \frac{2}{\dfrac{\sin x}{\cos x}} = \frac{\sin x}{\cos x} + \frac{2 \cos x}{\sin x}$$

$$= \frac{\sin^2 x + 2 \cos^2 x}{\sin x \cos x} = \frac{\sin^2 x + 2(1 - \sin^2 x)}{\sin x \cos x} = \frac{2 - \sin^2 x}{\sin x \cos x},$$

where $\cos^2 x = 1 - \sin^2 x$ from (III) on page 188.   Hence (1) is an identity, because the left-hand side has the same value as the right-hand side at all admissable values of x.

EXAMPLE 2.   Prove the identity:

$$\frac{1 - \tan x}{1 + \tan x} = \frac{1 - \sin 2x}{\cos 2x}. \tag{2}$$

Solution.   *By Method* B: On each side of (2), express all function values in terms of $\sin x$ and $\cos x$. In (2), use (II) of page 188 on the left; (IX) and (X) on the right:

Left-hand side:
$$\frac{1 - \tan x}{1 + \tan x} = \left(1 - \frac{\sin x}{\cos x}\right) \div \left(1 + \frac{\sin x}{\cos x}\right)$$

$$= \frac{\cos x - \sin x}{\cos x} \div \frac{\cos x + \sin x}{\cos x} = \frac{\cos x - \sin x}{\cos x} \cdot \frac{\cos x}{\cos x + \sin x}$$

$$= \frac{\cos x - \sin x}{\cos x + \sin x}. \tag{3}$$

Right-hand side:
$$\frac{1 - \sin 2x}{\cos 2x} = \frac{\sin^2 x + \cos^2 x - 2 \sin x \cos x}{\cos^2 x - \sin^2 x}$$

$$= \frac{(\cos x - \sin x)^2}{(\cos x - \sin x)(\cos x + \sin x)} = \frac{\cos x - \sin x}{\cos x + \sin x}. \tag{4}$$

Notice that, above, we replaced 1 by $(\sin^2 x + \cos^2 x)$. From (3) and (4), each side of (2) has the same value for all admissable values of x. Hence (2) is an identity.

Note 1.   In Example 2 we illustrated the following attitude: In case of uncertainty as to the method to use in proving an identity, express all function values on the two sides in terms of the sine and cosine of some convenient argument, and try to show the equality.

EXERCISE 43

By use of identities (I)–(XV), change the expression to a new form involving no trigonometric functions except the sine and cosine of x or y. Insert known function values.

1. $2 \cot x + \tan x.$ **2.** $\sec x + \tan x.$ **3.** $\csc^2 x \tan x.$

4. $\cos 2x - \sin 2x.$ **5.** $\sin \left(2x - \frac{1}{4}\pi\right).$ **6.** $\sec^2 x + 2 \tan x.$

7. $\tan \left(x - \frac{3}{4}\pi\right).$ **8.** $\sin x \tan x - \cot x.$ **9.** $(\tan x + 1)^2.$

10. $\dfrac{1 - \tan^2 x}{1 + \tan^2 x}.$ **11.** $\dfrac{1 - \cot x}{1 + \cot x}.$ **12.** $\dfrac{\sec x}{\tan x + \cot x}.$

13. $\dfrac{1 + \csc x}{\sec x}.$ **14.** $\dfrac{1 + \tan^2 x}{\sec^2 x}.$ **15.** $\dfrac{\tan x + \tan y}{1 - \tan x \tan y}.$

Prove the identity.

16. $\tan x = \sec x \sin x.$ **17.** $(\sin x - \cos x)^2 = 1 - \sin 2x.$

18. $\sec^2 x \cot^2 x = 1 + \cot^2 x.$ **19.** $\tan x \sin 2x = 2 \sin^2 x.$

20. $\cot x \sin 2x = 1 + \cos 2x.$ **21.** $2 \cos x = \csc x \sin 2x.$

22. $\dfrac{1 - \tan^2 x}{1 + \tan^2 x} = \cos 2x.$

23. $\dfrac{\sin^2 x}{1 - \cos x} = 1 + \cos x.$

24. $\dfrac{\cos y - \sin y}{\cos y + \sin y} = \dfrac{1 - \tan y}{1 + \tan y}.$

25. $1 + \cos x = \dfrac{\sin x}{\csc x - \cot x}.$

26. $\sin\left(x - \tfrac{1}{4}\pi\right) + \sin\left(x + \tfrac{1}{4}\pi\right) = \sqrt{2}\sin x.$

27. $2\cos x - \dfrac{\cos 2x}{\cos x} = \sec x.$

28. $(\cos x - \sin x)^2 = 1 - \sin 2x.$

29. $\csc^2 x = \dfrac{2}{1 - \cos 2x}.$

30. $\cot 2x = \dfrac{\csc x - 2\sin x}{2\cos x}.$

31. $\tan x = \dfrac{\sin 2x}{1 + \cos 2x}.$

32. $1 + \sin x = \dfrac{\cos x}{\sec x - \tan x}.$

33. $\dfrac{\cos x}{\sec 2x} - \dfrac{\sin 2x}{\csc x} = \cos 3x.$

34. $\sec^4 x - \tan^4 x = \dfrac{1 + \sin^2 x}{\cos^2 x}.$

★**35.** Prove identity (XVI) below by use of both parts of (V); (XVII) and (XVIII) by use of (VI).

$$2\sin x \cos y = \sin(x + y) + \sin(x - y). \qquad \text{(XVI)}$$

$$2\cos x \cos y = \cos(x + y) + \cos(x - y). \qquad \text{(XVII)}$$

$$2\sin x \sin y = \cos(x - y) - \cos(x + y). \qquad \text{(XVIII)}$$

★**36.** In (XVI), let $u = x + y$ and $v = x - y$. Then prove (XIX) below. Similarly, prove (XX) and (XXI). Also prove (XXII) by use of (XIX).

$$\sin u + \sin v = 2\sin\tfrac{1}{2}(u + v)\cos\tfrac{1}{2}(u - v). \qquad \text{(XIX)}$$

$$\cos u + \cos v = 2\cos\tfrac{1}{2}(u + v)\cos\tfrac{1}{2}(u - v). \qquad \text{(XX)}$$

$$\cos u - \cos v = -2\sin\tfrac{1}{2}(u + v)\sin\tfrac{1}{2}(u - v). \qquad \text{(XXI)}$$

$$\sin u - \sin v = 2\cos\tfrac{1}{2}(u + v)\sin\tfrac{1}{2}(u - v). \qquad \text{(XXII)}$$

70. TRIGONOMETRIC EQUATIONS OF SIMPLE TYPES

In this text, only trigonometric equations in one variable will be considered. Also, in equations of this type, the variable, x, will occur only in the arguments of trigonometric functions, with the typical argument being of the form kx, where k is an integer or a simple rational number.

Let T represent any trigonometric function, and let c be a given number. Then, the most simple variety of trigonometric equation is of the form $T(x) = c$. For any T, the domain for x is all real numbers, except for certain quadrantal values in the case of the tangent, cotangent, secant, and cosecant. However, unless otherwise mentioned, we agree to ask for the solutions of $T(x) = c$ only on the interval $\{0 \le x \le 2\pi\}$.

If c is a quadrantal value of $T(x)$, then $T(x) = c$ should be solved by inspection, as on page 186, perhaps with the aid of a rapid, qualitatively accurate graph of $y = T(x)$ to aid the memory.

ILLUSTRATION 1.   To solve $\sin x = 0$, we visualize a graph of $y = \sin x$ (given in Figure 123, although with experience the graph should not be required). The solutions, which are the x-intercepts of the graph, are $\{0, \pi, 2\pi\}$. The only solution of $\sin x = 1$ is $x = \frac{1}{2}\pi$.

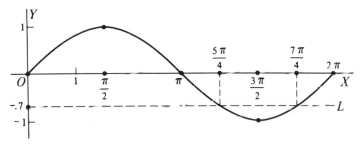

FIGURE 123

Suppose that c is not a value of $T(x)$ for some quadrantal values of x. Then, to solve $T(x) = c$, we may proceed as follows.

To solve $T(x) = c$, where any solution is not a quadrantal value.

A. *First obtain z such that $0 < z < \frac{1}{2}\pi$ and $T(z) = |c|$.*

B. *Locate all values of x on $\{0 < x < 2\pi\}$ such that z radians is the reference angle for x radians.*

In applying A and B, suppose that $|c|$ is one of the values for the trigonometric functions of convenient angles in quadrant I as given on page 165. Then z is found by inspection of $T(z) = |c|$. Otherwise, a trigonometric table is needed to obtain z.

ILLUSTRATION 2.   To solve $\sin x = -\frac{1}{2}\sqrt{2}$: First consider the equation $\sin z = \frac{1}{2}\sqrt{2} = .71$, with $0 \le z \le \frac{1}{2}\pi$. We recall that $z = \frac{1}{4}\pi$. The horizontal line $L: \{y = -\frac{1}{2}\sqrt{2}\}$ is shown in Figure 123. The values of x at the intersections of L and the sine curve are the desired solutions. Or, since $\sin x < 0$, the solutions are the values of $x > \pi$ such that $\frac{1}{4}\pi$ radians is the reference angle for x radians. The solutions are $x = \pi + \frac{1}{4}\pi$ and $x = 2\pi - \frac{1}{4}\pi$, or $\{\frac{5}{4}\pi, \frac{7}{4}\pi\}$, as seen in Figure 123.

ILLUSTRATION 3.   If θ is an angle, and if we desire the solutions of $\tan \theta = -1$ with θ expressed in degree measure, first we solve $\tan \phi = 1$, with ϕ in quadrant I. The result is $\phi = 45°$. Then, θ is an angle in quadrant II or quadrant IV with the reference angle $45°$. The solutions are $\theta = 135°$ and $\theta = 315°$.

If a trigonometric equation is not of the simple form $T(x) = c$ as just considered, by various devices we attempt to obtain one or more equations of the simple type $T(x) = c$ such that the union of their solution sets is the solution set of the given equation.

EXAMPLE 1.   Solve: $2 \cos^2 x - 1 = 0.$ (1)

Solution.   1. From (1), $\cos^2 x = \frac{1}{2}$, or $\cos x = \pm\sqrt{\frac{1}{2}}$. Hence,

$$\cos x = \tfrac{1}{2}\sqrt{2} \quad or \quad \cos x = -\tfrac{1}{2}\sqrt{2}. \quad (2)$$

2. The solution set of (1) is the union of the solution sets of the equations in (2). The solutions of $\cos x = \frac{1}{2}\sqrt{2}$ are $\{\frac{1}{4}\pi, \frac{7}{4}\pi\}$. The solutions of $\cos x = -\frac{1}{2}\sqrt{2}$ are $\{\frac{3}{4}\pi, \frac{5}{4}\pi\}$. Hence, the solutions of (1) are $\{\frac{1}{4}\pi, \frac{3}{4}\pi, \frac{5}{4}\pi, \frac{7}{4}\pi\}$.

EXAMPLE 2.   Solve: $2 \cos^2 x - \cos x - 1 = 0.$ (3)

Solution.   1. The equation is in the quadratic form in $\cos x$. We solve by factoring in (3):

$$(2 \cos x + 1)(\cos x - 1) = 0;$$

Hence, $2 \cos x + 1 = 0 \quad or \quad \cos x - 1 = 0.$ (4)

2. From (4), $\cos x = 1$ or $\cos x = -\frac{1}{2}$. The solutions of $\cos x = 1$ are $x = 0$ and $x = 2\pi$. The solution of $\cos z = \frac{1}{2}$ in quadrant I is $z = \frac{1}{3}\pi$. Hence, the solutions of $\cos x = -\frac{1}{2}$ are $x = \frac{2}{3}\pi$ and $x = \frac{4}{3}\pi$. The solution set of (3) is $\{0, \frac{2}{3}\pi, \frac{4}{3}\pi, 2\pi\}$.

ILLUSTRATION 4.   To solve $\sin x \cos x = \sin x$, rewrite with one member zero and factor:

$$\sin x \cos x - \sin x = 0, \quad or \quad (\sin x)(\cos x - 1) = 0.$$

Hence $\sin x = 0$ or $\cos x = 1$. The solution set is $\{0, \pi, 2\pi\}$, where the solutions 0 and 2π originate with each of $\sin x = 0$ and $\cos x = 1$.

EXAMPLE 3.   Solve: $\tan x = -1.881.$ (5)

Solution.   1. Consider $\tan z = 1.881$ with $0 < z < \frac{1}{2}\pi$. By use of Table IV, $z = 1.082$.

2. Any solution of (5) is on an interval in quadrant II or quadrant IV. With x interpreted in (5) as an angle x radians, if the solution is in quad-

rant II then $x = \pi - 1.082 = 3.142 - 1.082$, or $x = 2.060$. If the solution is in quadrant IV, then $x = 2\pi - 1.082 = 6.283 - 1.082$, or $x = 5.201$. The solutions of (5) are $\{2.060, 5.201\}$. A rough figure showing angles in standard positions would clarify this solution.

Comment. If we desired the angles in degree measure to satisfy $\tan \theta = -1.881$ instead of x as in (5), the solutions also are obtained by use of Table IV. We find $\theta = 180° - 62°$ and $\theta = 360° - 62°$, or $\theta = 118°$ and $\theta = 298°$.

EXERCISE 44

Solve the equation or verbal statement. Use $\pi = 3.142$ and Table IV if necessary.

1. $\cos x = \frac{1}{2}\sqrt{2}$.   2. $\tan x = \sqrt{3}$.   3. $\sin x = -\frac{1}{2}$.
4. $\cos x = -\frac{1}{2}\sqrt{3}$.   5. $\tan x = -1$.   6. $\tan x = -\sqrt{3}$.
7. $\sin x = -\frac{1}{2}\sqrt{2}$.   8. $\csc x = 1$.   9. $\cos x = 0$.
10. $\sec x = 0$.   11. $\tan x = \frac{1}{3}\sqrt{3}$.   12.* $\tan x$ is infinite.
13. $\cos^2 x = 1$.   14. $2\cos^2 x - 1 = 0$.   15. $\tan^2 x = 1$.
16. $\sin x = .602$.   17. $\cos x = -.857$.   18. $\tan x = -.625$.
19. $4\sin^2 x - 3 = 0$.   20. $4\cos^2 x - 3 = 0$.   21. $\cot^2 x - 3 = 0$.
22. $\sec x = 1$.   23. $\csc x = \frac{1}{2}$.   24. $\cot x = 1$.
25. $\csc x = 2$.   26. $\sec x = -2$.   27. $\csc x = .5$.
28. $2\sin^2 x + \sin x - 1 = 0$.   29. $2\sin^2 x - \sin x = 0$.
30. $2\cos^2 x + \cos x = 0$.   31. $2\cos^2 x + 3\cos x + 1 = 0$.
32. $\tan^3 x - \tan x = 0$.   33. $\sec^2 x - 2\sec x = 0$.
34. $\tan x \sin x - \tan x = 0$.   35. $\cos x \sin x = 2\cos x$.
36. $\sin x \sec x - 2\sin x = 0$.   37. $\cot x \sin x - \cot x = 0$.
38. $\sec^2 x + \sec x = 2$.   39. $2\csc^2 x + 3\csc x = 2$.

71. VARIOUS TYPES OF TRIGONOMETRIC EQUATIONS

If more than one trigonometric function of x is involved in an equation, it may be desirable to solve by first expressing all functions in terms of the sine and cosine, or all functions in terms of just one of those present in the equation.

EXAMPLE 1.   Solve: $\sin^2 x + 3\cos x + 3\cos^2 x = 0$.   (1)

Solution.   1. From the identity $\sin^2 x + \cos^2 x = 1$, we obtain

$$\sin^2 x = 1 - \cos^2 x; \qquad \cos x = \pm \sqrt{1 - \sin^2 x}.$$

*We prefer NOT to write "$\tan x = \infty$," which gives the undesirable appearance of considering "*infinity*" as if it were a number.

We could change to an equation involving $\sin x$ alone by using $\cos x = \pm \sqrt{1 - \sin^2 x}$, but the new equation would be complicated. Hence, instead, we decide to change to an equation in $\cos x$ alone by using $\sin^2 x = 1 - \cos^2 x$ in (1), which gives

$$1 - \cos^2 x + 3 \cos x + 3 \cos^2 x = 0; \ or,$$

$$2 \cos^2 x + 3 \cos x + 1 = 0; \ or,$$

$$(2 \cos x + 1)(\cos x + 1) = 0. \tag{2}$$

2. From (2), as in the preceding section, we obtain the solutions $\{\frac{2}{3}\pi, \pi, \frac{4}{3}\pi\}$.

EXAMPLE 2.   Solve: $\cos 2x - \sin x = 0.$ (3)

Solution.   It is desirable to have all function values expressed in terms of functions with the same argument.   Hence, we use (X) on page 188 to express $\cos 2x$ in terms of $\sin x$.   This gives $\cos 2x = 1 - 2 \sin^2 x$.   Then (3) becomes

$$1 - \sin x - 2 \sin^2 x = 0, \quad or \quad (1 + \sin x)(1 - 2 \sin x) = 0.$$

Therefore, $1 + \sin x = 0$ or $1 - 2 \sin x = 0$.   The solutions are $\{\frac{1}{6}\pi, \frac{5}{6}\pi, \frac{3}{2}\pi\}$.

EXAMPLE 3.   Solve: $\cos x - 1 = \sin x.$ (4)

Solution.   1. From $\sin^2 x = 1 - \cos^2 x$, we have $\sin x = \pm \sqrt{1 - \cos^2 x}$, which would be inconvenient as a substitution in (4). Hence, as an equivalent but more desirable method, we decide to square both sides in (4) before substituting. This gives

$$\cos^2 x - 2 \cos x + 1 = \sin^2 x. \tag{5}$$

2. Use $\sin^2 x = 1 - \cos^2 x$ in (5):

$$\cos^2 x - 2 \cos x + 1 = 1 - \cos^2 x, \ or$$

$$2 \cos^2 x - 2 \cos x = 0, \quad or \quad (\cos x)(\cos x - 1) = 0. \tag{6}$$

Hence, $\cos x = 0$ or $\cos x - 1 = 0$.   The solutions thus obtained are $\{0, \frac{1}{2}\pi, \frac{3}{2}\pi, 2\pi\}$.

3. When both sides of (4) were squared, we recognized that this might lead to obtaining *extraneous solutions* (recall the similar algebraic situation on page 41). Hence, each "*solution*" must be tested in (4) to determine whether or not the value of x actually deserves to be named a *solution*.

Test of $x = \frac{1}{2}\pi$ in (4): we obtain $0 - 1 = 1$, which is *not* true. Hence, $\frac{1}{2}\pi$ is not a solution. The student may check the fact that $\{0, \frac{3}{2}\pi, 2\pi\}$ are solutions of (4).

ILLUSTRATION 1.   *To solve* $\sec x + \tan x = 1$: Then $\sec x = 1 - \tan x$. We recall the identity $1 + \tan^2 x = \sec^2 x$. We would square both sides of $\sec x = 1 - \tan x$ and proceed as in Example 3.

EXAMPLE 4.   Solve: $\qquad\qquad\qquad \sin 2x = \tfrac{1}{2}.$ (7)

Solution.   1. In (7), let $2x = w$. Then $\sin w = \tfrac{1}{2}$. To obtain all x on $\{0 \le x \le 2\pi\}$ satisfying (7), first we shall find all w on $\{0 \le w \le 4\pi\}$ for which $\sin w = \tfrac{1}{2}$.

2. Recall that $\sin \tfrac{1}{6}\pi = \tfrac{1}{2}$ and $\sin \tfrac{5}{6}\pi = \tfrac{1}{2}$. The sine is periodic with the period 2π. Hence we also have

$$\sin\left(2\pi + \tfrac{1}{6}\pi\right) = \sin\left(2\pi + \tfrac{5}{6}\pi\right) = \tfrac{1}{2}.$$

Therefore, the solutions of $\sin w = \tfrac{1}{2}$ are $\{\tfrac{1}{6}\pi, \tfrac{5}{6}\pi, \tfrac{13}{6}\pi, \tfrac{17}{6}\pi\}$.   Since $w = 2x$, the desired solutions of (7) for x are $\{\tfrac{1}{12}\pi, \tfrac{5}{12}\pi, \tfrac{13}{12}\pi, \tfrac{17}{12}\pi\}$.

ILLUSTRATION 2.   To obtain all solutions of $\tan 3x = 1$ on $\{0 \le x \le 2\pi\}$, we would first find all solutions of $\tan w = 1$ on $\{0 \le w \le 6\pi\}$. In doing this, we would use the fact that the tangent function has the period π.

EXERCISE 45

Obtain all solutions where $\{0 \le x \le 2\pi\}$.
1. $2 \sin^2 x - 2 \cos^2 x = 1.$ 2. $2 \sin^2 x + \cos x = 1.$
3. $2 + 3 \sin x + \sin^2 x - \cos^2 x = 0.$
4. $\sin x = \cos x.$ 5. $\tan x = 3 \cot x.$

Hint for Problem 4. Divide by $\cos x$.

6. $\tan^2 x + \sec^2 x = 7.$ 7. $\sqrt{3} \tan x + 1 = \sec^2 x.$
8. $3 \sin x = -\sqrt{3} \cos x.$ 9. $\cos x + \cos 2x + 1 = 0.$
10. $\sin 2x - \cos x = 0.$ 11. $\cos 2x - 3 \cos x + 2 = 0.$
12. $\cos 2x = \sin x.$ 13. $\sin 2x - 2 \sin^2 x = 0.$
14. $\cos 2x = 2 \sin^2 x - 2.$ 15. $\sin x - 1 = \cos x.$
16. $\sec x = 1 + \tan x.$ 17. $\csc x = \cot x - 1.$
18. $3 \sec x + 2 = \cos x.$ 19. $2 \sin x = \csc x.$
20. $\cos 2x = 1.$ 21. $\sin 2x = 0.$ 22. $\tan 3x = 0.$
23. $\tan 2x = -1.$ 24. $\sin 3x = \tfrac{1}{2}\sqrt{3}.$ 25. $\cos 2x = -\tfrac{1}{2}\sqrt{2}.$
26. $\tan^2 2x - 1 = 0.$ 27. $\tan 2x = \tan x.$
28. $\sin \tfrac{1}{2}x = 1.$ 29. $\sin x = \sin \tfrac{1}{2}x.$
30. $\cos x = \sin \tfrac{1}{2}x.$ 31. $\sec x + \tan x = 1.$

72. THE INVERSE TRIGONOMETRIC FUNCTIONS

A graph of $y = \sin x$, with x unrestricted in size, is given in Figure 124. For each value of y on the interval $R:\{-1 \leq y \leq 1\}$, there are infinitely many values of x such that $y = \sin x$. Thus, if $y = \frac{1}{2}$ then $\frac{1}{2} = \sin x$, which is satisfied by $x = \frac{1}{6}\pi$, $x = \frac{5}{6}\pi$, and also by either of these numbers plus or minus any integral multiple of 2π (because of the periodicity of the sine function). If the line $y = \frac{1}{2}$ is drawn on the xy-plane, as in Figure 124, this line intersects the graph of $y = \sin x$ at the infinite number of points having as the horizontal coordinates those values of x which were mentioned above.

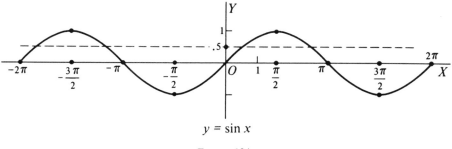

$$y = \sin x$$

FIGURE 124

Now consider just that part of the graph of $y = \sin x$, as in Figure 125, where x is on $D:\{-\frac{1}{2}\pi \leq x \leq \frac{1}{2}\pi\}$. Notice that y, or $\sin x$, is an *increasing function* of x when x is restricted to D. That is,

$$\text{if } x_1 < x_2 \quad \text{then} \quad \sin x_1 < \sin x_2, \tag{1}$$

and the graph of $y = \sin x$ rises steadily as x increases from $x = -\frac{1}{2}\pi$ to $x = \frac{1}{2}\pi$. Hence, a horizontal line through any point y on the interval $R:\{-1 \leq y \leq 1\}$ of the y-axis intersects the graph at just *one point P*

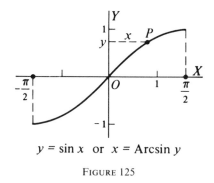

$$y = \sin x \quad \text{or} \quad x = \text{Arcsin } y$$

FIGURE 125

whose coordinate x falls in D. Thus, for each number y in R, there is *just one* corresponding number x in D so that $y = \sin x$. This correspondence *defines x as a function of y*. For any y in R:$\{-1 \leq y \leq 1\}$, let "**Arcsine** y," read "*arcsine of y*," represent the *single value of x on* D:$\{-\frac{1}{2}\pi \leq x \leq \frac{1}{2}\pi\}$, such that (x, y) satisfies $y = \sin x$. Hereafter we shall abbreviate "*Arcsine*" by writing "*Arcsin.*" Then, the equations

$$y = \sin x \quad and \quad x = \text{Arcsin } y \tag{2}$$

are equivalent and thus have the same graph, in Figure 125. On account of the equivalence in (2), the sine function and the Arcsine function are called *inverse functions,* where each one is said to be the **inverse** of the other.

To study the Arcsine, it is convenient to interchange the roles played by x and y in (2). Then we have the following basis:

$$\left\{\begin{array}{l} with\ -1 \leq x \leq 1, \\ -\frac{1}{2}\pi \leq y \leq \frac{1}{2}\pi \end{array}\right\} \quad \left\{\begin{array}{l} y = \text{Arcsin } x \\ is\ equivalent\ to \\ x = \sin y \end{array}\right\}. \tag{3}$$

It is important to remember that, for all x on $\{-1 \leq x \leq 1\}$,

$$-\tfrac{1}{2}\pi \leq \text{Arcsin } x \leq \tfrac{1}{2}\pi. \tag{4}$$

To graph $y = \text{Arcsin } x$, we may graph the equivalent equation $x = \sin y$, where y is restricted to $\{-\frac{1}{2}\pi \leq y \leq \frac{1}{2}\pi\}$. We make up a table of solutions (x, y) by assigning values to y in $x = \sin y$.

If $y = -\frac{1}{4}\pi$, *then* $x = \sin\left(-\frac{1}{4}\pi\right) = -\frac{1}{2}\sqrt{2} = -.707$.

If $y = -\frac{1}{6}\pi$, *then* $x = \sin\left(-\frac{1}{6}\pi\right) = -\frac{1}{2}$.

If $y = -\frac{1}{2}\pi$, *then* $x = \sin\left(-\frac{1}{2}\pi\right) = -1$.

Similarly, solutions (x, y) for positive values of x and y are obtained as in the following table, which is a basis for the graph AB in Figure 126. The graph of $x = \sin y$ with y *unrestricted* would be a complete sine graph along the y-axis as in Figure 127. The arc AB in Figure 126 corresponds to arc AB in Figure 127.

$x = \sin y$	-1	$-.7$	$-.5$	0	$.5$	$.7$	1
$y = \text{Arcsin } x$	$-\frac{1}{2}\pi$	$-\frac{1}{4}\pi$	$-\frac{1}{6}\pi$	0	$\frac{1}{6}\pi$	$\frac{1}{4}\pi$	$\frac{1}{2}\pi$
y (dec. form)	-1.6	$-.8$	$-.5$	0	$.5$	$.8$	1.6

On account of the equivalence in (3),

$$x = \sin(\text{Arcsin } x); \quad y = \text{Arcsin}(\sin y). \tag{5}$$

Let $y_1 = \text{Arcsin } x_0$ and $y_2 = \text{Arcsin}(-x_0)$, as seen in Figure 126. Geometrically, we decide that $y_2 = -y_1$, or $\text{Arcsin}(-x_0) = -\text{Arcsin } x_0$.

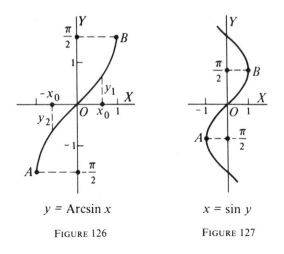

$y = \text{Arcsin } x$ $x = \sin y$

FIGURE 126   FIGURE 127

Later, we shall prove that this fact is a consequence of the identity $\sin (-z) = -\sin z$, obtained from (IV) on page 188. Thus, we have the identity

$$\text{Arcsin } (-x) = -\text{Arcsin } x, \tag{6}$$

or, the function Arcsin x is an *odd function*.

To obtain Arcsin x for any assigned value of x, let $y = \text{Arcsin } x$. By (3), $x = \sin y$. Then y can be found either by inspection, and memory of functions of convenient angles, or from a trigonometric table.

ILLUSTRATION 1.   To find Arcsin $\frac{1}{2}$, let $y = \text{Arcsin } \frac{1}{2}$.   Because of (3), $\sin y = \frac{1}{2}$ and y is on $\{-\frac{1}{2}\pi \leq y \leq \frac{1}{2}\pi\}$. Hence $y = \frac{1}{6}\pi$.

EXAMPLE 1.   Find all trigonometric functions of Arcsin $(-\frac{1}{2})$.

Solution.   Let $y = \text{Arcsin } (-\frac{1}{2})$. By (6),

$$y = -\text{Arcsin } \tfrac{1}{2} = -\tfrac{1}{6}\pi.$$

By (3), $\sin y = -\frac{1}{2}$; $\cos y = \cos(-\frac{1}{6}\pi) = \frac{1}{2}\sqrt{3}$; $\tan y = -1/\sqrt{3}$; $\csc y = -2$; $\sec y = 2/\sqrt{3}$; $\cot y = -\sqrt{3}$.

ILLUSTRATION 2.   To find Arcsin .643, let $y = \text{Arcsin } .643$. Then $\sin y = .643$. From Table IV, $y = .698$. By (6),

$$\text{Arcsin } (-.643) = -\text{Arcsin } .643 = -.698.$$

EXAMPLE 2.   Find $\tan [\text{Arcsin } (-\frac{3}{5})]$.

Solution.   1. Let $y = \text{Arcsin } (-\frac{3}{5})$. Then $\sin y = -\frac{3}{5}$ and $-\frac{1}{2}\pi \leq y \leq \frac{1}{2}\pi$. Hence, y is in quadrant IV and $-\frac{1}{2}\pi < y < 0$.

2. Let z be the acute reference angle for y radians. Then $\sin z = \frac{3}{5}$.

3. Draw right $\triangle ABC$ as in Figure 128, with $\overline{BC} = 3$ and $\overline{AB} = 5$; $\angle BAC = z$. Then $\overline{AC}^2 = 25 - 9 = 16$ and $\overline{AC} = 4$. Hence $\tan z = \frac{3}{4}$. Then $\tan y = -\frac{3}{4}$ because $-\frac{1}{2}\pi < y < 0$. Or, $\tan\left[\text{Arcsin}\left(-\frac{3}{5}\right)\right] = -\frac{3}{4}$.

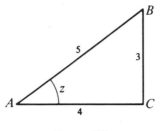

FIGURE 128

To obtain an inverse for the tangent function, we could start with $y = \tan x$, obtain $x = \text{Arctan } y$, and then interchange x and y, as was done when the Arcsine was introduced. Instead, we shall interchange x and y in the first stage as just described, and proceed as follows.

A complete graph of the equation $x = \tan y$ consists of the branch seen in Figure 129, and endless repetitions of this branch, upward and downward. Now, let us restrict y to the numbers on the interval $R:\left\{-\frac{1}{2}\pi < y < \frac{1}{2}\pi\right\}$. In $x = \tan y$, the range for x is the interval $D:\{-\infty < x < +\infty\}$. The complete graph of $x = \tan y$ with y restricted to R is the curve, C, as shown in Figure 129.

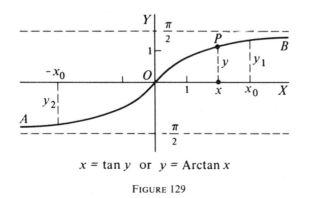

$x = \tan y$ or $y = \text{Arctan } x$

FIGURE 129

If a line L is drawn perpendicular to the x-axis at any point with coordinate x, then L intersects the graph C in just one point $P:(x,y)$. That is, to each number x there corresponds *just one number y* on the interval $R:\left\{-\frac{1}{2}\pi < y < \frac{1}{2}\pi\right\}$ such that the pair (x,y) is a solution of the equation $x = \tan y$. This correspondence defines *y as a function of x*.

For any x on $D:\{-\infty < x < +\infty\}$, let "**Arctangent** x," read "*Arctangent of x,*" represent the single value of y on $R:\{-\frac{1}{2}\pi < y < \frac{1}{2}\pi\}$ such that (x,y) satisfies $x = \tan y$. Hereafter we shall abbreviate "*Arctangent*" by writing "*Arctan.*" Then we have the following basis:

$$\left\{\begin{array}{c} with -\infty < x < +\infty, \\ -\frac{1}{2}\pi < y < \frac{1}{2}\pi \end{array}\right\} \quad \left\{\begin{array}{c} y = \text{Arctan } x \\ is\ equivalent\ to \\ x = \tan y. \end{array}\right\} \tag{7}$$

As a consequence of (7), the equations $y = $ Arctan x and $x = \tan y$ have the same graph, shown in Figure 129. It is important to remember that, for all values of x,

$$-\tfrac{1}{2}\pi < \text{Arctan } x < \tfrac{1}{2}\pi. \tag{8}$$

From (7) we obtain

$$\text{Arctan } (\tan y) = y; \qquad \tan (\text{Arctan } x) = x. \tag{9}$$

In $y = $ Arctan x, let $y_1 = $ Arctan x_0 and $y_2 = $ Arctan $(-x_0)$, as in Figure 129. Geometrically, we observe that $y_2 = -y_1$, or Arctan $(-x_0) = -$Arctan x_0. Later, we shall prove that this fact is a consequence of the identity $\tan (-z) = \tan z$, obtained from (IV) on page 188. Thus, we have the identity

$$\text{Arctan } (-x) = -\text{Arctan } x. \tag{10}$$

ILLUSTRATION 3.   To find solutions (x,y) of $y = $ Arctan x in order to graph the equation, we obtain $x = \tan y$ from (7) and then assign values to y on $\{-\frac{1}{2}\pi \leq y \leq \frac{1}{2}\pi\}$. Thus, $\tan \frac{1}{2}\pi$ and $\tan (-\frac{1}{2}\pi)$ are undefined, and the lines $y = \frac{1}{2}\pi$ and $y = -\frac{1}{2}\pi$ are asymptotes of the graph. If $y = -\frac{1}{4}\pi$ then $x = \tan (-\frac{1}{4}\pi) = -1$; etc. The student will obtain the graph of $y = $ Arctan x in the next exercise.

ILLUSTRATION 4.   To find Arctan (-2.246): By use of (10), we obtain Arctan $(-2.246) = -$Arctan 2.246. Let $y = $ Arctan 2.246. Then $\tan y = 2.246$, and $y = 1.152$ from Table IV. Hence Arctan $(-2.246) = -1.152$.

ILLUSTRATION 5.   To find Arctan $\sqrt{3}$, we write $y = $ Arctan $\sqrt{3}$ and obtain $\tan y = \sqrt{3}$. Hence $y = \frac{1}{3}\pi$.

ILLUSTRATION 6.   To find Arctan $(-\frac{1}{3}\sqrt{3})$, first use (10):

$$\text{Arctan } (-\tfrac{1}{3}\sqrt{3}) = -\text{Arctan } \tfrac{1}{3}\sqrt{3}.$$

Let $y = $ Arctan $\frac{1}{3}\sqrt{3}$. Then $\tan y = \frac{1}{3}\sqrt{3}$ and $y = \frac{1}{6}\pi$. Hence we obtain Arctan $(-\frac{1}{3}\sqrt{3}) = -\frac{1}{6}\pi$.

EXAMPLE 3.   Find $\sin$ (Arctan (-1)).

Solution.   Let y = Arctan (-1); tan y = -1 and hence y = $-\frac{1}{4}\pi$. Then, sin y = sin $\left(-\frac{1}{4}\pi\right)$ = $-\frac{1}{2}\sqrt{2}$.

EXAMPLE 4.   Find tan $(2$ Arctan $z)$.

Solution.   Let y = Arctan z. Then tan y = z and we desire tan $2y$. From identity (XIII) on page 188,

$$\tan (2 \text{ Arctan } z) = \frac{2z}{1 - z^2}.$$

THEOREM V.   *The function* Arcsin x *is an odd function. That is, the following equation is an identity:*

$$\text{Arcsin} (-x) = -\text{Arcsin } x. \tag{11}$$

Proof.   1. For any fixed number x, let y = Arcsin $(-x)$. By use of (3), we obtain

$$-x = \sin y, \quad or \quad x = -\sin y.$$

2. From (IV) on page 188, sin $(-y)$ = $-\sin y$, and thus sin $(-y)$ = x. Hence, by use of (3), as applied to x = sin $(-y)$,

$$\text{Arcsin } x = -y = -\text{Arcsin } (-x),$$

which proves (11).

Similarly as in Theorem V, the student may prove identity (10) for Arctan x in the next exercise.

Note 1.   In contrast to the *inverse* trigonometric functions, the {*sine, cosine, tangent,* $\cdots$} sometimes are called the *direct* trigonometric functions. Hereafter, if merely a *trigonometric function* is mentioned, assume that it is a *direct trigonometric function,* unless otherwise stated.

Note 2.   Inverses are introduced for the cosine and cotangent functions in the next exercise. In calculus, the arcsine and arctangent are indispensable in obtaining certain integrals. Thus

$$\int \frac{dx}{1 + x^2} = \text{Arctan } x + C, \quad and \quad \int \frac{dx}{\sqrt{1 - x^2}} = \text{Arcsin } x + C, \tag{12}$$

where C is any constant, and "$\int$" is read "*indefinite integral of.*" If a minus sign is inserted before each integral, the results can be written $(-\text{Arctan } x + C)$ and $(-\text{Arcsin } x + C)$, respectively, without any inconvenience. Also, however, the results could be written (Arccot $x + C$) and (Arccos $x + C$), respectively. Hence, by a mere change of signs in (12) we avoid all use of the arccotangent and arccosine functions in integration. For this reason, we shall not consider these functions except in optional problems. Inverses can be introduced for the secant and cosecant functions. However, the definitions involved are relatively complicated. In calculus, the arcsecant and arccosecant functions can be used in expressing the values of certain integrals, but optional forms involving the logarithm function are available. Hence, the arcsecant and arccosecant will not be introduced in this text. Their definitions are presented in calculus in any text which desires to use the functions.

EXERCISE 46

Find the value of the function by memory of convenient angles.

1. Arcsin $\frac{1}{2} \sqrt{3}$.

2. Arcsin 1.

3. Arcsin $\frac{1}{2} \sqrt{2}$.

4. Arcsin (-1).

5. Arcsin $\left(-\frac{1}{2}\right)$.

6. Arcsin $\left(-\frac{1}{2} \sqrt{3}\right)$.

7. Arctan 1.

8. Arctan 0.

9. Arcsin 0.

10. Arctan $\sqrt{3}$.

11. Arctan $\frac{1}{3} \sqrt{3}$.

12. Arctan (-1).

13. Arctan $(-\sqrt{3})$.

14. Arctan $\left(-\frac{1}{3} \sqrt{3}\right)$.

15. Arcsin $\left(-\frac{1}{2} \sqrt{2}\right)$.

Graph the equation. Use equal scale units on the coordinate axes.

16. $y = $ Arcsin x.

17. $y = $ Arctan x.

Find the value of the expression.

18. sin (Arcsin .6).

19. tan (Arctan .9).

20. sin $\left(\text{Arcsin} \left(-\frac{5}{6}\right)\right)$.

21. csc $\left(\text{Arcsin} \frac{1}{4}\right)$.

22. cot (Arctan 3).

23. sin (2 Arcsin x).

24. cos (2 Arcsin x).

25. Arcsin $\left(\sin \frac{1}{3} \pi\right)$.

26. Arctan $\left(\tan \frac{1}{4} \pi\right)$.

27. Arcsin $\left(\sin \frac{3}{4} \pi\right)$.

28. Arcsin $\left(\sin \frac{3}{2} \pi\right)$.

29. Arctan $\left(\tan \frac{3}{4} \pi\right)$.

Obtain the value of the function, by use of Table IV if necessary.

30. Arcsin .342.

31. Arcsin .921.

32. Arctan .404.

33. Arctan 3.078.

34. Arcsin $(-.588)$.

35. Arctan $(-.700)$.

36. Arctan (-1.483).

37. Arcsin $(-.946)$.

38. Arcsin $(-.848)$.

39. tan $\left(\text{Arcsin} \frac{1}{2}\right)$.

40. cos $\left(\text{Arcsin} \left(-\frac{1}{2} \sqrt{3}\right)\right)$.

41. sin (Arctan (-1)).

42. cos (Arctan $\sqrt{3}$).

Find all trigonometric functions of y without use of a table.

43. $y = $ Arcsin $\left(-\frac{1}{2}\right)$.

44. $y = $ Arctan $\left(-\frac{1}{3} \sqrt{3}\right)$.

45. $y = $ Arcsin $\frac{4}{5}$.

46. $y = $ Arctan $\frac{5}{12}$.

47. $y = $ Arctan $\left(-\frac{15}{8}\right)$.

48. $y = $ Arcsin $\left(-\frac{12}{13}\right)$.

49. Prove that the function Arctan x is an odd function.

50. Prove that the graph of $y = $ Arcsin x is symmetric to the origin. Prove the same fact for $y = $ Arctan x.

★**51.** Obtain a graph of the equation $x = $ cos y with the domain for y as the interval R below. Repeat the discussion of Section 72 with respect to the arcsine in order to show that the equation $x = $ cos y, with y on R and x on $D:\{-1 \leqq x \leqq 1\}$, defines y as a function of x, to be denoted by $y = $ Arccos x. Thus,

$$\left[\begin{matrix} x \text{ on } D : \{-1 \leq x \leq 1\}; \\ y \text{ on } R : \{0 \leq y \leq \pi\} \end{matrix}\right] \quad x = \cos y \text{ is equivalent to } y = \text{Arccos } x.$$

★**52.** Obtain a graph of $x = \cot y$ with the domain for y as the interval R below. Discuss why the equation $x = \cot y$, with y on R and x on $\{-\infty < x < +\infty\}$, defines y as a function of x, to be denoted by $y = \text{Arccot } x$. Thus,

$$\left[\begin{matrix} x \text{ on } D : \{-\infty < x < \infty\}; \\ y \text{ on } R : \{0 < y < \pi\} \end{matrix}\right] \quad x = \cot y \text{ is equivalent to } y = \text{Arccot } x.$$

73. COMPOSITE TRIGONOMETRIC FUNCTIONS

Let T represent any trigonometric function, and let $y = T(z)$. Now, suppose that $z = f(x)$, where the values of $f(x)$ are in the domain of T. Then, we have $y = T(f(x))$. In such a case, it is said that y is a *composite trigonometric function* of x.

ILLUSTRATION 1.   The function $\sin 3x$ is a composite trigonometric function of x.

For contrast with the name "*composite trigonometric function,*" a function $T(x)$, where the argument is simply x, may be called a *standard trigonometric function,* as was done at the beginning of this chapter. When no ambiguity results, it is customary to refer to either a standard or a composite trigonometric function simply as a trigonometric function, with the context making the nature of the function clear.

EXAMPLE 1.   Graph the function $\sin 2x$ for $\{0 \leq x \leq 2\pi\}$.

Solution.   1. *Periodicity.*   Let $f(x) = \sin 2x$.   Then, we have $\sin (2x + 2\pi) = \sin 2x$, or $\sin 2(x + \pi) = \sin 2x$. That is, $f(x + \pi) = f(x)$. Hence, f is periodic with the period π.

2. If x varies from $x = 0$ to $x = 2\pi$, then $2x$ will vary from $2x = 0$ to $2x = 4\pi$, and $\sin 2x$ will pass *twice* through all values of $\sin z$. Thus, we expect two "*waves*" in the graph. The "*amplitude*" of each wave will be 1 because the maximum value of $|\sin 2x|$ is 1.

3. The x-intercepts of the graph are the solutions of $\sin 2x = 0$. To obtain all solutions on $\{0 \leq x \leq 2\pi\}$, we require all values of $2x$ on $\{0 \leq 2x \leq 4\pi\}$ for which $\sin 2x = 0$. We obtain:

for $2x$*, the values* $\{0, \pi, 2\pi, 3\pi, 4\pi\}$;
for x*, the values* $\{0, \frac{1}{2}\pi, \pi, \frac{3}{2}\pi, 2\pi\}$.

4. *The crests of the waves.* The maximum value, 1, of y occurs when $\sin 2x = 1$. We obtain: *for* $2x$*, the values* $\{\frac{1}{2}\pi, \frac{5}{2}\pi\}$; *for* x*, the values* $\{\frac{1}{4}\pi, \frac{5}{4}\pi\}$.

5. *The troughs of the waves.* The minimum value, -1, of y occurs when $\sin 2x = -1$. We obtain, *for* $2x$, the values $\{\frac{3}{2}\pi, \frac{7}{2}\pi\}$; *for* x, the values $\{\frac{3}{4}\pi, \frac{7}{4}\pi\}$.

6. The preceding details yield the following table of coordinates for points on the graph of $y = \sin 2x$. The graph is the continuous curve showing two waves with amplitude 1 in Figure 130. For convenience, we use the approximation $\pi = 3.2$ in locating points.

y	0	1	0	-1	0	1	0	-1	0
x	0	$\frac{1}{4}\pi$	$\frac{1}{2}\pi$	$\frac{3}{4}\pi$	π	$\frac{5}{4}\pi$	$\frac{3}{2}\pi$	$\frac{7}{4}\pi$	2π
$x(dec.)$	0	.8	1.6	2.4	3.2	4.0	4.8	5.6	6.4

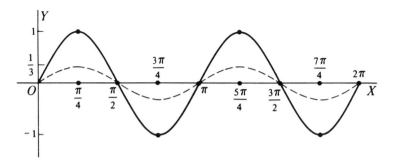

FIGURE 130

ILLUSTRATION 2.   To obtain a graph of the function $\frac{1}{3}\sin 2x$, first we draw the graph of $y = \sin 2x$ as in Figure 130. Let $y_1 = \frac{1}{3}\sin 2x$. For any value of x, the ordinate y_1 is one-third of the ordinate of the graph of $y = \sin 2x$. Hence, the graph of $y_1 = \frac{1}{3}\sin 2x$ consists of two waves, with the same x-intercepts as for the graph of $y = \sin 2x$, and with one-third of the amplitude of the waves of that graph. The graph of the equation $y_1 = \frac{1}{3}\sin 2x$ is the broken curve in Figure 130. We call this curve a sine curve with amplitude $\frac{1}{3}$ and period π.

EXAMPLE 2.   Graph the function $\sin (x - 2)$.

Solution.   Let $y = \sin (x - 2)$. In the xy-plane, transform coordinates by the substitution

$$x' = x - 2 \qquad and \qquad y' = y, \tag{1}$$

which translates the axes of the xy-plane to the new origin $O':(x = 2, y = 0)$, as shown in Figure 131. The graph of $y = \sin (x - 2)$ is the graph of $y' = \sin x'$ in the $x'y'$-system. This graph is the standard sine curve in Figure 131. Hence, the graph of $y = \sin (x - 2)$ in the xy-plane is a standard sine curve shifted 2 units to the right. To draw the graph, coordinates (x', y') for points on it were obtained from $y' = \sin x'$, and the points were plotted in the $x'y'$-system of coordinates.

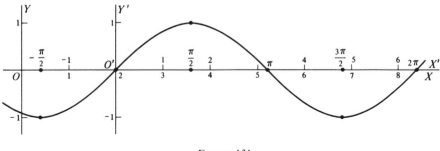

FIGURE 131

The following fact can be proved by remarks similar to those in the solution of Example 2.

> *If T is any standard trigonometric function, the graph of the function $T(x - w)$ is the result of shifting the graph of the function $T(x)$ a directed distance of w units horizontally.* (2)

EXAMPLE 3.   Graph the function $3 \sin (2x - .8)$.

Solution.   1. Let $y = 3 \sin (2x - .8)$. Then

$$y = 3 \sin 2(x - .4). \tag{3}$$

In (3), transform coordinates by the substitution

$$x' = x - .4 \quad and \quad y' = y, \tag{4}$$

which translates the axes of the xy-plane to the new origin $O':(x = .4, y = 0)$. Then (3) becomes $y' = 3 \sin 2x'$. Observe that the function $\sin 2x'$ is periodic with $2\pi/2$ or π as a period.

2. In Figure 132, the xy-axes and the $x'y'$-axes are shown. The graph of $y' = 3 \sin 2x'$ in the $x'y'$-plane is a sine curve with amplitude 3 and period π. Hence, the graph of (3) in the xy-plane is a sine curve with amplitude 3 and period π, shifted .4 unit to the right. The following table of coordinates was used to obtain the graph in Figure 132. For convenience, the approximation $\pi = 3.2$ was employed.

y'	-2.1	0	3	0	-3	0	3	0	-3	0
x'	$-\frac{1}{8}\pi$	0	$\frac{1}{4}\pi$	$\frac{1}{2}\pi$	$\frac{3}{4}\pi$	π	$\frac{5}{4}\pi$	$\frac{3}{2}\pi$	$\frac{7}{4}\pi$	2π
$x'(dec.)$	$-.4$	0	.8	1.6	2.4	3.2	4.0	4.8	5.6	6.4
$x(dec.)$	0	.4	1.2	2.0	2.8	3.6	4.4	5.2	6.0	6.8

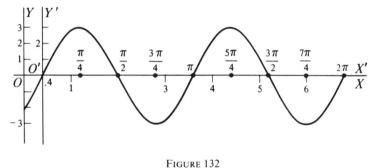

Observe that $\sin (hx - b) = \sin h(x - b/h)$. Then, by details similar to those in the solution of Example 3, we arrive at the following conclusions.

> *If $h > 0, K > 0$, and $F(x) = K \sin (hx - b)$, the graph*
> *of $y = F(x)$ is a sine graph with amplitude K and*   (5)
> *period $2\pi/h$, shifted horizontally through the directed*
> *distance b/h.*

In $\sin (hx - b)$ in (5), the number b is called the *phase constant,* because it determines the distance through which the sine wave is shifted to the right ($b > 0$) or to the left ($b < 0$).

Note 1.   Let the position of a physical particle P of mass m at time t seconds, in motion on an s-axis, be given by $s = K \sin (ht - b)$, where $\{K, h, b\}$ are constants with $K > 0$ and $h > 0$. Then, the motion is called *simple harmonic motion.* In such a case, the force acting on the particle P at any instant t is proportional to the distance of P from the s-origin, and is always directed at the origin. Thus, with x replaced by t, statement (5) describes the behavior of $s = K \sin (ht - b)$. From Figure 132, with y interpreted as s, in a special case it is seen that the particle continually oscillates, in periodic motion, back and forth ($s > 0, s = 0, s < 0$) through the origin on the s-axis.

In this section, all illustrations involved the sine function. We could have discussed similar facts about the cosine function. On account of the reduction formula $\cos x = \sin \left(x + \tfrac{1}{2}\pi\right)$, the composite *sine* function $K \sin (hx - b)$ of (5) could be rewritten as a *cosine* function with the phase constant reduced by $\tfrac{1}{2}\pi$. That is,

$$\sin (hx - b) = \cos \left(hx - b - \tfrac{1}{2}\pi\right).$$

Hence, if desired, all problems relating to the sine or cosine may be phrased in terms of the sine alone, as we have done, or in terms of the cosine alone if desired. Suppose that T is any standard trigonometric function. Then transformation of coordinates by translation of the axes, as in Examples 2 and 3, can be applied in obtaining the graph of $T(hx - b)$, where h and b are constants and $h > 0$. Except for the remark about amplitude, the statements in (5) would apply to $T(hx - b)$ when T is not the sine or cosine. However, no interesting application, such as mentioned in Note 1, is available except for the sine or, equally well, for the cosine.

EXERCISE 47

Graph each function, with the domain for x as an interval of length at least 3π. Only a few convenient points on any graph need be located accurately. Each graph should be qualitatively correct, with all asymptotes shown. Use transformation of coordinates by translation of the axes when appropriate. State the period of each function. Use equal scale units on the coordinate axes. To reduce the arithmetic in locating points on the graph, use $\pi = 3.2$ as an approximation where convenient.

1. $\sin\left(x - \frac{1}{4}\pi\right)$.   **2.** $\cos(x + 2)$.   **3.** $\frac{1}{2}\tan\left(x + \frac{1}{4}\pi\right)$.

4. $\frac{1}{2}\sec\left(x - \frac{1}{4}\pi\right)$.   **5.** $\frac{1}{3}\cot\left(x - \frac{3}{4}\pi\right)$.   **6.** $\frac{1}{2}\csc\left(x + \frac{1}{4}\pi\right)$.

7. $3\cos 2x$.   **8.** $\frac{1}{2}\sin 3x$.   **9.** $\frac{1}{2}\tan 2x$.

10. $\sin\frac{1}{2}x$.   **11.** $\cos\frac{1}{4}x$.   **12.** $3\cos(2x - \pi)$.

13. $3\sin\left(2x + \frac{1}{2}\pi\right)$.   **14.** $2\sin(3x - 1.2)$.

15. $\sin\left(\frac{1}{2}x - \frac{1}{4}\pi\right)$.   **16.** $\cos\left(\frac{1}{2}x + \frac{1}{2}\pi\right)$.

74. ADDITION OF ORDINATES

Suppose that $F(x) = g(x) + h(x)$ for all numbers x on an interval where g and h are both defined. Suppose that the graphs of g and h are familiar and easily drawn. To obtain a graph of F, first we may draw graphs of g and h on the same xy-plane. Then, by geometrically adding corresponding ordinates of the graphs of g and h for selected values of x, we can locate points on the graph of F, in order to draw this graph. In particular, for any value of x where $h(x) = 0$, we have $F(x) = g(x)$, and the graph of F intersects the graph of g. At any value of x where $g(x) = 0$, the graph of F intersects the graph of h.

EXAMPLE 1.   Graph the function $f(x) = \sin x - \cos x$ on the interval $\left\{-\frac{1}{2}\pi \le x \le \frac{9}{4}\pi\right\}$.

Solution.   1. Let $y = \sin x - \cos x$; $y_1 = \sin x$; $y_2 = -\cos x$. Then, at any value of x, we have $y = y_1 + y_2$.

2. In Figure 133, first construct the graphs of $y_1 = \sin x$ and $y_2 = -\cos x$. Then, at any value for x, a point on the graph of $y = f(x)$

can be located by geometrically adding the ordinates y_1 and y_2 for that value of x.

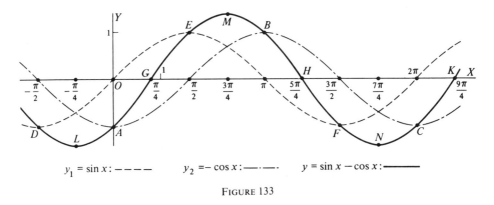

$y_1 = \sin x:$ $- - - -$ $y_2 = -\cos x:$ $-\cdot-\cdot-$ $y = \sin x - \cos x:$ $\underline{\quad\quad}$

FIGURE 133

3. If $y_1 = 0$ then $y = y_2$. Thus, corresponding to any x-intercept of the graph of $y_1 = \sin x$, the graph of $y = f(x)$ intersects the graph of $y_2 = -\cos x$. This gives points $\{A, B, C\}$ on the graph of $y = f(x)$ in Figure 133. If $y_2 = 0$ then $y = y_1$. Hence, corresponding to any x-intercept of the graph of $y_2 = -\cos x$, the graph of $y = f(x)$ intersects the graph of $y_1 = \sin x$. This gives points $\{D, E, F\}$ on the graph. When $y_1 = -y_2$ we have $y = 0$; this gives points $\{G, H, K\}$ on the graph. A few more points, $\{L, M, N\}$, on the graph were located by doubling the ordinate y_1 wherever $y_1 = y_2$. The graph of $y = f(x)$ is shown as the continuous curve in Figure 133.

Suppose that $f(x) = g(x)h(x)$. Then, by geometrically multiplying corresponding ordinates of the graphs of g and h, points can be located on the graph of f. If $x = c$ is a value such that $g(c) = 1$, then $f(c) = h(c)$; if $g(c) = 0$ or $h(c) = 0$, then $f(c) = 0$. The student will have an opportunity to obtain a few graphs in this manner in the next exercise.

Suppose that $y = F(x)$, where $F(x) = A \sin hx + B \cos hx$, with A and B not both zero, and $h > 0$. Then, numbers k and u can be found, with $k > 0$ and $-\pi < u \leqq \pi$, such that $y = k \sin (hx - u)$. Thus, the graph of F is a sine curve as described in (5) on page 216.

EXAMPLE 2.   If $f(x) = \sin x - \cos x$, find u and k in the form $f(x) = k \sin (hx - u)$ where $k > 0$ and $\{-\pi < u \leqq \pi\}$.

Solution.   1. We desire $k > 0$ and u, with $-\pi < u \leqq \pi$, so that

$$\sin x - \cos x = k \sin (x - u), \ or$$

$$\sin x - \cos x = k(\sin x \cos u - \cos x \sin u). \tag{1}$$

Statement (1) is true if

$$1 = k \cos u \quad \text{and} \quad 1 = k \sin u. \tag{2}$$

2. From (2), $k^2(\sin^2 u + \cos^2 u) = 1 + 1$, or $k^2 = 2$, because $\sin^2 u + \cos^2 u = 1$. Hence we choose $k = \sqrt{2}$.

3. From (2), $\quad \sin u = \dfrac{1}{\sqrt{2}} \quad \text{and} \quad \cos u = \dfrac{1}{\sqrt{2}}. \tag{3}$

Also, we desire $-\pi < u \leqq \pi$. Interpret u in (3) as the radian measure of an angle. Then u must be in either quadrant I or quadrant II because $\sin u > 0$. Also, u must be in quadrant I or quadrant IV because we have $\cos u > 0$. Hence, u is in quadrant I. Therefore $u = \frac{1}{4}\pi$. Thus,

$$\sin x - \cos x = \sqrt{2} \sin (x - \tfrac{1}{4}\pi). \tag{4}$$

On examining the graph of f in Figure 133, we observe that the graph checks with (4). Thus, the amplitude of the wave in Figure 133 is the ordinate of M, which is $\sqrt{2} \sin \frac{1}{2}\pi = \sqrt{2}$. The student should compare other features of Figure 133 with (4).

EXERCISE 48

On one xy-plane, separately draw a graph of each function in the sum defining $f(x)$. Then obtain a graph of f by geometrically adding ordinates of the component graphs. Let the domain of x be an interval of length at least 2π. Use relatively few accurate points but obtain a qualitatively correct graph. For convenience, use the approximation $\pi = 3.2$, or $\pi = 3$ in some problems.

1. $f(x) = x + \sin x$.
2. $f(x) = x + \cos x$.
3. $f(x) = \sin x - x$.
4. $f(x) = \sin x + \cos x$.
5. $f(x) = \sin 2x - \cos 2x$.
6. $f(x) = \sin 2x + \sin x$.
7. $f(x) = 2 \sin 3x + 3 \cos 2x$.
8. $f(x) = \cos 2x - \cos x$.

9. Construct points geometrically on the graph of the function $f(x) = x \sin x$, by first obtaining graphs of $y_1 = x$ and $y_2 = \sin x$ on the same xy-plane. Then, draw an approximate graph of $y = f(x)$, with $\{0 \leqq x \leqq 2\pi\}$.

Obtain an expression for $f(x)$ in the form $f(x) = k \sin (hx - u)$, with $k > 0$ and $-\pi < u \leqq \pi$. Also, where requested, draw a graph of f by use of the new form.

10. $f(x) = \sin x + \cos x$. Check the result with the graph for Problem 4.
11. $f(x) = 2 \sin x + 2\sqrt{3} \cos x$. Then obtain a graph of f.

12. By drawing graphs of the equations $y = .5x$ and $y = \sin x$ on the same xy-plane, solve the equation $.5x = \sin x$ graphically.

13. Solve the system of equations graphically:

$$y = x^2 \quad \textit{and} \quad y = 3 \sin 2x.$$

★14. Obtain a graph of $y = x \sin (1/x)$, for x on the domain $\{0 < x \leq 2\pi\}$. Calculate the coordinates of several points where x is near zero, by use of Table IV.

chapter 8

Certain Trigonometric Topics

75. AGREEMENTS CONCERNING THE CHAPTER

When a trigonometric function T is mentioned, it will be understood that T is one of the *standard* trigonometric functions. Also, in this chapter, T will be thought of as a function of angles, as in Chapter 6. We shall use either degree measure or radian measure for angles, as dictated by convenience. Any symbol, such as θ (*theta*) or ψ (*psi*) for an angle may be thought of as a number specifying the measure of the angle in radians. Then, for instance, we may write $\theta = w°$, indicating the corresponding measure in degrees.

In any xy-coordinate plane, the scale units on the axes will be equal, unless otherwise specified.

76. INCLINATION OF A LINE IN AN *XY*-PLANE

Let L be a nonhorizontal line in an xy-plane. Then, the inclination of L is defined as that positive angle ψ, less than 180°, through which the x-axis must be revolved about its intersection with L in order to coincide with L. If L is horizontal, its inclination is defined as $\psi = 0$. Thus, for any line L,

$$0 \leqq \psi < \tfrac{1}{2}\pi. \tag{1}$$

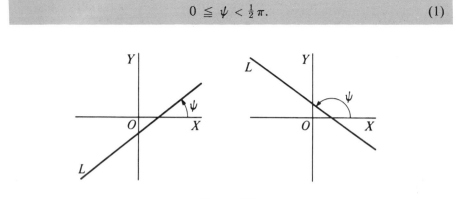

FIGURE 134

In Figure 134 we observe that $0 < \psi < \frac{1}{2}\pi$ if and only if L *rises* to the right, and thus has positive slope, as defined on page 74. Also, $\psi > \frac{1}{2}\pi$ if and only if L *falls* to the right, and thus has negative slope. Line L is vertical if the inclination of L is $\psi = \frac{1}{2}\pi$. By elementary geometry, two distinct lines L_1 and L_2, with inclinations ψ_1 and ψ_2, respectively, are parallel if and only if $\psi_1 = \psi_2$, as in Figure 135.

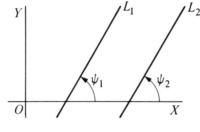

FIGURE 135

THEOREM I.   *If L is a nonvertical line in an xy-plane with slope m and inclination ψ, then*

$$m = \tan \psi. \tag{2}$$

Proof.   1. If L is horizontal, then $\psi = 0$, $\tan \psi = 0$, and also $m = 0$. Hence, (2) is true when $\psi = 0$.

2. In Figure 136, where $\psi \neq \frac{1}{2}\pi$ and $\psi \neq 0$, create a new $x'y'$-system of coordinates by translating the axes to a new origin located at the point Q where L intersects the x-axis. Then, in the $x'y'$-system, ψ is in its standard position for the definition of the trigonometric functions of ψ, as on page 157. Let $P:(x', y')$ be any point on the terminal side of ψ as represented in Figure 136, with Q having the coordinates $(x'=0, y'=0)$.

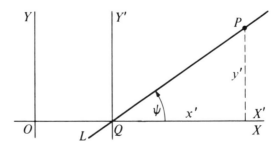

FIGURE 136

3. By the definition of the tangent function on page 158,

$$\tan \psi = \frac{y'}{x'}. \tag{3}$$

4. The slope of PQ in the $x'y'$-system is the same as in the xy-system. By use of the definition of slope for a line on page 74, the slope m of PQ is

$$m = \frac{y' - 0}{x' - 0} = \frac{y'}{x'}. \tag{4}$$

From (3) and (4), it is seen that $m = \tan \psi$.

ILLUSTRATION 1.   If line L has the inclination $\psi = \frac{1}{6}\pi$, then the slope is $m = \tan \frac{1}{6}\pi = \frac{1}{3}\sqrt{3}$. If $\psi = \frac{1}{4}\pi$, then $m = \tan \frac{1}{4}\pi = 1$. If $\psi = 120°$, then $m = \tan 120° = -\sqrt{3}$. If $\psi = \frac{1}{2}\pi$, then L is vertical and has no slope.

We have seen that every line in an xy-plane has an inclination, but we recall that *slope* is defined only for *nonvertical lines*.

Note 1.   Theorem I involved use of the definition of the tangent function of the inclination in an xy-plane. Hence, the equality of the scale units on the axes is a basic assumption in the proof of (2). This relation does not hold if the scale units on the axes are unequal, but we recall that the notion of slope applies in such a coordinate system.

EXAMPLE 1.   Obtain the inclination of the line through the points $A:(1, -4)$ and $B:(-2, 6)$.

Solution.   1. Let m be the slope of AB. Then

$$m = \frac{6 - (-4)}{-2 - 1} = -\frac{10}{3}, \quad or \quad m = -3.333.$$

2. From (2), $\tan \psi = -3.333$. Hence ψ is an obtuse angle, in quadrant II. Let ϕ (read *phi*) be the acute reference angle for ψ. Then we have $\tan \phi = 3.333$. By interpolation in Table IV, we find $\phi = 73.3°$. Hence $\psi = 180° - 73.3° = 106.7°$.

77. ANGLES FORMED BY INTERSECTING LINES

We shall have occasion to use the following reduction formula:

$$\tan (\pi - \phi) = -\tan \phi. \tag{1}$$

Proof of (1).   Recall that the tangent function is periodic, with the period π. Hence

$$\tan (\pi - \phi) = \tan (-\phi + \pi) = \tan (-\phi) = -\tan \phi,$$

where we used identity (IV) on page 188.

If two lines in a plane are not parallel, the lines intersect to form four positive angles, which are equal in pairs. Suppose that the equations of the lines are given. Then the angles formed by the lines can be found as in the following example, by first calculating the inclinations of the lines, and observing the graphs of the lines on the xy-plane.

EXAMPLE 1.   Find the angles formed by the following lines:

$$L_1: \{2x - 3y = 12\}; \qquad L_2: \{5x + 2y = 10\}.$$

Solution.   1. Write the lines in their slope-intercept forms:

$$L_1: \{y = \tfrac{2}{3}x - 4\}; \qquad L_2: \{y = -\tfrac{5}{2}x + 5\}.$$

The slopes are $m_1 = \tfrac{2}{3}$ and $m_2 = -2.5$.

2. Lines L_1 and L_2, and their inclinations ψ_1 and ψ_2, respectively, are shown in Figure 137. To obtain ψ_1 and ψ_2, use (2) on page 222:

$$\tan \psi_1 = .667 \qquad and \qquad \tan \psi_2 = -2.500. \tag{2}$$

By interpolation in Table IV, $\psi_1 = 33.7°$. Let $\phi = 180° - \psi_2$, where ϕ is the acute reference angle for ψ_2. By (2), $\tan \phi = 2.500$; from Table IV, $\phi = 68.2°$. Hence $\psi_2 = 180° - 68.2° = 111.8°$.

3. In Figure 137, let α be the acute angle, and θ be the obtuse angle formed by L_1 and L_2. With the aid of Figure 137, we obtain

$$\alpha = 111.8° - 33.7° = 78.1°; \qquad \theta = 180° - 78.1° = 101.9°.$$

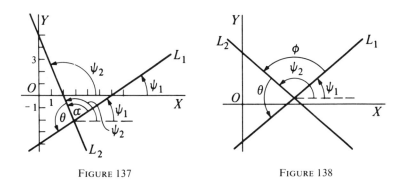

FIGURE 137   FIGURE 138

In an xy-plane, consider two nonvertical lines L_1 and L_2 with slopes m_1 and m_2, and inclinations ψ_1 and ψ_2, respectively, where $m_1 \neq m_2$. The lines intersect to form two positive angles, labeled θ and ϕ in Figure 138, where we take the notation so that $\psi_1 < \psi_2$. Then

$$\phi = \psi_2 - \psi_1 \qquad and \qquad \theta = \pi - \phi; \tag{3}$$

$$\tan \psi_1 = m_1; \qquad \tan \psi_2 = m_2. \tag{4}$$

Let α represent that one of the angles θ and ϕ which is *acute*. Then we shall prove that

$$\left\{\begin{array}{l} \textit{acute angle } \alpha, \\ \textit{formed by } L_1 \textit{ and } L_2 \end{array}\right\} \qquad \tan \alpha = \left| \frac{m_2 - m_1}{1 + m_1 m_2} \right|. \qquad (5)$$

Proof.   1. By use of identity (XII) on page 188 with $\phi = \psi_2 - \psi_1$,

$$\tan \phi = \frac{\tan \psi_2 - \tan \psi_1}{1 + \tan \psi_2 \tan \psi_1} = \frac{m_2 - m_1}{1 + m_1 m_2}. \qquad (6)$$

2. If $\phi < \frac{1}{2}\pi$, then ϕ is the desired acute angle, or $\phi = \alpha$. In this case, the fraction in (6) is positive, and $\tan \alpha$ is given by (5) where the absolute value rulings are redundant.

3. If $\phi > \frac{1}{2}\pi$, then $\alpha = \pi - \phi$. By use of (1),

$$\tan \alpha = \tan (\pi - \phi) = -\tan \phi.$$

Or, tan α is the *negative* of the right-hand side of (6), and hence is the absolute value of the fraction as seen in (5). Thus, in all cases, (5) is true.

ILLUSTRATION 1.   With L_1 and L_2 as in Example 1, where $m_1 = \frac{2}{3}$ and $m_2 = -\frac{5}{2}$, from (5) we obtain

$$\tan \alpha = \left| \frac{-\frac{5}{2} - \frac{2}{3}}{1 - \frac{5}{3}} \right| = 4.750$$

By interpolation in Table IV, $\alpha = 78.1°$, which agrees with the results in Example 1.

The result in (5) does not apply when $1 + m_1 m_2 = 0$, or $m_1 m_2 = -1$. In this case we know that L_1 and L_2 are perpendicular. Also, (5) does not apply if one of the lines is vertical and hence does not have a slope. In such a case, the method of Example 1 can be used to find the angles formed by the lines.

EXERCISE 49

Without use of a table, find the inclination of a line with the given slope m.

1. $m = 1$.   **2.** $m = -1$.   **3.** $m = -\sqrt{3}$.   **4.** $m = \frac{1}{3}\sqrt{3}$.

By use of Table IV, with interpolation when necessary, find the inclination of the line with the given slope m, or with the given equation.

5. $m = 2.145$. **6.** $m = -2.747$.   **7.** $m = -.552$.   **8.** $m = 1.620$.
9. $3x - 4y = 5$.   **10.** $2x - y = 3$.   **11.** $2y + 3x = 4$.

Find the acute angle formed by two lines with the given slopes.

12. $m_1 = -2; m_2 = 2.$ **13.** $m_1 = \frac{1}{2}; m_2 = -\frac{3}{2}.$ **14.** $m_1 = 4; m_2 = -\frac{1}{4}.$

Find the two angles formed by the lines, (a) without using (5) on page 225, by first obtaining inclinations if necessary; (b) by use of (5) on page 225, if (5) applies.

15. $\begin{cases} 2x - 5y = 5. \\ 4x + y = 2. \end{cases}$ **16.** $\begin{cases} x + y = 2. \\ 2y - 3x = 7. \end{cases}$ **17.** $\begin{cases} 3x - 2y = 0. \\ 2x + 3y = 0. \end{cases}$

18. $\begin{cases} 3x + 5 = 0. \\ 2x - y = 1. \end{cases}$ **19.** $\begin{cases} 3y - x = 6. \\ x - y = 5. \end{cases}$ **20.** $\begin{cases} 2y + 3x = 4. \\ 2x = 7. \end{cases}$

78. TERMINOLOGY CONCERNING TRIANGLES

In any triangle ABC, let $\{\alpha, \beta, \gamma\}$ be the angles at the vertices $\{A, B, C\}$, respectively. Let $\{a, b, c\}$ be the sides (or, the lengths of the sides) opposite $\{A, B, C\}$, respectively. The angles and sides of a triangle will be called its *parts*. A triangle is called an *oblique triangle* if it is not a right triangle. Many identities exist involving the parts of a general triangle. Hereafter, when we mention a *triangle,* it may be *any* triangle, either a right triangle or an oblique triangle.

From plane geometry, it is known that a triangle ABC can be constructed if three of its parts, including at least one side, are given. With any data of the preceding type, in numerical trigonometry the student learned that it is possible to compute the unknown parts of the triangle by use of trigonometric identities involving its parts. This computation is referred to as the *solution* of the triangle. We shall not consider the solution of triangles systematically. Instead, we shall discuss just two of the sets of trigonometric identities for a triangle. These relations will be considered because they are useful in mathematical discussion not related to the solution of triangles in numerical trigonometry. Also, we shall merely illustrate the use of these identities numerically, instead of investigating all of their applications in the solution of triangles.

79. THE LAW OF COSINES

We shall infer that the standard notation for the parts of a triangle ABC applies as given in Section 78.

THEOREM II. (**Law of cosines.**) *In any triangle ABC, the square of any side is equal to the sum of the squares of the other sides minus twice the product of these sides and the cosine of the angle included by them. Thus*

$$a^2 = b^2 + c^2 - 2bc \cos \alpha; \qquad (1)$$

$$b^2 = a^2 + c^2 - 2ac \cos \beta; \tag{2}$$

$$c^2 = a^2 + b^2 - 2ab \cos \gamma. \tag{3}$$

From the preceding equations, we obtain

$$\cos \alpha = \frac{b^2 + c^2 - a^2}{2bc}; \quad \cos \beta = \frac{a^2 + c^2 - b^2}{2ac}; \quad \cos \gamma = \frac{a^2 + b^2 - c^2}{2ab}. \tag{4}$$

Proof of (1).   1. Place $\triangle ABC$ on an xy-coordinate plane, as in Figure 139, with A at the origin, B at the right on the x-axis, and C above the x-axis. Then α is in its standard position as in the definition of the trigonometric functions on page 157. The coordinates of B are $(c,0)$, and $\overline{OC} = b$. Let (h,k) be the coordinates of C. By use of Definition II on page 157,

$$\cos \alpha = \frac{h}{b}; \quad \sin \alpha = \frac{k}{b}; \quad h = b \cos \alpha; \quad k = b \sin \alpha.$$

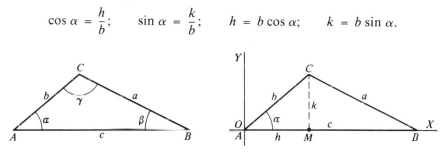

FIGURE 139

2. We have $\overline{BC} = a$. When the distance formula of page 57 is applied to find $\overline{CB}$, with $B:(c,0)$ and $C:(b \cos \alpha, b \sin \alpha)$, we obtain

$$\overline{BC}^2 = a^2 = (b \cos \alpha - c)^2 + (b \sin \alpha - 0)^2$$
$$= b^2 (\cos^2 \alpha + \sin^2 \alpha) - 2bc \cos \alpha + c^2$$
$$= b^2 + c^2 - 2bc \cos \alpha. \tag{5}$$

In (5) we arrived at (1). The student may draw a substitute for Figure 139 in case $\alpha > \frac{1}{2}\pi$, and verify that all of the preceding details remain unchanged in form.

On referring to "a" in (1) as *any* side of $\triangle ABC$, it is seen that (2) and (3) may be obtained from (1) by mere changes in notation. The results in (4) are obtained by solving for the cosine in each of (1), (2), and (3).

EXAMPLE 1.   Obtain the angles in a triangle ABC if $a = 4$, $b = 8$, and $c = 10$.

Solution.   1. By use of (4),

$$\cos \alpha = \frac{64 + 100 - 16}{160}; \quad \cos \beta = \frac{52}{80}; \quad \cos \gamma = -\frac{20}{64};$$

$$\cos \alpha = .925; \quad \cos \beta = .650; \quad \cos \gamma = -.312.$$

2. By interpolation in Table IV, we obtain $\alpha = 22.3°$ and $\beta = 49.5°$. Since $\cos \gamma < 0$, angle γ is obtuse. Let ϕ be the acute reference angle for γ. Then $\cos \phi = .312$ and $\phi = 71.8°$. Hence $\gamma = 180° - 71.8° = 108.2°$.

Summary of results: $\alpha = 22.3°$; $\beta = 49.5°$; $\gamma = 108.2°$.

Check: $22.3° + 49.5° + 108.2° = 180.0°$.

In Example 1, we illustrated the fact that a triangle ABC can be solved by use of the law of cosines if all sides of the triangle are given. We shall need the law of cosines in the proof of a fundamental theorem in Chapter 11.

EXERCISE 50

The problems apply to a triangle ABC in the standard notation introduced on page 226.

Find the specified part of a triangle ABC with the given parts.

1. $c = 3, b = 2, \alpha = 60°$; find a.
2. $a = 5, c = \sqrt{2}, \beta = 45°$; find b.
3. $b = 5, c = \sqrt{3}, \alpha = 30°$; find a.
4. $a = \sqrt{2}, b = 4, \gamma = 135°$; find c.

Solve triangle ABC by use of Table IV. Check with $\alpha + \beta + \gamma = 180°$.

5. $a = 6, b = 7, c = 5$.   6. $a = 8, b = 6, c = 9$.
7. $a = 3, b = 7, c = 6$.   8. $a = 4, b = 2, c = 5$.
9. $a = 2, b = 4, c = 3$.   10. $a = 3, b = 9, c = 8$.

80. LAW OF SINES

To state that three numbers $\{u, v, w\}$ are proportional, respectively, to three numbers $\{r, s, t\}$ means that there exists a constant $k \neq 0$ such that

$$u = kr; \quad v = ks; \quad w = kt. \tag{1}$$

Then, if none of $\{r, s, t\}$ is zero, from (1) we obtain

$$\frac{u}{r} = k; \quad \frac{v}{s} = k; \quad \frac{w}{t} = k. \tag{2}$$

Hence, from (2), $\qquad\qquad \dfrac{u}{r} = \dfrac{v}{s} = \dfrac{w}{t}. \tag{3}$

To state that $\{u, v, w\}$ are proportional to $\{r, s, t\}$, sometimes we write "$u:v:w = r:s:t$," which is read "u is to v is to w as r is to s is to t."

THEOREM III.   In any triangle, the lengths of the sides are pro-portional to the sines of the opposite angles. That is,

$$a:b:c = \sin \alpha:\sin \beta:\sin \gamma, \; or \tag{4}$$

$$\frac{a}{\sin \alpha} = \frac{b}{\sin \beta} = \frac{c}{\sin \gamma}. \tag{5}$$

The equalities in (5) abbreviate the following three equations:

$$\frac{a}{\sin \alpha} = \frac{b}{\sin \beta}; \qquad \frac{b}{\sin \beta} = \frac{c}{\sin \gamma}; \qquad \frac{c}{\sin \gamma} = \frac{a}{\sin \alpha}. \tag{6}$$

Proof of (6).   1. Let a and b be any two sides of $\triangle ABC$. At least one of α and β is acute; let it be assigned the notation β. Then place $\triangle ABC$ on an xy-plane with A at the origin, B on the x-axis where $x = c$, and C above the x-axis, as in Figure 140. Let the coordinates of C be (h, k).

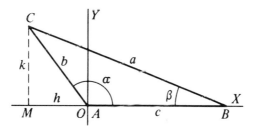

FIGURE 140

2. In Figure 140, angle α is in its standard position for use of the definitions of the trigonometric functions on page 157. Hence,

$$\sin \alpha = \frac{k}{b}, \qquad or \qquad k = b \sin \alpha. \tag{7}$$

From right $\triangle MBC$, by the special formulas for the trigonometric func-tions of acute angles on page 168,

$$\sin \beta = \frac{k}{a}, \qquad or \qquad k = a \sin \beta. \tag{8}$$

3. From (7) and (8),

$$a \sin \beta = b \sin \alpha, \qquad or \qquad \frac{a}{\sin \alpha} = \frac{b}{\sin \beta}, \tag{9}$$

where we divided both sides of the equation at the left by sin α sin β. Hence, the equation at the left in (6) is true. Since a and b represented *any* two sides, the other equations in (6) can be written by mere changes in notation.

EXAMPLE 1.   In a triangle ABC, if $c = 6$, $\beta = 50°$, and $\gamma = 30°$, solve the triangle.

Solution.   1. $\alpha = 180° - (30° + 50°) = 100°.$

2. By the law of sines,

$$\frac{b}{\sin \beta} = \frac{c}{\sin \gamma}, \quad or \quad b = \frac{c \sin \beta}{\sin \gamma}. \tag{10}$$

By use of (10) and Table IV,

$$b = \frac{6 \sin 50°}{\sin 30°} = \frac{6(.766)}{\frac{1}{2}} = 9.2.$$

3. By the law of sines,

$$\frac{a}{\sin \alpha} = \frac{c}{\sin \gamma}, \quad or \quad a = \frac{c \sin \alpha}{\sin \gamma}. \tag{11}$$

Since $80°$ is the reference angle for $100°$, we have sin α = sin $100°$ = sin $80°$. From (11) and Table IV,

$$a = \frac{6 \sin 100°}{\sin 30°} = \frac{6 \sin 80°}{\sin 30°} = \frac{6(.985)}{\frac{1}{2}} = 11.8. \tag{12}$$

Summary of results: $\alpha = 100°$; $a = 11.8$; $b = 9.2$.

The solution of Example 1 illustrates the fact that a triangle ABC may be solved by use of the law of sines in case two angles and a side of the triangle are given.

EXAMPLE 2.   In a triangle ABC, if $a = 20$, $c = 30$, and $\alpha = 30°$, find all angles of the triangle.

Solution.   1. *Construction of the triangle.* In Figure 141, an angle of $30°$ is constructed with A as the vertex. Let side AC be horizontal, with the location of C unknown. Locate B on the upper side of α with $\overline{AB} = 30 = c$. With B as a center and radius $a = 20$, strike an arc. The unknown vertex C is at the point, or points, if any, where this arc intersects the horizontal side of α. In Figure 141, two locations for vertex C arise. Hence, the analytic solution of this problem should produce *two solutions.*

2. From the law of sines,

$$\frac{\sin \gamma}{c} = \frac{\sin \alpha}{a}, \qquad or \qquad \sin \gamma = \frac{c \sin \alpha}{a}; \qquad (13)$$

$$\sin \gamma = \frac{30 \sin 30°}{20} = \frac{30(\frac{1}{2})}{20} = .750. \qquad (14)$$

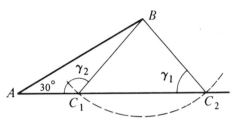

FIGURE 141

From Table IV, an acute angle satisfying (13) is $\gamma_1 = 48.6°$. An obtuse angle γ_2 also satisfies (13), where the reference angle for γ_2 is γ_1. Hence we obtain $\gamma_2 = 180° - 48.6°$, or $\gamma_2 = 131.4°$. These solutions are in harmony with Figure 141.

3. From $\alpha + \beta + \gamma = 180°$ with $\alpha = 30°$, when $\gamma = \gamma_1 = 48.6°$ we find $\beta_1 = 101.4°$; when $\gamma = \gamma_2 = 131.4°$ we find $\beta_2 = 18.6°$. Hence the angles of triangle ABC are either one of the following sets:

$$\{\alpha = 30°, \beta_1 = 101.4°, \gamma_1 = 48.6°\}; \quad \{\alpha = 30°, \beta_2 = 18.6°, \gamma_2 = 131.4°\}.$$

For each set of angles, a corresponding value could be found for side b by use of the law of sines. Thus the triangle corresponding to the data could be solved completely by use of the law of sines, and two solutions exist. The student may desire to complete a solution of the triangle. It would be found that side $b_1 = 39.2$, corresponding to angles $\{\beta_1, \gamma_1\}$, and $b_2 = 12.8$, corresponding to $\{\beta_2, \gamma_2\}$.

In case two sides and an angle opposite one of them are given for a triangle ABC, the corresponding problem of the solution of the triangle is referred to as the *ambiguous case*. This name is appropriate because, corresponding to the data, there may be *no* triangle (inconsistent data), or just *one* triangle, or *two* triangles, as found in Example 2.

Note 1. All cases which arise in the solution of triangles can be solved by use of the laws of cosines and sines. However, in many situations, the details of such solutions are less convenient than when other identities for a triangle are applied. In this text, only moderate use will be made of the laws in numerical applications.

EXERCISE 51

Solve the triangle ABC. Use Table IV.

1. $a = 4, \gamma = 70°, \alpha = 30°.$ **2.** $b = 10, \alpha = 40°, \beta = 30°.$
3. $c = 56, \gamma = 16°15', \alpha = 25°20'.$ **4.** $a = 8, \alpha = 150°, \beta = 15°.$
5. $b = 34, \beta = 149°20', \alpha = 9°40'.$
6. $c = 65, \gamma = 114°30', \alpha = 20°30'.$

*Find the angles of the triangle ABC with the given angle and sides.
If desired, complete the solution of the triangle.*

7. $b = 40, c = 30, \beta = 30°.$ **8.** $a = 10, b = 15, \alpha = 30°.$
9. $b = 10, c = 30, \beta = 150°.$ **10.** $a = 20, c = 10, \alpha = 150°.$
11. $a = 17.5, c = 15, \alpha = 20°30'.$
12. $a = 28, b = 26, \beta = 65°30'.$

chapter 9

Exponential and Logarithm Functions

81. THE GENERAL EXPONENTIAL FUNCTION

Suppose that $a > 0$. Then, if m and n are integers, with $n > 0$, and m/n in lowest terms, $a^{m/n}$ was defined as follows on page 27:

$$a^{m/n} = \sqrt[n]{a^m}. \tag{1}$$

Also, it was observed that

$$a^{m/n} = \left(\sqrt[n]{a}\right)^m. \tag{2}$$

Now let x be any real number, rational or irrational. Then, when x is rational, $x = m/n$ as in (1), we have defined a^x. It is essential also to have meaning for a^x when x is irrational.

ILLUSTRATION 1.   Consider the irrational number $\sqrt{2} = 1.414\cdots$, where the decimal part is endless, and is not of the repeating type. If $a > 0$, we may consider the sequence

$$a^1, a^{1.4}, a^{1.41}, a^{1.414}, \cdots. \tag{3}$$

Each exponent in (3) is a rational number; thus $1.41 = 141/100$. Hence, each power in (3) has meaning as defined in (2). A discussion beyond the scope of this text would show that, as closer and closer decimal approximations to $\sqrt{2}$ are used as exponents, or as we proceed to the right in the sequence of powers in (3), these powers approach a limit. This limit is defined as $a^{\sqrt{2}}$. Thus, $a^{1.414}$ is an approximation (very close) to $a^{\sqrt{2}}$.

If x is *any* irrational number, positive or negative, it is known that x can be expressed as an endless, nonrepeating decimal. Then, as in (3), if $a > 0$ we define a^x as the *limit* of the sequence of rational powers obtained by using as exponents the successive decimal approximations to x. Let $E(x) = a^x$, where a^x is defined by (2) if x is a *rational number,* and a^x is defined as specified in the preceding sentence when x is *irrational.* With $E(x) = a^x$, we refer to E as the **exponential function** with the **base*** a. A discussion above the level of this text would show that the function a^x is

---

*In considering $E(x) = a^x$, we do not permit $a \leq 0$, in order to avoid difficulties whose explanation is beyond the scope of this text.

a *continuous function,* or the graph of $y = a^x$ is a *continuous curve.* In obtaining the graph for any particular value of a, we may use rational values of x in computing coordinates (x, y) for points on the curve.

ILLUSTRATION 2.   To obtain coordinates for points on the graph of the exponential function 2^x in Figure 142, we let $y = 2^x$ and compute solutions (x, y) by substituting values of x as in the following table.

$x =$	-3	-2	-1	0	1	2	3
$y = 2^x$	$\frac{1}{8}$	$\frac{1}{4}$	$\frac{1}{2}$	1	2	4	8

We accept the fact that various properties of 2^x, as shown by Figure 142, are typical for a^x with $a > 1$, as follows.

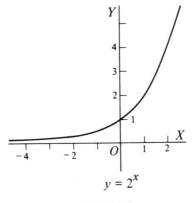

$y = 2^x$

FIGURE 142

Properties of a^x with $a > 1$.

1. *If $y = a^x$, the domain for x consists of* **all real numbers,** *and the range for y is the set of* **all positive numbers.**

2. *a^x increases if x increases, or the function a^x is an increasing function for all values of x. That is,*

$$\text{if } x_1 < x_2 \quad \text{then} \quad a^{x_1} < a^{x_2}. \tag{4}$$

3. *If x is negative and $|x|$ grows large without bound, then $a^x \to 0$, or*

$$\lim_{x \to -\infty} a^x = 0. \tag{5}$$

4. *If $x \to +\infty$, then a^x grows large without bound, or $a^x \to +\infty$ when $x \to +\infty$.*

As a consequence of (5), the graph of $y = a^x$ has the x-axis as an *asymptote,* which the graph approaches from above as $x \to -\infty$. This fact is illustrated in Figure 142.

The function a^x can be considered similarly when $0 < a < 1$. If $0 < a < 1$, the behavior of a^x when $x < 0$ is of the same nature as the behavior of a^x when $a > 1$ and $x > 0$. The behavior of a^x when $0 < a < 1$ and $x > 0$ is of the same nature as the behavior of a^x when $a > 1$ and $x < 0$.

Without discussion, we accept the fact that the familiar laws of exponents apply to powers of the type a^x with $a > 0$, for all values of x. A proof of this result could be based on the fact that each power a^x can be approximated as closely as we please by a power with a rational exponent. Then, the laws of exponents for a^x for any x are found to be consequences of the laws as they apply to powers with rational exponents. Thus, we have the following results, with the understanding that each base for a power is positive, and the exponents are any real numbers.

$$a^{-x} = \frac{1}{a^x}; \qquad a^x a^y = a^{x+y}; \qquad (a^x)^u = a^{ux}. \tag{6}$$

$$\frac{a^x}{a^y} = a^{x-y}; \qquad (ab)^x = a^x b^x; \qquad \left(\frac{a}{b}\right)^x = \frac{a^x}{b^x}. \tag{7}$$

In more advanced mathematics, and particularly in calculus, it is found that the most important base for exponential functions is a certain positive irrational number, represented almost always by e, where

$$e = 2.71828 \cdots.$$

Without added comment, we remark that e can be defined as the following limit:

$$e = \lim_{h \to 0} (1 + h)^{1/h} = 2.71828 \cdots. \tag{8}$$

Let $F(h) = (1 + h)^{1/h}$. Then, an appreciation of the limit in (8) is gained by observing the values of $F(h)$ in the following table, and the corresponding graph of $y = F(h)$ in Figure 143 on page 236. Since $F(h)$ is undefined at $h = 0$, the graph has no y-intercept, and hence there is a hole at the point $(h = 0, y = e)$ in Figure 143.

$y = F(h)$	4.000	2.868	2.732	2.718	$\cdots$	2.717	2.705	2.594	2.250
$h =$	$-.5$	$-.1$	$-.01$	$-.001$	$\cdots$	.001	.01	.1	.5

Later, we shall employ certain exponential functions as a means for introducing the related *inverse* functions, called *logarithm functions*. However, apart from that use, exponential functions are of great importance in many fields of application as well as in more advanced mathematics. Later problems in this text will suggest the wide range of usefulness for the functions.

Consider any exponential function $f(x) = a^x$. If we replace the

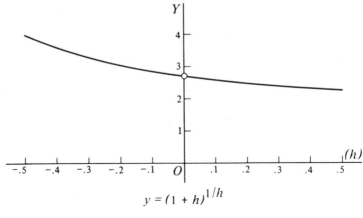

$$y = (1 + h)^{1/h}$$

FIGURE 143

argument x by $g(x)$, where g is any suitable function, then the function $a^{g(x)}$ also is called an *exponential function*. When the argument of f is *just* x as in $f(x) = a^x$, sometimes we may refer to f as being the *simple* exponential function with base a.

ILLUSTRATION 3.   The following table of solutions for $y = e^x$ is made up by use of Table VIII. To obtain e^x with $x = -.7$, use the column for e^{-x} with $x = .7$. With $x = -.7$, then $e^{-x} = e^{.7}$, and is obtained from the column for e^x in Table VIII. Graphs of $y = e^{-x}$ and $y = e^x$ are given in Figure 144. Notice that the graph of either equation is the "*reflection in the y-axis*" of the graph of the other equation.

$y = e^{-x}$	7.4	2.7	1	.4	.14
$y = e^x$	.14	.4	1	2.7	7.4
$x =$	-2	-1	0	1	2

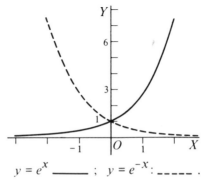

$y = e^x$ _____ ; $y = e^{-x}$: ------ .

FIGURE 144

ILLUSTRATION 4.   Let $f(x) = e^{kx}$. Then, if $k > 0$, the graph of f has the same general nature as the graph of $y = e^x$. If $k > 1$, the graph of $y = e^{kx}$ rises *faster* than the graph of $y = e^x$ when $x > 0$, and approaches the asymptote $x = 0$ *more rapidly* than the graph of $y = e^x$ when x is negative and decreases.

ILLUSTRATION 5.   A graph of $y = e^{x-2}$ is seen in Figure 145. Introduce a transformation of coordinates by translating the axes to the new origin $O':(x = 2, y = 0)$ by the equations $x' = x - 2$ and $y' = y$. Then the equation $y = e^{x-2}$ becomes $y' = e^{x'}$, whose graph is the curve, C, in Figure 145. Notice that C is the result of *translating* the graph of $y = e^x$ a distance of 2 units to the *right*. Similarly, the graph of $y = e^{x-h}$ would be the result of translating the graph of $y = e^x$ horizontally through a directed distance h.

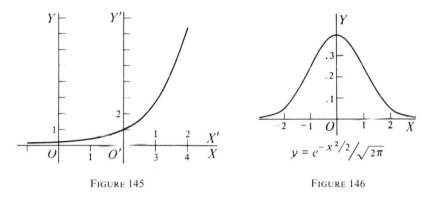

FIGURE 145   FIGURE 146

ILLUSTRATION 6.   In statistics, the exponential function $e^{-x^2/2}/\sqrt{2\pi}$ is called the *standard normal probability density function*, and forms the basis for extensive statistical theory. The student will obtain the graph of a function of this type in the next exercise. A graph of $y = e^{-x^2/2}/\sqrt{2\pi}$ is in Figure 146.

EXERCISE 52

Graph the equation. Choose unequal scale units on the coordinate axes so that a representative part of the graph will appear on the coordinate system. Use Table VIII where it applies. Employ transformation of coordinates by translating the axes where convenient.

1. $y = 3^x$.   2. $y = 10^x$.   3. $y = 10^{-x}$.   4. $y = .5^x$.
5. On the same xy-plane: $y = e^{2x}$ and $y = e^{-2x}$.
6. On the same xy-plane: $y = 10^{x-2}$; $y = 10^{2-x}$.
7. On the same xy-plane: $y = e^{x-3}$; $y = e^{3-x}$.

8. $y = 10^{-x^2}$ **9.** $y = e^{-x^2/2}$.   **10.** $y = e^{-(x-2)^2}$.

11. $y = x + e^x$, by first graphing the equations $y_1 = x$ and $y_2 = e^x$ on the same xy-plane, and then adding ordinates geometrically. What line is an asymptote of the graph?

82. APPLICATIONS OF EXPONENTIAL FUNCTIONS

Consider a biological experiment where a colony (*population*) of insects is being allowed to increase without interference. Let time be measured as t units of time before ($t < 0$) or after ($t > 0$) some assigned zero instant of time. Then, usually it is found that the *time rate of increase* of the population at any time t is proportional to *the size of the population at that time*. In particular, if the population has H_0 individuals at time t_0, the length of time for the population to double (to become $2H_0$) is a *constant,* which does not depend on H_0 or t_0. This principle of population growth is of fundamental importance in the biological and social sciences, and also in analogous situations in the physical sciences. On the basis of the principle, if H_0 is the population at time t_0, in calculus it is proved that the population, y, at any time t is given by

$$y = H_0 e^{k(t-t_0)}. \tag{1}$$

In (1), k is a positive constant depending on the type of population and not depending on t_0 or H_0. If a variable y is defined by (1), it is said that y increases in accordance with **the law of exponential growth.**

EXAMPLE 1.   Suppose that the number, y, of individuals in an increasing population of insects is given by

$$y = 50 e^{.5(t-2)}, \tag{2}$$

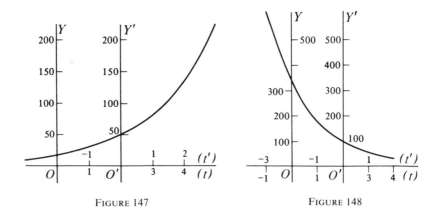

FIGURE 147   FIGURE 148

where the time unit is one day. By use of a graph, find approximately how long it takes for the population to double from the size it has (a) at $t = 2$; (b) at $t = 3$.

Solution. 1. Transform coordinates in (2) by use of $y = y'$ and $t' = t - 2$, which translates the axes to the new origin $O':(t = 2, y = 0)$. The graph of (2) is the graph of $y' = 50e^{.5t'}$ in the $t'y'$-system of coordinates in Figure 147. The following table for the graph of (2) was made up by use of Table VIII.

$y = y' =$	18	30	50	83	136	224
$t' =$	-2	-1	0	1	2	3
$t =$	0	1	2	3	4	5

2. (a) With a ruler edge placed horizontally on Figure 147 where $y = 100$, we find that the edge intersects the graph where $t' = 1.4$. Hence, the time necessary for the population to increase from 50 at $t' = 0$ to 100 is 1.4 days.

3. (b) At $t = 3$, $y = 83$. From the graph as in (a), $y = 166$ at approximately $t = 4.4$. The elapsed time is 1.4 days, which agrees with (a).

Consider a given mass consisting of a certain number of atoms of some radioactive element E. It has the property of continually disintegrating, due to some of the atoms breaking up by emitting radiation and changing to atoms of a different element. In physics it has been found that the rate at which the atoms of E in the mass disintegrate is proportional to the number of atoms which remain. Suppose that H_0 is the number of atoms of E in the given mass at some assigned time t_0. Then, by use of calculus, it can be demonstrated that the remaining number, y, of atoms at any time t is given by

$$y = h_0 e^{-k(t - t_0)}, \tag{3}$$

where $k > 0$. Let T be the time necessary for a mass consisting of a given number of atoms of E to disintegrate to one-half* of the original number of atoms. Then it can be proved that T is a *constant* which does not depend upon the original number of atoms in the mass. The time T is called the **half-life** of E. If any variable y is specified as a function of t by an equation of type (3), then it is said that y decreases according to **the law of exponential decay**.

EXAMPLE 2. Suppose that at time t the number, y, of atom units† in a certain mass of a radioactive element E is given by

$$y = 100e^{-.6(t-2)}, \tag{4}$$

*Or, to *any specified fraction* of the original number. The proof can be based on the properties of exponents.

†A unit might consist of 10^{20} atoms, or any number of atoms.

where the unit for t is 4,000 years. Graph (4). Find the half-life of E approximately from the graph.

Solution.   1. The graph of (4) in Figure 148 on page 238 was obtained as in Example 1 by first transforming coordinates to obtain $y' = 100e^{-.6t'}$, and then using Table VIII.

2. The number of atom units is $y = 100$ at $t' = 0$. Because of the small slope of the graph where $t' > 0$, a graphical determination of the point where $y = 50$ would be very inaccurate. Hence, instead, we find the point where y is 200. This occurs when $t = .8$. Then y decreases to $y = 100$ at $t = 2$ in 1.2 units of time, which therefore is the half-life. In years, this would be 4,800 years, which is near the half-life of the element carbon 14, one of the most troublesome contaminants in the fallout from atomic bombs.

EXAMPLE 3.   Consider the motion of a physical particle P which is oscillating along a y-axis, with the y-coordinate of P at any instant t seconds given by

$$y = e^{-.3t} \cos t. \tag{5}$$

Obtain a graph of (5).

Comment.   Assume that a particle P is in simple harmonic motion on a y-axis. Then, let this axis be immersed in a medium where the resistance to the motion of P is proportional to its velocity. In such a setting, P might move as described by (5).

Solution.   1. Let $y_1 = e^{-.3t}$ and $y_2 = \cos t$. From (5), $y = y_1 y_2$. Figure 149 shows a graph of $y_1 = e^{-.3t}$ as a broken curve *above* the t-axis,

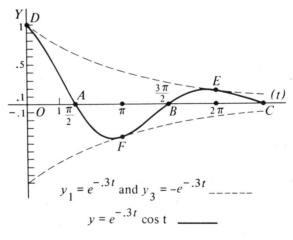

FIGURE 149

and a graph of $y_3 = -e^{-.3t} = -y_1$ as a broken curve *below* the t-axis. We visualize a graph of $y_2 = \cos t$, or draw this graph on scratch paper.

2. If $y_2 = 0$ then $y = y_1 y_2 = 0$, or the graph, W, of (5) has a t-intercept at each value of t where $\cos t = 0$. This gives points $\{A, B, C\}$ on W, at the values $\{\frac{1}{2}\pi, \frac{3}{2}\pi, \frac{5}{2}\pi\}$ of t.

3. If t is such that $\cos t = 1$, or $y_2 = 1$, then $y = y_1 y_2 = y_1$ and W touches the graph of $y_1 = e^{-.3t}$. This determines the points $\{D, E\}$ on W, at the values $\{0, 2\pi\}$ of t.

4. If t is such that $y_2 = \cos t = -1$ then $y = y_1 y_2 = -y_1$, and W touches the graph of $y_3 = -e^{-.3t}$. This gives the point F on W, where $t = \pi$. In calculus, it is proved that W is tangent to each of the curves $y_1 = e^{-.3t}$ and $y_3 = -e^{-.3t}$ wherever W intersects either one of them. This situation occurs at $\{D, E, F\}$ in Figure 149. On the basis of the preceding facts, a qualitatively accurate graph of (5) is drawn in Figure 149.

In the motion specified by (5), the particle P oscillates back and forth through $y = 0$ on the y-axis,* with the amplitudes (maxima and minima of y) decreasing and having the limit 0. We speak of the amplitudes being *damped* down by the damping factor $e^{-.3t}$ in (5). This motion is referred to as a *damped vibration*. Illustrations of such a phenomenon occur in many physical situations.

★*Note 1.*   Suppose that a sum of money, P, is invested at compound interest at an annual rate j (a small decimal, say .06, meaning 6%), compounded m times per year. Then, it is found that, at the end of t years, the resulting accumulated amount, A_m, of money is given by

$$A_m = P\left(1 + \frac{j}{m}\right)^{mt}. \tag{6}$$

Usually, interest is compounded semiannually ($m = 2$), or quarterly ($m = 4$), or monthly ($m = 12$), or even daily ($m = 365$) in recent arrangements for savings deposits in certain banks in the United States. It proves useful in the theory of life insurance to consider the result of interest being compounded *continuously,* which is interpreted to mean the *limit* of the result in (6) as $m \to \infty$. It is found that

$$\lim_{m \to \infty} A_m = P\left[\lim_{m \to \infty}\left(1 + \frac{j}{m}\right)^{mt}\right] = P\left[\lim_{m \to \infty}\left(1 + \frac{j}{m}\right)^{\frac{m}{j}}\right]^{jt}$$

$$(with\ h = j/m) \qquad = P\left[\lim_{h \to 0}(1 + h)^{1/h}\right]^{jt} = Pe^{jt},$$

*Not the y-axis of Figure 149, but the y-axis where the motion is occurring.

where the definition of e on page 235 was used. Thus, if A is the amount at the end of t years when interest is compounded *continuously* at the annual rate j, then $A = Pe^{jt}$. Hence, money increases in accordance with the law of exponential growth. On account of this background, the law of exponential growth sometimes is referred to as the "*compound interest law of growth.*"

★*Note 2.* In preceding sections, the exponential functions with the base e have been emphasized. Wherever a power of e has been used, it would have been possible to use a power of any other number $b > 0$ with $b \neq 1$. Use of e is preferred, first, because extensive tables for its powers are available. Table VIII in this text is a very abbreviated illustration of the tables which exist. Second, and most important, the function e^x is the most simple exponential function from the standpoint of calculus. Thus the derivative and integral of e^x are as follows, where C is any constant.

$$\frac{de^x}{dx} = e^x; \qquad \int e^x dx = e^x + C. \tag{7}$$

Or, the derivative of e^x is e^x itself; the indefinite integral of e^x is simply e^x itself, plus a constant. On the other hand, if $0 < a, a \neq 1$, and $a \neq e$, we have

$$\frac{da^x}{dx} = a^x \log_e a; \qquad \int a^x dx = \frac{a^x}{\log_e a} + C. \tag{8}$$

The greater complexity of (8) as compared with (7) is sufficient evidence that, if calculus is to be employed with exponential functions, *the most convenient base for the functions is e.*

EXERCISE 53

In each of these problems, choose generously large scale units, probably unequal, on the coordinate axes, and arrange to spread the figure vertically.

A variable y is increasing according to the law of exponential growth in the problem. Obtain a graph of y as a function of the time t. Use relatively few accurately located points on the graph. By use of the graph find the length of time it takes for y to double in value.

1. $y = 50e^{.3(t-3)}$.   2. $y = 100e^{.5(t+2)}$.   3. $y = 50e^{.2(t-2)}$.

A certain mass of a radioactive element E is decaying according to the given law of exponential decay. Draw a graph showing the number, y, of atom units left at any value of t. From the graph, find the half-life of E.

4. $y = 100e^{-.4(t-1)}$.   5. $y = 50e^{-.2(t+2)}$.   6. $y = 50e^{-.5(t-2)}$.

First obtain a graph of the exponential factor on the right in the equation, and draw a graph on a different coordinate plane for the trigonometric *factor. By considering these two graphs, and perhaps obtaining a few solutions of the equation by use of tables, construct a graph of the equation on* $\{-\pi \leq x \leq 2\pi\}$.*

7. $y = e^{-.2t} \sin t$.   **8.** $y = e^{-.4t} \cos t$.

Graph the equation by geometrically adding ordinates for two graphs. Restrict x to the interval $\{-\pi \leq x \leq 2\pi\}$.

9. $y = \sin x + e^{-x}$.   **10.** $y = \cos x + e^x$.

11. By drawing the graphs of two functions of x on the same xy-plane, find the solution of the equation $\sin x = e^{-x}$ with $\{0 \leq x \leq \frac{1}{2}\pi\}$. Use large scale units on the coordinate axes.

83. THE LOGARITHM FUNCTION

Let $E(x) = a^x$ with $a > 1$. A typical graph of $y = E(x)$ is seen in Figure 150, with $a = 10$. The *domain* of E is the interval $D:\{-\infty < x < +\infty\}$, the *range* of E is the interval $R:\{0 < y < +\infty\}$, and $E(x)$ *increases* as x increases. Hence for each number y in R there exists *just one* corresponding value of x in D such that $y = E(x)$, as illustrated in Figure 150. This correspondence *defines x as a function of y.* For any y in R, let $L(y)$ represent the corresponding value of x in D such that (x, y) is a solution of $y = a^x$. Then, the equation

$$y = E(x) \qquad is \ equivalent \ to \qquad x = L(y). \qquad (1)$$

Or, the functions $E(x)$ and $L(y)$ are *inverse functions* defined by the equation $y = E(x)$. With $E(x) = a^x$, instead of using $L(y)$ in (1) we shall write "$\log_a y$," read "*logarithm of y to the base a.*" Then, from (1),

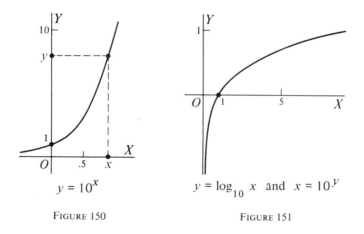

$y = 10^x$

FIGURE 150

$y = \log_{10} x$ and $x = 10^y$

FIGURE 151

*The approximation $\pi = 3.2$ may be used if desired in graphing.

$$y = a^x \quad \text{is equivalent to} \quad x = \log_a y. \tag{2}$$

Or, the exponential and logarithm functions to the *base a* are a pair of *inverse functions,* where each one is called the *inverse* of the other function.

In order to study the logarithm function, it proves convenient to interchange the roles of x and y in (2). Then

$$y = \log_a x \quad \text{is equivalent to} \quad x = a^y, \tag{3}$$

and the two equations in (3) have the same graph. To obtain the graph of $y = \log_a x$ for any assigned value of a, instead we may graph $x = a^y$.

ILLUSTRATION 1.   To graph the logarithm function to the base 10, let $y = \log_{10} x$. Since

$$y = \log_{10} x \quad \text{is equivalent to} \quad x = 10^y, \tag{4}$$

we use $x = 10^y$ and assign values for y in order to obtain solutions (x, y) for $y = \log_{10} x$ in the following table. Thus, if $y = 1$ in (4) then $x = 10$; if $y = 0$ then $x = 10^0 = 1$; if $y = -1$ then $x = 10^{-1} = .1$; if $y = .5$ then $x = 10^{\frac{1}{2}} = \sqrt{10} = 3.2$. The graph is in Figure 151. It was desirable to use a much larger scale unit vertically than horizontally in Figure 151. The y-axis is an asymptote of the graph because, in (4),

$$\left.\begin{array}{l} \lim\limits_{y \to -\infty} x = \lim\limits_{y \to -\infty} 10^y = 0, \text{ or} \\[2mm] \lim\limits_{x \to 0+} y = \lim\limits_{x \to 0+} \log_{10} x = -\infty. \end{array}\right\} \tag{5}$$

$x =$	.01	.1	1	3.2	10
$y = \log_{10} x$	-2	-1	0	.5	1

The graph in Figure 151, with $a = 10$, is typical of the graph of $y = \log_a x$.* On the basis of Figure 151, we accept the following facts.

I. *The domain of the function* $\log_a x$ *consists of all positive numbers, or the infinite interval* $D: \{0 < x < +\infty\}$. *Thus, negative numbers and zero do not have logarithms.*

II. *The function* $\log_a x$ *is an increasing function. That is,*

$$x_1 < x_2 \quad \text{is equivalent to} \quad \log_a x_1 < \log_a x_2. \tag{6}$$

III. $\log_a x \to -\infty$ *as* $x \to 0+$, *or the y-axis is an asymptote of the graph of* $y = \log_a x$.

IV.
$$\begin{array}{llll} & \textit{If } x < 1 & \textit{then} & \log_a x < 0. \tag{7}\\[1mm] & \textit{If } x > 1 & \textit{then} & \log_a x > 0. \end{array} \tag{8}$$

*With $a > 1$ in all cases in this text.

Note 1. In the preceding discussion of $\log_a x$, we have assumed that $a > 1$. Similar content can be discussed with $a < 1$, but is not valuable for future purposes in this text, and hence will be omitted.

Inspection of (3) justifies the following statement, which often is used in elementary introductions to the logarithm function:

$$\left.\begin{array}{l} \textit{the logarithm of a number } x > 0 \textit{ to a base a is the ex-} \\ \textit{ponent of the power of a which is equal to x.} \end{array}\right\} \qquad (9)$$

In (3), we may refer to "$x = a^y$" as the *exponential form*, and to "$y = \log_a x$" as the *logarithmic form* of the relationship between x and y. If one of these forms is given, the other form can be written immediately by use of (3), or (9).

ILLUSTRATION 2.   If $H = 2^3$, then $\log_2 H - 3$ or, the logarithm of H to the base 2 is 3.

ILLUSTRATION 3.   We read "$\log_3 27$" as "*the logarithm of 27 to the base 3.*" If $y = \log_3 27$ then, by (3), $27 = 3^y$ and $y = 3$, because $3^3 = 27$. Thus, $\log_3 27 = 3$.

ILLUSTRATION 4.   Since $\sqrt[3]{2} = 2^{\frac{1}{3}}$, from (9) we obtain $\log_2 \sqrt[3]{2} = \frac{1}{3}$.

Since $\qquad \dfrac{1}{16} = \dfrac{1}{2^4} = 2^{-4}, \qquad$ *then* $\qquad \log_2 \dfrac{1}{16} = -4.$

ILLUSTRATION 5.   To obtain $\log_{10} .001$, write .001 as a power of 10:

$$.001 = \frac{1}{1000} = \frac{1}{10^3}, \qquad \textit{or} \qquad .001 = 10^{-3}.$$

Hence $\log_{10} .001 = \log_{10} 10^{-3} - -3$.

ILLUSTRATION 6.   To obtain the base a such that $\log_a 81 = 4$, we write the exponential form $81 = a^4$. Hence $a = \sqrt[4]{81} = \sqrt[4]{3^4} = 3$.

ILLUSTRATION 7.   For any base a,

$$\log_a a = 1 \qquad \textit{because} \qquad a^1 = a;$$
$$\log_a 1 = 0 \qquad \textit{because} \qquad a^0 = 1.$$

EXERCISE 54

1. Later, we shall emphasize the use of e as a base for logarithms. If $y = \log_e x$ then $x = e^y$. Obtain a graph of the function $\log_e x$, with the x-axis horizontal as usual. Use Table VIII.

Write a logarithmic equation equivalent to the given exponential equation.

2. $W = 3^6$.   **3.** $H = 4^3$.   **4.** $N = 2^{-3}$.   **5.** $K = 6^{\frac{1}{4}}$.

6. $N = 10^{-3}$.   **7.** $N = 10^4$.   **8.** $H = 10^{\frac{1}{2}}$.   **9.** $W = 1.5^4$.

10. $\frac{1}{32} = 2^{-5}$.   **11.** $\frac{25}{4} = \left(\frac{5}{2}\right)^2$.   **12.** $\frac{1}{81} = 3^{-4}$.   **13.** $.01 = 10^{-2}$.

Find the number with the given logarithm by writing the equivalent exponential form.

14. $\log_{10} H = 4$.   **15.** $\log_5 N = 2$.   **16.** $\log_4 W = 3$.

17. $\log_{10} N = 0$.   **18.** $\log_{10} N = -1$.   **19.** $\log_{10} N = 3$.

20. $\log_{10} N = 1$.   **21.** $\log_h T = 2$.   **22.** $\log_h K = 0$.

23. $\log_b N = -1$.   **24.** $\log_{64} H = \frac{1}{3}$.   **25.** $\log_{16} K = \frac{1}{2}$.

26. $\log_4 N = -\frac{1}{2}$.   **27.** $\log_{36} N = \frac{3}{2}$.   **28.** $\log_{27} H = \frac{2}{3}$.

Find the logarithm.

29. $\log_{10} 10{,}000$.   **30.** $\log_5 125$.   **31.** $\log_5 \frac{1}{5}$.   **32.** $\log_2 32$.

33. $\log_{10} .0001$.   **34.** $\log_{100} 10{,}000$.   **35.** $\log_2 \frac{1}{8}$.   **36.** $\log_3 \frac{1}{27}$.

Find the unknown base.

37. $\log_a 36 = 2$.   **38.** $\log_a 9 = 2$.   **39.** $\log_a 64 = 6$.

40. $\log_a 6 = \frac{1}{2}$.   **41.** $\log_a 2 = \frac{1}{3}$.   **42.** $\log_a .0001 = -2$.

84. FUNDAMENTAL PROPERTIES OF LOGARITHMS

I. *The logarithm of a product is equal to the sum of the logarithms of its factors. Thus,*

$$\log_a HK = \log_a H + \log_a K. \tag{1}$$

Proof.   Let $x = \log_a H$ and $y = \log_a K$. By use of (3) on page 243, we obtain $H = a^x$ and $K = a^y$; hence

$$HK = a^x a^y, \qquad or \qquad HK = a^{x+y}.$$

Therefore, by use of (3) on page 243, $\log_a HK = x + y$, which proves (1).

ILLUSTRATION 1. $\log_{10} 35(59) = \log_{10} 35 + \log_{10} 59$.

II. *The logarithm of a fraction H/K is equal to the logarithm of the numerator minus the logarithm of the denominator:*

$$\log_a \frac{H}{K} = \log_a H - \log_a K. \tag{2}$$

Proof.   As in proving (1), we have $H = a^x$ and $K = a^y$. Hence

$$\frac{H}{K} = \frac{a^x}{a^y} \qquad or \qquad \frac{H}{K} = a^{x-y}.$$

Therefore, by the definition of a logarithm, $\log_a (H/K) = x - y$, which proves (2).

ILLUSTRATION 2. $\log_{10} \frac{79}{87} = \log_{10} 79 - \log_{10} 87$.

III. *The logarithm of the* hth *power of a number* $K > 0$ *is equal to* h *times the logarithm of* K, *or*

$$\log_a K^h = h \log_a K. \tag{3}$$

Proof. Let $\log_a K = x$. Then

$$K^h = (a^x)^h, \quad or \quad K^h = a^{hx}.$$

Hence, $\log_a K^h = hx$, which proves (3).

Suppose that n is a positive integer. Then $\sqrt[n]{K} = K^{1/n}$ and, as a special case of (3) with $h = 1/n$, we obtain

$$\log_a \sqrt[n]{K} = \frac{1}{n} \log_a K. \tag{4}$$

ILLUSTRATION 3. $\log_a H^3 = 3 \log_a H.$

$$\log_a \sqrt[3]{H} = \log_a H^{\frac{1}{3}} = \tfrac{1}{3} \log_a H.$$

Logarithms to the base 10 are called **common logarithms.** They are the most convenient variety when logarithms are used as an aid in computation. Hereafter in this chapter, if we mention a *logarithm* it will be a *common logarithm,* unless otherwise stated. We shall write simply "log N" for "$\log_{10} N$." The student should verify the following common logarithms.

$N =$	.001	.01	.1	1	10	100	1000	10,000
$\log N =$	-3	-2	-1	0	1	2	3	4

ILLUSTRATION 4.   From a table of logarithms which will be met later, we obtain log 4 = .6021 and log 7 = .8451. By use of these logarithms, the logarithms of powers of 10, and (1) extended to a product of three factors, we obtain

$$\log 2800 = \log 4(7)(100)$$
$$= \log 4 + \log 7 + \log 100 = 2 + .6021 + .8451 = 3.4472.$$

EXERCISE 55

Find the common logarithm of the number by use of the logarithms of powers of 10 *and the following logarithms:*

$$\log 3 = .4771; \quad \log 4 = .6021; \quad \log 7 = .8451.$$

1. 12.   **2.** 21.   **3.** $\frac{3}{4}$.   **4.** $\frac{3}{7}$.   **5.** $\frac{1}{28}$.

6. 300.   **7.** 40.   **8.** 7000.   **9.** 30.   **10.** .3.

11. .7.   **12.** .04.   **13.** .003.   **14.** 28,000.   **15.** .21.

16. $\sqrt{3}$.   **17.** $\sqrt[3]{4}$.   **18.** $\sqrt[5]{7}$.   **19.** $\sqrt{21}$.   **20.** $\sqrt[3]{12}$.

★85. REVIEW OF CHARACTERISTIC AND MANTISSA

Every number, and hence every logarithm, can be written in just one way as the sum of an *integer* and a *decimal* which is positive or zero and less than 1. When log N is written in this fashion, we call the integer the **characteristic** and the decimal the **mantissa** of log N.

$$\log N = \text{(an integer)} + \text{(decimal} \geq 0, <1); \tag{1}$$

$$\log N = \text{characteristic} + \text{mantissa}. \tag{2}$$

ILLUSTRATION 1.   If log N = 3.216, the characteristic of log N is 3 and the mantissa is .216.

ILLUSTRATION 2.   If log N = -2.168, then log N lies between -3 and -2, and hence is equal to -3 plus a decimal, <1. We find that $3 - 2.168 = .832$. Hence

$$\log N = -2.168 = -3 + .832;$$

the characteristic is -3 and the mantissa is .832.

Note 1.   In computation in this chapter, in referring to any number N we shall suppose that N is represented in decimal form. Also, unless otherwise specified, we shall assume that N is a terminating decimal. If this is not the case, any given number immediately is understood to be rounded off to a terminating decimal.

Any positive number N can be expressed as the product of a number K such that $1 \leq K < 10$, and some power 10^h where h is an integer. Thus we obtain $N = K(10^h)$, which is called the **scientific notation** for N.

ILLUSTRATION 3.   Since $1,000,000 = 10^6$ and $.00001 = 1/100,000 = 10^{-5}$, in scientific notation we obtain

$$3,789,000 = 3.789(10^6); \qquad .000079 = 7.9(10^{-5}). \tag{3}$$

The following facts were illustrated in (3).

When $N \geq 1$, and N has h digits to the left of the decimal point, then $N = K(10^{h-1})$, where $1 \leq K < 10$. $\qquad$ (4)

When $0 < N < 1$, and the first significant digit of N appears in the hth decimal place, then $N = K(10^{-h})$, where $1 \leq K < 10$. $\qquad$ (5)

Suppose that $1 \leq K < 10$ as in (4) and (5). Then

$$\log 1 \leq \log K < \log 10, \text{ or}$$

$$0 \leq \log K < 1, \tag{6}$$

because $\log x$ is an *increasing function of* x (as emphasized in (6) on page 244). By (6) above, $\log K$ is a *nonegative decimal* which is less than 1.

THEOREM I.   *Suppose that* $N > 0$ *and* $N = K(10^w)$ *in scientific notation, where* $1 \leq K < 10$ *and* w *is an integer. Then the characteristic of* $\log N$ *is* w *and the mantissa of* $\log N$ *is* $\log K$.

Proof.   We have $\log N = \log K(10^w)$, or

$$\log N = \log K + \log 10^w = w + \log K. \tag{7}$$

In (7), $\log K$ is a nonnegative decimal less than 1 and w is an integer. Hence, by (1), w is the characteristic and $\log K$ is the mantissa of $\log N$.

THEOREM II.   *The mantissa of* $\log N$ *depends only on the sequence of significant* digits in* N. *That is, if two numbers* N_1 *and* N_2 *differ only in the position of the decimal point, then* $\log N_1$ *and* $\log N_2$ *have the same mantissa.*

Proof.   In scientific notation, $N_1 = K(10^h)$ and $N_2 = K(10^w)$, where K appears in both cases since N_1 and N_2 have the same significant digits. Hence, by Theorem I, $\log K$ is the mantissa for both $\log N_1$ and $\log N_2$.

From the study of logarithms in algebra, we recall the following facts about the characteristic of the logarithm of a number N. The facts are consequences of the statement about the characteristic in Theorem I, and about the exponents in statements (4) and (5).

> If $N \geq 1$, the characteristic of $\log N$ is a nonnegative
> integer which is one less than the number of digits in N (8)
> to the left of the decimal point.
> If $0 < N < 1$, and the first significant digit of N is in
> the hth decimal place, then the characteristic of $\log N$ (9)
> is $-h$.

By use of (8) and (9), we obtain the characteristic of $\log N$ in any particular case by merely inspecting N.

ILLUSTRATION 4.   By (8), the characteristic of $\log 56{,}700$ is 4. By (9), the characteristic of $\log .000739$ is -4.

---

*The **significant digits** of N are its digits, in sequence from the left, starting with the first one not zero and ending with the last digit which is specified definitely.

★86. REVIEW OF INTERPOLATION FOR LOGARITHMS

Mantissas for the logarithms of numbers can be computed by methods met in calculus and, as a rule, are endless decimals. By Theorem I, if N is any number, the mantissa of log N is the logarithm of some number K where $1 \leq K < 10$. Table V gives log K rounded off to four decimal places if $1 \leq K < 10$ and K has at most three significant digits, aside from any definitely specified zeros at the right. Thus, the table gives log K from $K = 1.00$ to $K = 9.99$, at intervals of .01 in K. The table may be called a table of *mantissas* for logarithms, or a table of the *actual logarithms* of numbers K such that $1 \leq K < 10$. A decimal point is understood at the left of each mantissa in Table V.

To obtain log N by use of Table V, the characteristic is found by inspection of N. The mantissa for log N is determined by finding log K from Table V, where $1 \leq K < 10$ and K has the same significant digits as N.

Note 1. It is assumed that the student has studied logarithms before, and has had experience with interpolation in tables of logarithms. Hence, only a few examples of use of Table V will be considered.

ILLUSTRATION 1. To obtain log 295,000: The characteristic is 5. From Table V, the mantissa (which is log 2.950) is .4698. Hence we have log 295,000 = 5.4698.

ILLUSTRATION 2. To obtain log .00378: By (9) on page 249, the characteristic is −3. From Table V, the mantissa (which is log 3.780) is .5775. Hence, log .00378 = −3 + .5775.

For convenience, if the characteristic of log N is −h where $h > 0$, usually we change −h to

$$[(10 - h) - 10], \quad or \quad [(20 - h) - 20].$$

Thus, in Illustration 2, we write

$$\log .00378 = -3 + .5775 = 7.5775 - 10.$$

It is said that a number N is the **antilogarithm** of a number L in case log $N = L$. Then we may write $N = $ antilog L, which is read "*N is equal to the antilogarithm of L.*"

ILLUSTRATION 3. 100 = antilog 2 because log 100 = 2.

EXAMPLE 1. Find N if log $N = 4.8893$.

Solution. 1. The mantissa of log N is .8893. We find .8893 in Table V in the row with 77 at the left, in the column headed by 5. Hence, .8893 = log 7.75, and "775" are the significant digits of N.

2. The characteristic of log N is 4. Hence N has $(4 + 1)$ or 5 digits to the left of the decimal point, so that $N = 77,500$.

Comment. Since only a four-place table of mantissas is being used, we cannot guarantee the accuracy of the zeros in 77,500. Hence, we prefer to write the result in scientific notation as $7.75(10^4)$, to indicate certainty of only three significant digits.

EXAMPLE 2. Find N if log N = 7.6170 − 10.

Solution. The mantissa is .6170. It is found as the mantissa in Table V for 414 (more elaborately: .6170 = log 4.14). The characteristic is $(7 - 10)$ or -3. Hence, the first significant digit of N is in the 3d decimal place, or N = .00414.

We interpolate in Table V by use of the following **principle of proportional parts.**

> *Suppose that values of a function $f(x)$ are listed in a table for values of x spaced at some regular interval. Then, we assume that, for small changes in x from any listed value, the corresponding changes in $f(x)$ are proportional to the changes in x.*   (1)

When interpolating in Table V, we act as if every number N whose logarithm is to be found has *just four significant digits.* If N initially has *less* than four significant digits, we adjoin zero as often as necessary at the right in N to create four digits. If N initially has *more* than four significant digits, we agree to round off N to just four significant digits before using Table V to obtain log N. Each number from 1.00 to 9.99 whose logarithm is given in Table V is thought of now as having an additional 0 at the right. Thus Table V is considered as listing all numbers from 1.000 to 9.990 at intervals of .010 in the numbers. If log N is given and N is to be found from Table V by interpolation, we agree to round off the result to just four significant digits. Use of any more digits would not be justified by the limited accuracy of interpolation.

EXAMPLE 3. Find log 17.36 from Table V.

Solution. 17.36 is *bracketed* by (is *between*) 17.30 and 17.40, whose logarithms can be found from Table V. Then we interpolate by principle (1): Since 17.36 is 6/10 of the way from 17.30 to 17.40, we assume that log 17.36 is 6/10 of the way from log 17.30 to log 17.40. The solution is compactly arranged as follows.

						Tabular difference is .0025.
10	6	log 17.30 = 1.2380	x	25		$\dfrac{x}{25} = \dfrac{6}{10}$; x = .6(25).
		log 17.36 = ?				
		log 17.40 = 1.2405				

x = .6(.0025) = .0015; log 17.36 = 1.2380 + .0015 = 1.2395.

EXAMPLE 4. Find log .003037.

Solution. 1. The *characteristic* is -3.

2. *The mantissa:* 3037 (where we drop the decimal point for brevity) is bracketed by 3030 and 3040. We obtain the mantissa as follows.

		Digits	Mantissa	Tabular difference is 15.
10	7	3030	.4814	$\dfrac{x}{15} = \dfrac{7}{10};$
		3037	?	
		3040	.4829	$x = .7(15) = 10.5.$

Hence, the mantissa is $.4814 + .0010 = .4824$. Finally, $\log .003037 = -3 + .4824 = 7.4824 - 10$. In the interpolation we could have used $x = 10$ or $x = 11$ with equal justification. In such a case, we agree to make that choice which causes the final result of interpolation to end with an *even integer.*

EXAMPLE 5.   Find N if $\log N = 8.2471 - 10$.

Solution.   1. The mantissa .2471 is bracketed by the tabulated mantissas .2455 and .2480, for the logarithms of 1.760 and 1.770. We obtain the significant digits of N as follows (with practice, the following work should be done mentally).

25	16	16	.2455, *mantissa for* 1760	x	10	$\dfrac{x}{10} = \dfrac{16}{25} = .6$
			.2471, *mantissa for N*			
			.2480, *mantissa for* 1770			$x = .6(10) = 6$
			Hence .2471 *is the mantissa for* 1760 $+$ 6 *or* 1766.			

2. The characteristic of $\log N$ is -2. Hence $N = .01766$.

ILLUSTRATION 4.   To find N if $\log N = 4.3867$: The mantissa is bracketed by the tabulated mantissas .3856, for 2430, and .3874, for 2440:

$$3874 - 3856 = 18; \qquad 3867 - 3856 = 11.$$

Hence, the significant part of N is

$$2430 + \tfrac{11}{18}(10) = 2430 + 6 = 2436.$$

Therefore $N = 24,360$, which is better written as $2.436(10^4)$ to show that only four digits are significant.

★EXERCISE 56

Find the logarithm of the number by use of Table V.

1. 1856.	**2.** 3625.	**3.** 4714.
4. 19.26.	**5.** 8.219.	**6.** .4792.
7. .7538.	**8.** .003184.	**9.** .003965.
10. .0002317.	**11.** .00001468.	**12.** 23,560.
13. $1.726(10^5)$.	**14.** $2.848(10^{-3})$.	**15.** $6.537(10^6)$.

Find the antilogarithm of the given logarithm by use of Table V.

16. 1.6947.	**17.** 2.4634.	**18.** 7.0564 − 10.

19. 9.2179 − 10.	**20.** 0.3629.	**21.** 1.5248.
22. 7.6108 − 10.	**23.** 8.5319 − 10.	**24.** 2.0648.
25. 7.4613 − 10.	**26.** 6.0753 − 10.	**27.** 4.0552.
28. 6.8415 − 10.	**29.** 7.3238.	**30.** 2.1689 − 10.

★87. REVIEW OF LOGARITHMIC COMPUTATION

The accuracy of any computation with logarithms depends on the number of decimal places to which mantissas are given in the table being used. The result frequently is subject to an unavoidable error which, under favorable circumstances, would be at most a few units in the last significant place given by the interpolation. However, no general statement can be made about the accuracy obtainable, because it depends also on the nature of the expression being computed. In problems in this text, we shall state results as obtained by interpolation.

EXAMPLE 1.   Compute: $P = (688)(.004314)(5.316)$.

Solution.   By Property I of logarithms, log P is the sum of the logarithms of the factors. We obtain these logarithms, then log P, and then find P as the antilogarithm of its logarithm. The computing form, given below in bold face type, was made up as the first stage of the solution.

$$
\begin{aligned}
\textbf{log 688} &= 2.8376 \\
\textbf{log .004314} &= 7.6349 - 10 \\
\textbf{log 5.316} &= 0.7256 \\
\hline
\textbf{(add) log } P &= 11.1981 - 10 = 1.1981.
\end{aligned}
$$

From Table V, $P = 15.78.$

EXAMPLE 2.   Compute: $W = \dfrac{321.8}{14.376}.$

Solution.   By Property II of logarithms, log W is equal to the logarithm of the numerator minus the logarithm of the denominator. We round off 14.376 to 14.38 before computing because only a four-place table is being used.

$$
\begin{aligned}
\textbf{log 321.8} &= 2.5076 \\
(-)\ \textbf{log 14.38} &= 1.1578 \\
\hline
\textbf{log } W &= 1.3498; \qquad W = 22.38.
\end{aligned}
$$

EXAMPLE 3.   Compute: $H = 469/1857.$

Solution.   Below, we noticed that $3.2688 > 2.6712$; hence log H will be negative, and thus will have a negative characteristic. In order to obtain log H in the standard form for a negative logarithm, first we added 10, and then subtracted 10, with log 469.

$$
\begin{aligned}
\log 469 &= 2.6712 = 12.6712 - 10 \\
(-)\log 1857 &= 3.2688 = 3.2688 \\
\hline
\log H = ? &= 9.4024 - 10 \\
H = .2526 & \quad \text{(From Table V)}
\end{aligned}
$$

EXAMPLE 4.   Compute: $W = \dfrac{65(188)(27)}{16.3(.084)}.$

Solution.   Separately, we obtain the logarithms of the numerator and denominator, and then use Property II of logarithms.

$$
\begin{aligned}
\log 65 &= 1.8129 \\
\log 188 &= 2.2742 \\
\log 27 &= 1.4314 \\
\hline
\textbf{(add) log num.} &= 5.5185 \\
(-)\textbf{ log den.} &= 0.1365 \\
\hline
\log W &= 5.3820;
\end{aligned}
\qquad
\begin{aligned}
\log 16.3 &= 1.2122 \\
\log .084 &= 8.9243 - 10 \\
\hline
\textbf{(add) log den.} &= 0.1365 \\[1em]
W &= 2.410(10^5).
\end{aligned}
$$

EXAMPLE 5.   Compute: $H = (.285)^3.$

Solution.   By Property III of logarithms, log $H = 3$ log .285.

$$\log .285 = 9.4548 - 10$$

$$\log H = 3 \log .285 = 28.3644 - 30 = 8.3644 - 10$$

Hence, $H = .02314.$ (From Table V)

EXAMPLE 6.   Compute: $W = \sqrt[3]{.0746}.$

Solution. $W = (.0746)^{\frac{1}{3}}.$ Hence log $W = \frac{1}{3}$ log .0746.

$$\log .0746 = 8.8727 - 10$$

$$\log W = \frac{8.8727 - 10}{3} = \frac{28.8727 - 30}{3} = 9.6242 - 10$$

Hence, $W = .4209.$

Comment.   In finding log W, we noticed that the logarithm would be negative. In order to obtain the result in the standard form for a negative logarithm, we subtracted 20 from -10 to get -30, and added 20 to 8.8727 to compensate for the subtraction. Then, division of 30 by 3 gives 10.

★EXERCISE 57

Compute by use of Table V.

1. 31.65(.768).   **2.** 03142(.7693).   **3.** 78(65.6)(.0716).
4. 425(−8.132)(.00167).   **5.** (−19.81)(−26.45)(.00871).

Hint. Remember that *a negative number does not have a logarithm.* Compute with all factors positive. Determine the sign of the result by inspection.

6. $\dfrac{869}{13.26}$.   **7.** $\dfrac{.0835}{.97643}$.   **8.** $\dfrac{31.48}{695.3}$.   **9.** $\dfrac{1}{49.618}$.

10. $\dfrac{(3.146)(12.8)(.4316)}{29.3(.08146)}$.   **11.** $\dfrac{5.36(10^{-5})(89.63)}{(.077)(346)(1.68)}$.

12. Find the reciprocal of 783; .00296.

13. 2.13^3.   **14.** $.564^4$.   **15.** $\sqrt{38.67}$.   **16.** $\sqrt{.0536}$.

17. $\sqrt[3]{81.13}$.   **18.** $\sqrt[3]{.647}$.   **19.** $\sqrt[4]{.0075}$.   **20.** $\sqrt[3]{-.0069}$.

21. $\sqrt{\dfrac{54.6}{315(.74)}}$.   **22.** $\sqrt[3]{\dfrac{18.3}{53.644}}$.   **23.** $\sqrt{\dfrac{10^{1.36}(32.4)}{527.63}}$.

Hint. For Problem 21, first obtain the logarithm of the fraction.

88. EXPONENTIAL AND LOGARITHMIC EQUATIONS

Suppose that a logarithm of some expression involving the variable or variables is involved in an equation. Then, we shall call it a *logarithmic equation.*

EXAMPLE 1.   Solve: $\log 12x = 5 - \log x.$ (1)

Solution.   By use of a property of logarithms, from (1) we obtain

$$\log 12 + \log x = 5 - \log x, \textit{ or}$$
$$2 \log x = 5 - \log 12. \qquad (2)$$

By use of Table V in (2), we find that

$$2 \log x = 5 - 1.0792 = 3.9208; \quad \log x = 1.9604; \quad x = 91.28,$$

where we obtained antilog 1.9604 from Table V by interpolation.

If the variable or variables appear in one or more exponents in an equation, we may refer to it as an *exponential equation.* If it is of simple form, as in the following problem, the equation may be solved by first taking the logarithm of both of its sides. In this operation, as a rule, we shall take logarithms to the base 10.

EXAMPLE 2.   Solve: $15^x = 38.$ (3)

Solution.   1. Take the logarithm of both sides of (3) to the base 10, with Property III of page 247 used on the left:

$$x \log 15 = \log 38; \qquad x = \frac{\log 38}{\log 15} = \frac{1.5798}{1.1761},$$ (4)

where the logarithms were obtained from Table V.

2. The fraction at the right in (4) should be computed by use of Table V. The numbers in the fraction are rounded off to four significant digits before Table V is used below. Interpolation was required.

$$\begin{aligned} \log 1.580 &= 0.1987 \\ (-) \log 1.176 &= 0.0704 \\ \hline \log x = 0.1283; \quad & x = 1.344. \end{aligned}$$

EXAMPLE 3.   Obtain $\log_{15} 38$.

Solution.   Let $x = \log_{15} 38$. By the definition of a logarithm on page 245, $38 = 15^x$. Hence, the present problem merely rephrases Example 2. From the previous solution, $\log_{15} 38 = 1.344$.

In Example 3, the following method was illustrated.

> *For any number $a > 0$, in order to find $\log_a N$, write the equation $N = a^x$, and obtain $x = \log_a N$ by solving $N = a^x$ for x by use of common logarithms.*   (5)

Any desired logarithm to the base e can be obtained by the method of Example 3 with the aid of a table of common logarithms. For this purpose, we list the following logarithms:

> $$\log_{10} e = .43429; \qquad \log_{10} .43429 = 9.63778 - 10.$$ (6)

EXAMPLE 4.   Find $\log_e 45$.

Solution.   Let $x = \log_e 45$. Then $e^x = 45$. On taking the common logarithm of both sides of $45 = e^x$, we obtain

$$x \log_{10} e = \log_{10} 45; \qquad x(.4343) = 1.6532. \qquad \text{(Table V)}$$

Hence, $x = 1.653/.4343$, which we compute by use of Table V below: $x = \log_e 45 = 3.806$.

$$\begin{aligned} \log 1.653 &= 0.2183 & = 10.2183 - 10 \\ (-) \log .4343 &= 9.6378 - 10 & = 9.6378 - 10 \\ \hline & & \log x = 0.5805; \quad x = 3.806. \end{aligned}$$

THEOREM III.   *If* $N > 0, a > 0, b > 0, a \neq 1$ and $b \neq 1$, then*

$$\log_a N = (\log_a b)(\log_b N). \tag{7}$$

Proof.   Let $y = \log_b N$. Then $N = b^y$. Hence, by Property III of logarithms on page 247,

$$\log_a N = \log_a b^y = y \log_a b = (\log_b N)(\log_a b).$$

In (7), the number $\log_a b$ is called the **modulus** of the system of logarithms to the base a with respect to the system to the base b. If we have available a table of logarithms to the base b, by use of (7) we could compute logarithms to the base a by multiplication with the modulus $\log_a b$, which can be expressed in terms of $\log_b a$ as follows.

THEOREM IV.   *If $a > 0, b > 0, a \neq 1$, and $b \neq 1$, then*

$$\log_a b = \frac{1}{\log_b a}. \tag{8}$$

Proof.   By the definition of a logarithm, $b = a^{\log_a b}$. Hence,

$$\log_b b - \log_b a^{\log_a b}, \ or$$

$$1 = (\log_a b)(\log_b a), \tag{9}$$

by Property III of logarithms on page 247. We obtain (8) from (9).

EXERCISE 58

Solve for x, or compute the specified logarithm, by use of the four-place Table V.

1. $\log x^2 - \log 3x = 2.6$.   **2.** $\log (2/x) + \log 3x^2 = 5$.
3. $12^x = 17$.   **4.** $25^x = 482$.   **5.** $3^x = 20(2^x)$.
6. $\log_5 30$.   **7.** $\log_{15} 48$.   **8.** $\log_e 100$.
9. $\log_e 56$; $\log_e 5.6$; $\log_e .56$.

Comment. Notice that the results in Problem 9 do *not* differ merely by an integer. Thus, the property of the mantissa for a common logarithm stated in Theorem II on page 000 does not hold in the case of a logarithm to the base e.

10. $\log_{.8} 50$.   **11.** $\log_{.5} 1000$.   **12.** $\log_e 1000$.

---

*The conditions are necessary so that a and b may serve as bases for logarithms, and so that $\log_a N$ and $\log_b N$ may exist.

89. THE FUNCTION ln *x*

Numerous tables of common logarithms are available with the logarithms given to a great many decimal places. Some of the most extensive tables were calculated by painstaking efforts hundreds of years ago. With the aid of modern computing machines and methods developed in calculus, a table of logarithms to any base $a > 0$ could be computed now with only a reasonable amount of effort. In such computation, (7) of page 257 and common logarithms [with $b = 10$ in (7)] would not have to be used, although a table of common logarithms could be applied conveniently in this way.

If logarithms are to be used as aids in computation* as in Section 87, then 10 is the most desirable base for logarithms. This is true because of the property of the mantissa of a common logarithm specified in Theorem II on page 249.

In connection with calculus, *e* is the most desirable base for logarithms. It was mentioned on page 242 that *e* is the most desirable base for exponential functions as used in calculus, because of the resulting simplicity of associated formulas for derivatives and integrals. Similarly, the formulas for derivatives and integrals involving logarithms are much more simple if the base for logarithms is *e*, than if any other base is employed. For instance, in calculus, the following simple formulas are met:

$$\frac{d(\log_e x)}{dx} = \frac{1}{x}; \qquad \int a^x \, dx = \frac{a^x}{\log_e a} + C. \tag{1}$$

If logarithms to any other base $b > 0$ are used instead of logarithms to the base *e*, then formulas (1) are replaced by the following more complicated results:

$$\frac{d(\log_b x)}{dx} = \frac{\log_b e}{x}; \qquad \int a^x \, dx = \frac{a^x}{(\log_b a)(\log_e b)}. \tag{2}$$

Hence, in calculus, logarithms to the base *e* are used whenever possible, if logarithms occur. Therefore, hereafter in this chapter, we shall deal exclusively with logarithms to the base *e*.

The logarithm of a number *N* to the base *e* is called the **natural logarithm** of *N*. From the standpoint of calculus applications, the name *"natural logarithm"* is justified by our previous comments. However, these facts probably were not the reason why the name originally was selected. Hereafter, we shall use "ln *N*" instead of "$\log_e N$" for the logarithm of *N* to the base *e*. We read "ln *N*" simply as *"the logarithm*

---

*The importance of this field of application has decreased radically because of the widespread use of modern electronic digital computing machines.

of N" when the context makes it clear that the base for logarithms is *e*. For emphasis, sometimes we may read "ln *N*" as "*the natural logarithm of N.*"

Table VII in this text is a brief table of natural logarithms. There is no convenient property for natural logarithms such as is stated in Theorem II on page 249 for the mantissas of common logarithms. Hence, we emphasize that the notions of a characteristic and a mantissa DO NOT APPLY for natural logarithms. Table VII provides ln *N* if $1 \leq N \leq 10.09$. To obtain ln *N* when $N > 10$ or $N < 1$, first we may write *N* in scientific notation as $N = K(10^h)$ where $1 \leq K < 10$, and ln *K* is found in Table VII. Then, we may obtain ln *N* as follows:

$$\ln N = \ln K + h \ln 10. \tag{3}$$

ILLUSTRATION 1.   To graph the function ln *x*: Recall that only positive numbers have logarithms.   Hence, the domain for *x* is $\{0 < x < +\infty\}$. Also, as seen on page 244, if $y = \ln x$ the range of values for *y* is $-\infty < y < +\infty$; the line $x = 0$ is an asymptote of the graph, or

$$\lim_{x \to 0+} y = \lim_{x \to 0+} \ln x = -\infty.$$

The *x*-intercept of the graph of $y = \ln x$ is $x = 1$. The following table of coordinates for points on the graph of $y = \ln x$ in Figure 152 was made up by use of Table VII. In doing this, for some entries we used (3), with ln $10 = 2.3026$ from Table VII. Thus

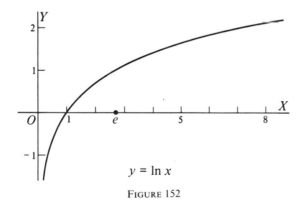

$$y = \ln x$$

FIGURE 152

$$\ln 15 = \ln 10(1.5) = \ln 10 + \ln 1.5 = 2.3026 + .4055 = 2.7081;$$

$$\ln .5 = \ln \frac{5}{10} = \ln 5 - \ln 10 = 1.609 - 2.303 = -.69.$$

The larger values of *x* in the following table were not used in Figure 152.

$y = \ln x$	$\downarrow - \infty$	-1.6	$-.7$	0	$.4$	$.7$	1	2.3	2.7	3
$x =$	$0 \leftarrow x$	$.2$	$.5$	1	1.5	2	e	10	15	20

EXAMPLE 1.   Obtain a graph of $y = \ln (2x - 3)$.

Solution.   1. Let $y = \ln (2x - 3)$.   The domain for x consists of those numbers x such that $0 < 2x - 3$, or $3 < 2x$. Hence, the domain is $\{\frac{3}{2} < x\}$. The function $\ln (2x - 3)$ is undefined when $2x - 3 = 0$ or $x = \frac{3}{2}$, and the vertical line $x = \frac{3}{2}$ is an asymptote of the graph.

2. The x-intercept of the graph is the solution of $\ln (2x - 3) = 0$. Since $\ln 1 = 0$, the x-intercept satisfies $2x - 3 = 1$, or $x = 2$.

3. A qualitatively accurate graph can be drawn, as in Figure 153, on the basis of a few accurate points (x,y), found by use of Table VII, and the following facts: the x-intercept is 2; the line $x = \frac{3}{2}$ is an asymptote; the domain for x is $\{\frac{3}{2} < x\}$.

$y = \ln (2x - 3)$	$\downarrow - \infty$	0	1.1	$\ln 10 = 2.3$
$x =$	$\rightarrow 1.5$	2	3	6.5

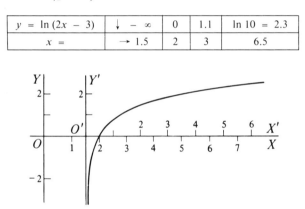

FIGURE 153

Comment.   From $y = \ln (2x - 3)$, we obtain $y = \ln 2(x - 1.5)$. Transform coordinates by translating the axes to the new origin $O':(x = 1.5, y = 0)$ by the equations $x' = x - 1.5$ and $y' = y$. Then the given equation becomes $y' = \ln 2x'$. We could obtain the graph of $y = \ln (2x - 3)$ by finding the graph of $y' = \ln 2x'$ in the $x'y'$-system, as shown in Figure 153. Similarly, for any number h, the graph of $y = \ln (x - h)$ is the result of shifting the standard graph of $y = \ln x$ a directed distance h horizontally in the xy-plane.

EXERCISE 59

Find the natural logarithm of the number by use of Table VII.

1. 8.2;   82;   820.   **2.** 7.5;   75;   .75;   .075.

3. Obtain a meticulously accurate graph of the function ln x for x on $\{0 < x \leq 50\}$, with large scale units.

Obtain a graph of the function by use of relatively few accurately located points and general properties. Perhaps use transformation by translation of the coordinate axes.

4. ln $(x - 3)$.   **5.** ln $2x$.   **6.** ln $(2x + 4)$.

Obtain a graph of the equation by the graphical addition of ordinates for two auxiliary graphs drawn on the same xy-plane. If desired, the approximation $\pi = 3.2$ *may be used where trigonometric functions occur.*

7. $y = x + \ln x$, for $\{0 < x \leq 10\}$.

8. $y = \ln x + \sin x$, for $\{0 < x \leq 4\pi\}$.

9. $y = \ln (x - 2) - \cos x$, for $\{2 < x \leq 11.2\}$.

10. Solve the equation 3 ln $x = x$ by obtaining the graphs of $y_1 = 3 \ln x$ and $y_2 = x$ on the same coordinate system.

Solve the equation graphically.

11. ln $x = \sin x$.   **12.** ln $x = \cos x$.   **13.** ln $2x = e^{-x}$.

Graph the equation.

14. $x = \ln y$.   **15.** $x = \ln (2y - 3)$.

16. Solve the system of equations graphically:

$$x = y^2 \qquad and \qquad x = \ln (2 - y).$$

In the following problems, if logarithms are to be used, employ natural logarithms.

17. A population, y, of insects is increasing in accordance with the law of exponential growth $y = He^{kt}$ where the unit for t is one day. (a) How large is the population when $t = 0$? (b) If the population doubles in three days, find k.

18. The radioactive chemical element strontium 90 has a half life of approximately twenty-eight years. The disintegration of a mass, y, of atoms of the element obeys the exponential law of decay $y = He^{-kt}$, where the unit for the time t is one year. Find k. (If $y = \frac{1}{2}H$ then $t = 28$.)

19. Repeat Problem 18 for the element iodine 131, whose half life is about 8 days, with the unit for time as one day.

20. Repeat Problem 18 for the element carbon 14, whose half life is approximately 5,600 years, with the unit for time as 1,000 years.

chapter 10

Miscellaneous Topics in Graphing

90. PARAMETRIC EQUATIONS FOR A LINE

Suppose that L is a nonvertical line through the distinct points $P_1:(x_1, y_1)$ and $P_2:(x_2, y_2)$ in an xy-plane, where the scale units on the coordinate axes need not be equal.* Let $P:(x, y)$ be any point on L. The projections of $\{P_1, P, P_2\}$ on OX, as in Figure 154, are $\{H_1, H, H_2\}$ with the x-coordinates $\{x_1, x, x_2\}$, respectively. From (4) on page 5,

$$\overline{H_1 H} = x - x_1 \qquad \text{and} \qquad \overline{H_1 H_2} = x_2 - x_1. \tag{1}$$

If $P:(x, y)$ is any assigned point on L, define a number t by

$$t = \frac{\overline{H_1 H}}{\overline{H_1 H_2}}, \qquad \text{so that} \qquad \overline{H_1 H} = t(\overline{H_1 H_2}). \tag{2}$$

From (1) and (2),

$$x - x_1 = t(x_2 - x_1), \qquad \text{or} \qquad x = x_1 + t(x_2 - x_1). \tag{3}$$

For any point $P:(x, y)$ on L, with x as in (3), we shall prove that $y = y_1 + t(y_2 - y_1)$ and, hence, that L consists of those points whose coordinates are

$$x = x_1 + t(x_2 - x_1) \qquad \text{and} \qquad y = y_1 + t(y_2 - y_1), \tag{4}$$

where the domain for t is all real numbers.

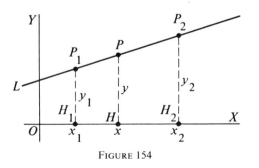

FIGURE 154

*Recall that the concept of *slope* of a line, as used in Section 29, is available under such circumstances.

Proof of (4) for any line L.   1. **If L is not vertical:** the slope of L is

$$m = \frac{y_2 - y_1}{x_2 - x_1}, \qquad so \ that \qquad y_2 - y_1 = m(x_2 - x_1). \qquad (5)$$

An equation for L is $y - y_1 = m(x - x_1).$ (6)

2. Suppose that $P:(x, y)$ is on L, and t is defined by (2). With this value for t, from (3) we have $(x - x_1) = t(x_2 - x_1)$. Then (6) gives

$$y - y_1 = tm(x_2 - x_1). \qquad (7)$$

With $m(x_2 - x_1) = (y_2 - y_1)$ from (5), in (7) we obtain

$$y - y_1 = t(y_2 - y_1), \qquad or \qquad y = y_1 + t(y_2 - y_1),$$

which proves (4) when L is not vertical.

3. **If L is vertical:** then $x_1 = x_2$ and $x = x_1$ for all points $P:(x, y)$ on L, and the equation for x in (4) is true. When L is vertical, the reasoning applied in (1), (2), and (3) may be repeated with the y-axis and y involved, which would give the equation for y in (4). Hence (4) is true for *any* line L in the xy-plane.

We refer to (4) as *parametric equations* for the line L with t as the *parameter*. For each value of t, from (4) we obtain the coordinates of just one point on L. For each point P on L, there exists just one value for t, defined in (2), such that the coordinates of P are given by (4). Later in the chapter, we shall consider parametric equations for curves of a general type.

ILLUSTRATION 1.   By use of (4), parametric equations for the line L through $P_1:(-2, 5)$ and $P_2:(3, 2)$ are

$$x = -2 + (3 + 2)t \qquad and \qquad y = 5 + (2 - 5)t, \ or$$
$$x = -2 + 5t \qquad and \qquad y = 5 - 3t. \qquad (8)$$

When $t = 1$, (8) gives $P_2:(3, 2)$ on L. When $t = 3$, (8) gives the point $(13, -4)$ on L.

In (4), if t is eliminated by use of the equations, an xy-equation is obtained for the line.

ILLUSTRATION 2.   From $x = -2 + 5t$ in (8), we obtain the equation $t = \frac{1}{5}(x + 2)$. If this is used for t in $y = 5 - 3t$ in (8), then

$$y = 5 - \tfrac{3}{5}(x + 2), \quad or \quad y = \tfrac{19}{5} - \tfrac{3}{5}x,$$

which is an xy-equation for the line (8).

In Figure 154, conceive of a number scale being imposed on line L, with any desired unit for length, and the positive direction to the *right* so that L now is a directed line. The vertical lines through $\{P_1, P, P_2\}$ in Figure 154 form a set of parallel lines intercepting line segments $\{P_1 P, P_1 P_2\}$ on L, and $\{H_1 H, H_1 H_2\}$ on OX. By elementary geometry, the following ratios of lengths of corresponding segments are equal:

$$\frac{|\overline{P_1 P}|}{|\overline{P_1 P_2}|} = \frac{|\overline{H_1 H}|}{|\overline{H_1 H_2}|}. \tag{9}$$

Since the positive direction is to the *right* on both $P_1 P_2$ and OX, the directed line segments in the numerators in (9) are both *zero*, both *positive*, or both *negative*, and similarly for the denominators. Hence, the equality in (9) persists when absolute value bars are removed, so that

$$\frac{\overline{P_1 P}}{\overline{P_1 P_2}} = \frac{\overline{H_1 H}}{\overline{H_1 H_2}}, \tag{10}$$

where the right-hand side is equal to t, from (2). Thus, we have the following new interpretation of t in (4), when L is a directed line:

$$\frac{\overline{P_1 P}}{\overline{P_1 P_2}} = t, \qquad or \qquad \overline{P_1 P} = t(\overline{P_1 P_2}). \tag{11}$$

In (11), $t = \frac{1}{2}$ corresponds to P being the *midpoint* of $P_1 P_2$. The points obtained from (11) with $t = \frac{1}{3}$ and $t = \frac{2}{3}$ are the *trisection points* of $P_1 P_2$. On account of the meaning for t in (11), sometimes equations (4) are referred to as the "*point of division formulas.*"

ILLUSTRATION 3.   To find the midpoint $P:(x, y)$ of segment $P_1 P_2$ with $P_1:(-2, 5)$ and $P_2:(6, -1)$, use $t = \frac{1}{2}$ in (4), and obtain

$$x = -2 + \tfrac{1}{2}(6 + 2); \qquad y = 5 + \tfrac{1}{2}(-1 - 5).$$

Thus $P:(2, 2)$ is the midpoint. If $P:(x, y)$ is the trisection point of $P_1 P_2$ nearest P_1, we use $t = \frac{1}{3}$ in (4) and obtain

$$x = -2 + \tfrac{1}{3}(6 + 2) = \tfrac{2}{3} \qquad and \qquad y = 5 + \tfrac{1}{3}(-1 - 5) = 3.$$

Hence $P:(\frac{2}{3}, 3)$ is the specified trisection point. The other one would be found by use of $t = \frac{2}{3}$ in (4).

With $t = \frac{1}{2}$ in (4), after simplification we obtain

$$(\text{\textbf{midpoint} } of\ P_1 P_2) \qquad x = \tfrac{1}{2}(x_1 + x_2); \ y = \tfrac{1}{2}(y_1 + y_2). \tag{12}$$

Note 1.   In obtaining (11), we agreed that the positive direction on the nonvertical line L was to the *right.* If, instead, the positive direction were to the *left,* this would multiply both numerator and denominator of $\overline{P_1 P}/\overline{P_1 P_2}$ in (11) by -1, and thus leave the equality unaltered. Hence, either direction on L might be considered as the positive direction.

For every pair of points P_1 and P_2 on a line L, a corresponding pair of parametric equations for L is obtained from (4). Thus, L has infinitely many different pairs of parametric equations of type (4).

EXERCISE 60

Write parametric equations for the line through the points. Then, find the midpoint and the points of trisection of the line segment joining the points.

1. $(2, 4)$; $(8, 10)$.   2. $(0, 6)$; $(12, 0)$.   3. $(-2, 3)$; $(-8, -3)$.
4. $(3, 5)$; $(9, 8)$.   5. $(-4, 5)$; $(-10, -2)$.   6. $(2, 3)$; $(6, 11)$.

In Problems 7–9, consider AB to be a directed line, as in the discussion leading to (11) *on page* 264. *With $A:(3, -4)$ and $B:(-5, 2)$, find the point C to satisfy the given condition. Solve by the point of division formulas.*

7. C divides segment AB internally in the ratio $2:3$.
8. $\overline{AC} = 3\overline{AB}$.   9. $\overline{AC} = -3\overline{AB}$.
10. If $(3, 4)$ is the center of a circle through $(-1, 2)$, find another point on the circle. (Consider using $t = -1$ in the parametric equations.)
11. Obtain an xy-equation for the line through the point $(3, 5)$ and the midpoint of the segment joining $(2, -4)$ and $(6, 2)$.
12. If $P:(-1, 2)$ is an endpoint and $Q:(3, 5)$ is the midpoint of a segment PC, find C.
13. If $P:(-2, 3)$ is the center of a circle through $Q:(4, 7)$, find the other endpoint of the diameter through P and Q.
14. If $(2, 6)$, $(1, 2)$, and $(4, 3)$ are consecutive vertices of a parallelogram, find the fourth vertex by use of the point of division formulas. (First find the intersection of the diagonals.)

Note 1.   A **median** of a triangle is a line segment from a vertex to the midpoint of the opposite side. A proof of the following theorem is requested in Problem 20. *"The medians of a triangle meet in a point two thirds of the way from any vertex to the midpoint of the opposite side."* The point of intersection of the medians is called the **centroid** of the triangle.

15. Obtain the equation of each median of the triangle with the vertices $A:(2, 1)$, $B:(8, 9)$, and $C:(4, 5)$. Then, prove that the medians are concurrent (meet in a point) and intersect as described in Note 1.

Obtain an equation for the perpendicular bisector of the line segment joining the points.

16. $(2, -4)$; $(-4, 2)$.   17. $(2, -6)$; $(4, 2)$.

For the triangle with the given vertices, obtain equations for the perpendicular bisectors of the sides. Then prove the lines concurrent.

18. $(-4, -3)$; $(4, 1)$; $(3, 4)$.   19. $(-4, 3)$; $(-2, 7)$; $(-6, 11)$.

★**20.** Prove the theorem stated in Note 1 for any triangle. [Locate the triangle on an xy-plane with one vertex at $(-b, 0)$ and another at $(b, 0)$, etc.]

91. GRAPHS OF RATIONAL FUNCTIONS

Recall that a function $R(x)$ is said to be a **rational function** of x in case there exist polynomials $P(x)$ and $Q(x)$ such that $R(x) = P(x)/Q(x)$, where $P(x)$ and $Q(x)$ have no common factor* (other than a constant) which is a polynomial in x.

ILLUSTRATION 1.   Any polynomial function $P(x)$ is a rational function because $P(x) = P(x)/1$. The function $5x^2/(x + 3)$ is a rational function of x.

Suppose that $R(x) = P(x)/Q(x)$, where we shall suppose that $Q(x)$ is not merely a constant (then $R(x)$ would be merely a polynomial). As a characteristic feature, the graph of $y = R(x)$ has a *vertical asymptote,* with the equation $x = c$, corresponding to each real solution $x = c$ of the equation $Q(x) = 0$. That is, $| R(x) | \rightarrow +\infty$ *as* $x \rightarrow c$. In such a case, it is said that $R(x)$ has a **pole** at $x = c$. Suppose that the equation $P(x) = 0$ has the real solution $x = k$. Then $R(k) = 0$ and the graph of $y = R(x)$ has $x = k$ as an x-intercept. Thus, with $R(x) = P(x)/Q(x)$, the *real zeros* of $P(x)$ are the *zeros of $R(x)$*; the *real zeros of $Q(x)$ are the poles of $R(x)$*.

ILLUSTRATION 2.   The graph of $y = 6/x$ is the hyperbola in Figure 155 with the line $x = 0$ as a vertical asymptote. Thus, the rational function $6/x$ has the pole $x = 0$. From $y = 6/x$, we obtain $xy = 6$ and $x = 6/y$. Hence, the equation also defines x as a rational function of y. With the roles of x and y now interchanged, the graph is seen to have the horizontal line $y = 0$ (the x-axis) as an asymptote. We have

$$\lim_{x \to 0} | y | = \lim_{x \to 0} \left| \frac{6}{x} \right| = +\infty, \textit{ and} \tag{1}$$

$$\lim_{y \to 0} | x | = \lim_{y \to 0} \left| \frac{6}{y} \right| = +\infty. \tag{2}$$

ILLUSTRATION 3.   If $y = 1/(x - 2)^2$, then y is not defined when $x = 2$, and the line $x = 2$ will be an asymptote of the graph of the rational function $1/(x - 2)^2$, as seen in Figure 156. If "x *approaches* 2 *from the right*," abbreviated by "$x \rightarrow 2+$," then $y \rightarrow +\infty$. Also, if "x *approaches* 2 *from the left*," or "$x \rightarrow 2-$," then $y \rightarrow +\infty$, because $(x - 2)^2$ is always

*This assumption will apply whenever the notation $R(x) = P(x)/Q(x)$ is used.

positive. Thus, the graph approaches the asymptote $x = 2$ *upward* from both sides. Also, if $|x| \rightarrow +\infty$ then $y \rightarrow 0$; hence the line $y = 0$, or the x-axis, is a *horizontal asymptote.*

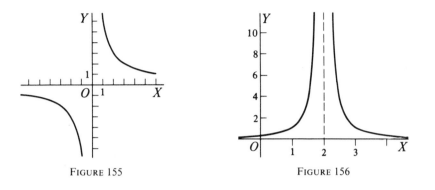

FIGURE 155   FIGURE 156

EXAMPLE 1.   Graph: $x^2y - 9y - 2x^2 + 8 = 0.$ (3)

Solution.   1. The graph will be symmetric to the y-axis because altering x to $-x$ does not affect the equation (it involves only *even* powers of x).

2. Notice that the equation is *linear* in y. Hence, (3) can be solved for y without introducing any radical, to obtain

$$y(x^2 - 9) = 2x^2 - 8, or$$

$$y = \frac{2(x^2 - 4)}{x^2 - 9}.$$ (4)

3. From (4), y is not defined at $x = \pm3$, where $x^2 - 9 = 0$. Thus, y has *poles* at $x = \pm3$; the graph has the lines $x = 3$ and $x = -3$ as vertical asymptotes.

4. From (4), $y = 0$ when $x^2 - 4 = 0$ or $x = \pm2$, which are the x-intercepts of the graph. From (3) with $x = 0$, we obtain $9y = 8$ or $y = 8/9$, which is the y-intercept of the graph.

5. *Horizontal asymptote.*   To investigate y as $|x| \rightarrow +\infty$, first divide the numerator and denominator in (4) by x^2, the *highest power of x in the denominator:*

$$\lim_{|x| \rightarrow +\infty} y = \lim_{|x| \rightarrow +\infty} \frac{2 - \dfrac{8}{x^2}}{1 - \dfrac{9}{x^2}} = \frac{2 - 0}{1 - 0} = 2.$$

Hence, $y \to 2$ as $|x| \to +\infty$, so that the line $y = 2$ is a *horizontal asymptote* of the graph.

6. To obtain the graph in Figure 157, first the asymptotes $x = 3$, $x = -3$, and $y = 2$ were drawn. The following table of coordinates was prepared, with values chosen for x between ± 3, and with $x < -3$ and $x > 3$. Then, the graph was drawn through points from the table to approach the asymptotes, and pass through the intercepts on the axes. The symmetry with respect to the y-axis was made use of in preparing the table of coordinates.

$x =$	$\downarrow -\infty$	-5	-4	-3	-2	0	2	3	4	5	$\uparrow +\infty$
$y =$	2	$\frac{21}{8}$	$\frac{24}{7}$	∞	0	$\frac{8}{9}$	0	∞	$\frac{24}{7}$	$\frac{21}{8}$	2

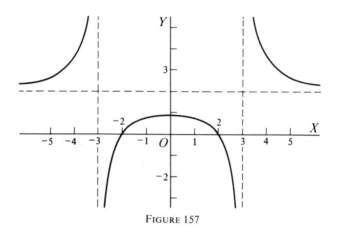

FIGURE 157

Summary. *To graph a rational function $R(x) = P(x)/Q(x)$.*

I. *Find the poles of $R(x)$ to locate the vertical asymptotes, by solving $Q(x) = 0$.*

II. *Divide both numerator and denominator of $R(x)$ by the highest power of x in $Q(x)$, and calculate $\lim_{|x| \to +\infty} R(x)$. If this limit exists and is equal to h, then the line $y = h$ is a horizontal asymptote of the graph of $y = R(x)$.*

III. *With $y = R(x)$, apply tests for symmetry from page 96. Also, obtain any x-intercept by solving $P(x) = 0$.*

IV. *Form a table of coordinates and draw the graph, with the branches approaching the asymptotes smoothly.*

Note 1.   The preceding summary can be applied with the roles of x and y interchanged, as in the next example.

EXAMPLE 2.   Graph: $xy^2 - 6y + 4x = 0.$ (5)

Solution.   1. Since (5) is linear in x, first solve (5) for x in terms of y, to obtain

$$x = \frac{6y}{y^2 + 4}.\tag{6}$$

There is no real solution of $y^2 + 4 = 0$. Hence, the graph of (6) has no horizontal asymptote $y = c$.

2. *To find the intercepts.*   If $y = 0$ then $x = 0$ is obtained as the x-intercept. If $x = 0$, then $y = 0$. Hence each intercept is zero, or the curve goes through the origin.

3. *Symmetry.*   If x is replaced by $-x$, and y by $-y$, then (6) is unaltered (except for multiplication by -1). Hence, the graph, C, of (6) is symmetric to the origin, as seen in Figure 158.

4. On dividing both numerator and denominator by y^2 in (6), we obtain

$$\lim_{|y| \to +\infty} x = \lim_{|y| \to +\infty} \frac{6y}{y^2 + 4} = \lim_{|y| \to +\infty} \frac{\dfrac{6}{y}}{1 + \dfrac{4}{y^2}} = \frac{0}{1} = 0.$$

Therefore the line $x = 0$ is an asymptote, approached as $y \to +\infty$ and also as $y \to -\infty$.

5. Figure 158 was based on the following table of coordinates, obtained from (6). The curve was drawn to approach the asymptote $x = 0$ smoothly.

$y =$	-8	-3	-2	-1	0	1	2	3	8
$x =$	$-\frac{12}{17}$	$-\frac{18}{13}$	$-\frac{3}{2}$	$-\frac{6}{5}$	0	$\frac{6}{5}$	$\frac{3}{2}$	$\frac{18}{13}$	$\frac{12}{17}$

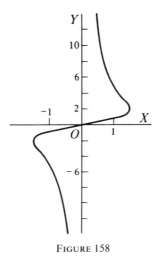

FIGURE 158

Note 2.   If a rational function $R(x) = P(x)/Q(x)$ is given, with the degree of P greater than or equal to the degree of Q, it may be desirable to divide $P(x)$ by $Q(x)$ by long division before graphing. However, the routine of the preceding Summary applies without carrying out the long division. By experience, it will be observed that a horizontal asymptote for the graph of R is obtained when the degree of P is less than or at most equal to the degree of Q.

EXERCISE 61

Graph the function.

1. $\dfrac{1}{x^2}$.   **2.** $\dfrac{1}{x^3}$.   **3.** $\dfrac{1}{x - 2}$.   **4.** $\dfrac{1}{(x + 1)^2}$.

Find the limit of the fraction as $|x| \to +\infty$.

5. $\dfrac{3x - 2}{2x + 5}$.   **6.** $\dfrac{x^2 + 3}{2x^2 - 5}$.   **7.** $\dfrac{2x - 1}{x^2 + 1}$.   **8.** $\dfrac{3x^2 + 2}{x^3 - 5}$.

Graph the equation. If necessary, start by solving for one variable in terms of the other, or possibly for each variable in terms of the other variable, without introducing any radical.

9. $xy = -8$.   **10.** $x^2 y = 1$.   **11.** $xy^2 = 1$.

12. $y = \dfrac{1}{(x - 1)(x + 2)}$.   **13.** $y = \dfrac{x^2 - 9}{4 - x^2}$.   **14.** $x = \dfrac{1}{(y + 1)(y - 2)}$.

15. $xy + x = 8 - 4y$.   **16.** $x^2 y - 4x + 4y = 0$.

17. $x + xy^2 = 4$.   **18.** $3x - xy = 5y$.

19. $yx^2 - 2y + xy = 2x^2$.   **20.** $y(x - 1)(x - 3)(x + 3) = 1$.

92. PARAMETRIC EQUATIONS OF A CURVE

With t as a variable, consider the system of equations

$$x = g(t) \quad \text{and} \quad y = h(t), \tag{1}$$

where the domain for t is some interval. For each value of t, a pair of numbers (x, y), and hence a point $P:(x, y)$ in the xy-plane is produced. The set of all points obtained in this way is referred to as the graph, C, of (1). Also, the equations in (1) are said to be *parametric equations* for C with t as the parameter.

In Section 90 we considered systems of type (1) where f and g were linear functions of t, and the graph of (1) was a line.

Sometimes it is possible to eliminate t in (1) by solving one of the equations for t in terms of x, or of y, and substituting the result for t in the other equation. If this is done, the result is an *xy*-equation* for the curve defined by (1). Other means for eliminating *t* will occur, as in the next example.

EXAMPLE 1.   Obtain an *xy*-equation for the curve defined as follows, with *t* as the parameter:

$$x = 3 \cos t \qquad and \qquad y = 2 \sin t. \tag{2}$$

Solution.   1. From (2), $\cos t = \frac{1}{3}x$ and $\sin t = \frac{1}{2}y$. Hence,

$$\sin^2 t + \cos^2 t = 1 = \tfrac{1}{9}x^2 + \tfrac{1}{4}y^2, or \tag{3}$$

$$4x^2 + 9y^2 = 36. \tag{4}$$

Therefore, (2) gives parametric equations for the ellipse whose *xy*-equation is in (4).

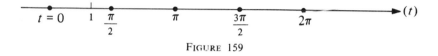

FIGURE 159

2. For any value of *t*, let P_t be the point whose coordinates are found from (2). By assigning selected increasing values to *t* in (2), corresponding coordinates for P_t are obtained as given in the following table, which is the basis for the graph of the ellipse in Figure 160. If *t* moves continuously on the *t*-scale in Figure 159 from $t = 0$ to $t = 2\pi$, then P_t moves

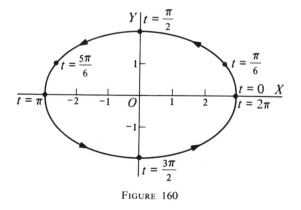

FIGURE 160

*Sometimes the graph of the resulting *xy*-equation includes not only the graph of (1), but also other curves. Hence, any result of elimination of *t* must be examined with care to see whether or not the equation is equivalent to (1).

continuously counterclockwise around the ellipse. In Figure 160, values of t from the table are beside corresponding points P_t, and arrowheads on the curve indicate how P_t moves as t increases. We say that equations (2) *map* the t-axis on the ellipse. By the periodicity of the sine and cosine, (2) produces all points of the ellipse if t has a domain which is an interval of length 2π. The preceding method for finding points on the ellipse has advantages over the method employing the xy-equation (4) to compute coordinates.

$t =$	0	$\frac{1}{6}\pi$	$\frac{1}{2}\pi$	$\frac{5}{6}\pi$	π	$\frac{3}{2}\pi$	2π
$y =$	0	1	2	1	0	-2	0
$x =$	3	2.6	0	-2.6	-3	0	3

Note 1.   It is not always convenient, or possible, to obtain an xy-equation for a curve C in an xy-plane with given parametric equations for C. Then, the method of Example 1 for obtaining C is the only method which applies in obtaining C.

If a curve C can be defined by one set of parametric equations as in (1), then C can be represented parametrically by infinitely many distinct pairs of parametric equations. With two different parametric representations as in (1), the domains of the corresponding parameters in general would be different. Thus, if we should replace t by $2t$ in (2), a new parametric form would result. A study of the effects of a change in the parameter will not be considered in this text.

EXAMPLE 2.   Graph the curve, W, defined parametrically by

$$x = 6t - \tfrac{1}{2}t^3 \qquad and \qquad y = \tfrac{1}{2}t^2 - 6 \ . \tag{5}$$

Solution.   1. For any value of t, let P_t be the point (x, y) on W given by (5).

2. *Values of t when $y = 0$*: from (5),

$$\tfrac{1}{2}t^2 - 6 = 0, \qquad or \qquad t^2 = 12; \qquad t = \pm 2\sqrt{3}.$$

From (5), $x = -ty$. Hence $x = 0$ when $t = \pm 2\sqrt{3}$; for each of these values, P_t is the origin, $(0, 0)$.

3. *Values of t when $x = 0$*: from (5),

$$6t - \tfrac{1}{2}t^3 = 0, \qquad or \qquad t\!\left(6 - \tfrac{1}{2}t^2\right) = 0; \, hence$$

$$t = 0 \qquad or \qquad 6 - \tfrac{1}{2}t^2 = 0, \qquad and \qquad t = \pm 2\sqrt{3}.$$

When $t = 0$, from (5) we find $y = -6$; P_t is the point $(0, -6)$ when $t = 0$.

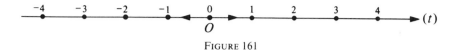

FIGURE 161

4. The domain for t is the set of all real numbers on the t-axis, as shown in Figure 161. If the point t moves continuously on the t-axis, the corresponding point P_t from (5) will move continuously on the curve W. First, in (5), we substitute selected positive values of t in increasing order, starting at $t = 0$, to obtain coordinates of points P_t. We repeat this process for negative values of t in decreasing order. The points thus obtained, as given in the following table, are the basis for W in Figure 162. Some values used for t are shown in Figure 162 beside the corresponding points on W, to indicate how (5) maps the t-axis on W.

$t =$	0	1	2	3	$2\sqrt{3}$	4	5
$y =$	-6	$-\frac{11}{2}$	-4	$-\frac{3}{2}$	0	2	6.5
$x =$	0	$\frac{11}{2}$	8	$\frac{9}{2}$	0	-8	-32.5
$t =$	0	-1	-2	-3	$-2\sqrt{3}$	-4	-5
$y =$	-6	$-\frac{11}{2}$	-4	$-\frac{3}{2}$	0	2	6.5
$x =$	0	$-\frac{11}{2}$	-8	$-\frac{9}{2}$	0	8	32.5

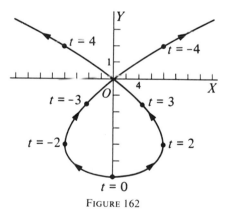

FIGURE 162

EXERCISE 62

By use of the given parametric equations for a curve W, obtain the coordinates of points on W and construct W. Place values of the parameter t at the corresponding points on W, to show how the given equations map the domain of t on W. Also find an xy-equation for W. Where trigonometric

*functions occur, t radians may be replaced by u° in order to use trigono-
metric tables.*

1. $x = 2 \sin t$ and $y = 2 \cos t$.   **2.** $x = 2 \sin t$ and $y = 5 \cos t$.
3. $x = 3 \cos t$ and $y = 4 \sin t$.   **4.** $x = 2 - \cos t$ and $y = 2 + \sin t$.
5. $x = 2 + 2 \sin t$ and $y = 1 + 3 \cos t$.
6. $x = t^2$ and $y = t^3$.   **7.** $x = t^3$ and $y = t^2$.
8. $x = \dfrac{2}{\sqrt{t}}$ and $y = 2\sqrt{t}$. (Is this curve the same as the whole graph of

the xy-equation obtained by eliminating t?)
9. $x = 12 - t^2$ and $y = 12t - t^3$.
10. $x = \cos^3 t$ and $y = \sin^3 t$. The curve is called a *hypocycloid* with four
cusps.
★11. With t having the domain $\{-2\pi \leq t \leq 2\pi\}$, obtain the curve
defined by

$$x = t - \sin t \qquad and \qquad y = 1 - \cos t.$$

The graph is called a *cycloid.* Any attempt to find an xy-equation
for the curve would be useless. On the graph, show how the equations map the domain of t on the curve. It is the path traced by any
point on the circumference of a circle if it rolls on a straight line in a
plane.

93. INEQUALITIES IN TWO VARIABLES

A solution of an inequality $f(x, y) < g(x, y)$ is a pair of numbers
(x, y) for which the inequality becomes a true statement. The graph of
the inequality in an xy-plane is *the graph of the solution set* of the inequality. Usually, it will have infinitely many solutions, and hence we
cannot expect to describe them analytically. As a rule, *to solve* an inequality $f(x, y) < g(x, y)$ will mean *to draw its graph,* which then serves
to describe the solutions geometrically.

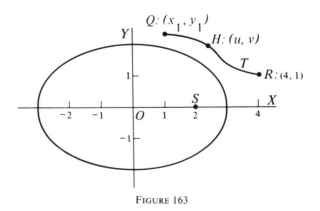

FIGURE 163

ILLUSTRATION 1.   Let $f(x,y) = 4x^2 + 9y^2 - 36$. Then $f(x,y) = 0$ becomes $4x^2 + 9y^2 = 36$, whose graph is the ellipse in Figure 163. Consider the inequality

$$4x^2 + 9y^2 - 36 < 0, \quad or \quad f(x,y) < 0. \tag{1}$$

In Figure 163, observe that $R:(4,1)$ is outside the ellipse, and

$$f(4,1) = 4(16) + 9 - 36 = 37 > 0. \tag{2}$$

Hence, at R, the inequality (1) is *not* satisfied.   Now suppose that $Q:(x_1, y_1)$ in Figure 163 is any point *outside* the ellipse.   Let T be a continuous* curve joining R and Q and NOT intersecting the ellipse.   Let $H:(u,v)$ be any point on T.   If H moves on T from R to Q, the value $f(u,v)$ at H changes continuously from $f(4,1) > 0$ when $(u = 4, v = 1)$, to $f(x_1, y_1)$ at Q.   Also, $f(u,v) \neq 0$ at any point H because T *does not meet the ellipse,* where $f(x,y) = 0$.   Therefore $f(x_1, y_1) > 0$, because $f(u,v)$ could not change continuously from a *positive* value $f(4,1)$ at R to a *negative* value $f(x_1, y_1)$ at Q without passing through the value 0. Hence, $f(x,y) > 0$ at each point $P:(x,y)$ *outside* the ellipse.   Or, the points outside the ellipse are in the graph of $f(x,y) > 0$.   Similarly, at $S:(2,0)$ *inside* the ellipse, we find $f(2,0) = 16 - 36 = -20$.   Hence, $(x = 2, y = 0)$ is a solution of $f(x,y) < 0$.   Then, by continuity reasoning as before, $f(x,y) < 0$ at each point $P:(x,y)$ *inside* the ellipse.   Thus, the graph of $f(x,y) > 0$ is the set of points *outside the ellipse;* the graph of $f(x,y) < 0$ is the set of points *inside the ellipse.*   The *boundary* between the graphs of $f(x,y) > 0$ and $f(x,y) < 0$ is the *ellipse* where $f(x,y) = 0$.

On the basis of continuity reasoning as in Illustration 1, we accept the following procedure for graphing an inequality $f(x,y) < g(x,y)$, or $f(x,y) > g(x,y)$.   We assume that the values of $f(x,y)$, and of $g(x,y)$, change continuously† if $P:(x,y)$ moves in a continuous fashion in the xy-plane.

Summary. *To graph $f(x,y) < g(x,y)$, or $f(x,y) > g(x,y)$.*

I. *Obtain the graph, T, of $f(x,y) = g(x,y)$, where we assume that T divides the xy-plane into disjoint sets‡ of points, say $\{W_1, W_2, W_3\}$, separated by boundaries on T.*

II. *Select any point $P:(x_0, y_0)$ in W_1, and substitute $(x = x_0, y = y_0)$ in $f(x,y) < g(x,y)$.   If $f(x_0, y_0) < g(x_0, y_0)$, then all of W_1 is part of the graph of $f(x,y) < g(x,y)$.   If $f(x_0, y_0) > g(x_0, y_0)$, then W_1 is in the graph of $f(x,y) > g(x,y)$.*

---

*That is, a curve without breaks.
†That is, f and g are "*continuous functions,*" in the sense described formally in calculus, and accepted intuitively here.
‡*If* P_1 and P_2 are points in any one of these sets, we assume that P_1 and P_2 can be connected by a continuous curve lying in the set.

III. *Repeat the action in* II *with* W_2 *and* W_3 *to find all parts of the graphs of* $f(x,y) < g(x,y)$, *and of* $f(x,y) > g(x,y)$.

An inequality $f(x,y) < g(x,y)$ is said to be a *polynomial inequality* in case each of $f(x,y)$ and $g(x,y)$ is a polynomial or is zero. Such an inequality is said to be *linear* in case it is equivalent to $ax + by + c < 0$ where a and b are not both zero.

If an inequality $f(x,y) < g(x,y)$ is linear, the graph of $f(x,y) = g(x,y)$ is a line, L, separating the xy-plane into two mutually exclusive sets of points with L as a common boundary. Each of these sets is called an **open half-plane.** The method of the Summary will show that one of these open half-planes is the graph of $f(x,y) < g(x,y)$, and the other open half-plane is the graph of $f(x,y) > g(x,y)$. The *union* of either one of these open half-planes and the boundary line L is called a **closed half-plane,** and it is the graph of $f(x,y) \leqq g(x,y)$, or of $f(x,y) \geqq g(x,y)$.

Note 1.   The Summary is valid for $f(x,y) < g(x,y)$ if it is equivalent to $H(x) < K(x)$, where y does not occur, or to $H(y) < K(y)$, where x does not occur.

EXAMPLE 1.   Obtain the graph of $3x + 2y > 6.$ (3)

Solution.   1. The line $L:\{3x + 2y = 6\}$ is shown in Figure 164.

2. *Test* (3) *at* $Q:(0,2)$. When $(x = 0, y = 2)$ in (3), we obtain $4 > 6$, which is *false,* but $4 < 6$ is *true.* Hence $3x + 2y < 6$ at each point $P:(x,y)$ in the open half-plane *below L.*

3. *Test* (3) *at* $P:(3,0)$. When $(x = 3, y = 0)$ in (3), we obtain $9 > 6$, which is *true.* Hence, (3) is satisfied at each point $P:(x,y)$ *above L.* Or, the graph of (3) is the open half-plane *above L,* as shown by the ruled region in Figure 164.

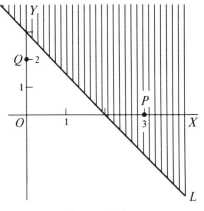

FIGURE 164

Comment. The graph of $3x + 2y \geq 6$ is the union of L, where $3x + 2y = 6$, and the graph of $3x + 2y > 6$. Or, the graph of the inequality $3x + 2y \geq 6$ is the closed half-plane consisting of L and the open half plane above L.

EXAMPLE 2. Obtain the graph of the system

$$\begin{cases} 3x - 2y \leq 6, \text{ and} & (4) \\ x + y > 3. & (5) \end{cases}$$

Solution. In Figure 165, the graph of $3x - 2y = 6$ is line AB. The graph of (4) is the closed half-plane *above* and including AB, and is indicated by horizontal rulings. The graph of $x + y = 3$ is CD, and the graph of (5) is the open half-plane *above* CD, indicated by vertical rulings. The graph of system [(4), (5)] is the *intersection* of the graphs of (4) and (5), or is the set of points which are *doubly ruled,* including a ray of the line $3x - 2y = 6$, because of "$=$" in (4), but not including any part of the line $x + y = 3$.

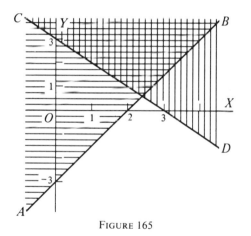

FIGURE 165

In preceding remarks and examples, the following facts about linear inequalities were emphasized.

Any open half-plane T is the graph of a linear inequality

$$ax + by + c < 0, \tag{6}$$

and the boundary of the half-plane is the line $L:\{ax + by + c = 0\}$. *The union of this line and the open half-plane T is a closed half plane which is the graph of*

$$ax + by + c \leq 0. \tag{7}$$

In a plane, the intersection, W, of two or more closed half-planes is called a **polygonal set** of points. If W has a finite area,* then the boundary of W is called a **convex†** **polygon,** and W is called a *bounded polygonal set of points.*

EXAMPLE 3.   Find the polygonal set which is the graph of the following system of inequalities:

$$y - x - 3 \leq 0, \tag{8}$$

$$y \leq 4, \tag{9}$$

$$x \leq 4, and \tag{10}$$

$$2y + x \geq 2. \tag{11}$$

Solution.   1. In Figure 166, the four lines are shown whose equations are obtained by using only the signs of equality in (8)–(11). These lines intersect at A, B, C, and D. Let W be the set of points which is the intersection of the closed half planes which are the graphs of (8)–(11).

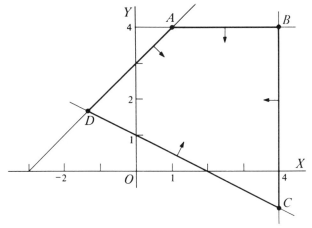

FIGURE 166

2. Temporarily disregard points in the xy-plane which lie on the lines satisfying (8)–(11) with equality signs. Then, W lies *below AB*, as shown by a short arrow in Figure 166, and to the *left* of BC. On substituting $(x = 0, y = 0)$ on the left in (8), we obtain $-3 \leq 0$, which is true. Thus, the graph of (8) is the half-plane on the side of the origin, or *below* the

*That is, if W lies within some rectangle, however large.
†It can be verified geometrically that, if P and Q are any two points in a polygonal set T, then all points of line segment PQ are in T. A set with this property is called a **convex set** of points.

line $y - x - 3 = 0$, which is AD in Figure 166. With $(x = 0, y = 0)$ in (11), we obtain $0 \geq 2$, which is *false*. Hence, (11) is true on the other side of DC, or *above* that line, as shown by an arrow. Therefore, the intersection of the graphs of (8)–(11) is the polygon $ABCD$ and its interior.

Note 2. Polygonal sets in a plane are of fundamental importance in the study of *linear programming* with two independent variables. A typical problem of linear programming in a simplified setting is as follows: "A drug company will produce a cough syrup containing an anti-histamine A, a barbiturate, B, and an aspirin compound, C. In the mixture, by weight, at least 20% but not more than 45% will be A; at least 30% will be B; the amount of B will be greater than the amount of A. The amount of C plus the amount of A must exceed 32% of the syrup. The costs of A, B, and C per ounce are $6, $4, and $5, respectively. What percentages of A, B, and C, respectively, should be used to minimize the cost?" With $\{x, y, z\}$ as the percents involved, the data produce a system of inequalities involving $\{x, y, z\}$, where $z = 100 - x - y$, defining a polygonal set in the xy-plane. The boundary of the set is a certain polygon. Then, in the theory of linear programming, it is proved that the desired minimum cost results from taking (x, y, z) as the numbers corresponding to some vertex of the polygon just mentioned.*

EXERCISE 63

Exhibit the solution set of the inequality, or system of inequalities, graphically. If the graph is a polygon and its interior, find the vertices of the polygon.

1. $x^2 + y^2 < 4.$ **2.** $4x^2 + 9y^2 > 36.$ **3.** $2x - 3y \leq 6.$
4. $x - 4y > 4.$ **5.** $2y + x \leq 6.$ **6.** $x - 3y > 0.$
7. $y \leq x, x \geq 0$, and $y + x - 2 \leq 0.$
8. $x \geq -2, y \geq 2$, and $y - x \geq 0.$
9. $x - y - 2 \leq 0, x + y - 2 \geq 0$, and $x - 2y + 1 \geq 0.$
10. $x \leq 6, x - y + 3 \geq 0, x + y \geq 3$, and $x + 2y \leq 12.$
11. $2x + y \leq 2, 2y + 1 \geq x, 2y - x \leq 4$, and $3y - 4x \leq 11.$
12. $2y - x \geq 0, y + x \geq 0, 4x + 7 \geq 3y$, and $2y + 3x \leq 16.$
13. $x^2 - 4y < 0.$ **14.** $y^2 - 2x > 0.$ **15.** $9x^2 + 25y^2 \leq 100.$
16. $4x^2 + y^2 \leq 16, and$ **17.** $y^2 \geq 12x, and$
   $x^2 + y^2 - 9 \geq 0.$ $9x^2 + y^2 < 9.$
★18. Solve the problem stated in Note 2 on this page.

*See William L. Hart, *Algebra, Elementary Functions, and Probability* (Lexington, Mass.: D. C. Heath and Company, 1965), Chap. 15.

94. A SYSTEM OF POLAR COORDINATES

In a given plane, the position of any point can be designated by specifying its direction and distance from a selected fixed point. This fact supplies a background for use of so-called **polar coordinates.**

In the plane, select a point, *O*, to be called the **pole,** and a fixed half-line or ray *OI*, to be called the **initial ray.** Let the whole line through *OI* be called the **polar axis,** shown as horizontal in the typical Figure 167. Let *P* be any point not the pole. Let *θ* be any angle whose initial side is *OI* and terminal side falls on line *OP*. Let *r* be the corresponding value of the directed distance $\overline{OP}$, considered positive or negative according as *P* is on the terminal side of *θ* or on the extension of this side back through the pole *O*. Then, we call *θ* a **polar angle** for *P*, and *r* the corresponding **radius vector** for *P*. The pair [*r, θ*], where we shall use square brackets, will be referred to as a set of **polar coordinates** for *P*. The pole *O* is assigned the coordinates [0, *θ*], where *θ* may be any angle. The notation *P*:[*r, θ*] can be read "*P with polar coordinates r and θ*," or simply "*P, r, θ*."

To plot an assigned point *P*:[*r, θ*], lay off *θ* by rotation from the initial ray *OI*, and then locate *P* on the terminal side of *θ* by use of the value of *r*.

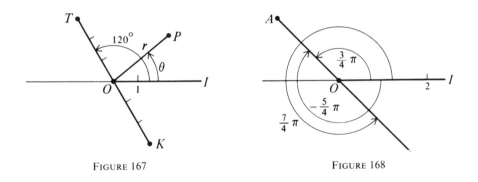

FIGURE 167   FIGURE 168

ILLUSTRATION 1.   Figure 167 shows *T*:[3, 120°] and *K*:[−3, 120°]. Figure 168 shows $A:[2,\frac{3}{4}\pi]$, with radian measure for the angle. Other coordinates for *A* are $[2, -\frac{5}{4}\pi]$; $[-2,\frac{7}{4}\pi]$. The following coordinates locate the same point: [3, 135°]; [3, −225°]; [−3, 315°]. In this way, we may give an unlimited number of sets of polar coordinates for any assigned point.

Observe that any point *P*, not the pole, has just one set of polar co-ordinates [*r, θ*] where 0° ≤ *θ* < 360° and *r* > 0. Then $r = |\overline{OP}|$, the length of *OP*. However, it is impossible to avoid use of *negative* values for *r* in the applications of polar coordinates [*r*, θ]. When allowable, usually we shall select [*r*, θ] for any point *P* so that 0 ≤ θ < 360° and *r* ≥ 0.

95. GRAPH* OF AN EQUATION IN *r* AND θ

The *graph* of an equation $G(r, \theta) = 0$ consists of the set of points where, for each point *P* in the set, there exists *at least* one pair of coordinates [*r*, θ] which form a solution of $G(r, \theta) = 0$. An equation for a set, *T*, of points in an *r*θ-plane is an equation $G(r, \theta) = 0$ whose graph is *T*.

ILLUSTRATION 1.   An equation for the circle with radius *h* and center at the pole, as in Figure 169, is *r* = *h*. For any angle, α, the point *P*:[*h*, α] is on the circle. Another equation for this circle is *r* = −*h*, because *P*:[−*h*, α] traces out the circle if α varies from α = 0 to α = 2π. This illustrates the fact that two equations, such as *r* = *h* and *r* = −*h*, may have the same graph in polar coordinates although the equations are *not algebraically equivalent*.

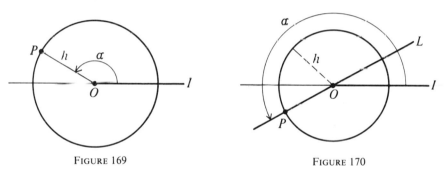

| FIGURE 169 | FIGURE 170 |

ILLUSTRATION 2.   If an angle from the polar axis *OI* to a line *L* through the pole is α, then an equation for *L* is θ = α, as in Figure 170. Another equation for *L* would be tan θ = tan α. In Figure 170, notice that the point *P*:[*h*, α] is an intersection of the circle *r* = *h* and the line θ = α.

The following fact will be proved later.

> *For any constant k ≠ 0, the graph of r = k sin θ is a* **circle** *tangent to the polar axis at the pole, with radius* | *k* | /2. *The circle is above the axis if k > 0, and below if k < 0.*   (1)

---

*For convenience in graphing, special coordinate paper for use with polar coordinates might be used. Otherwise, a protractor should be available for measuring angles. Usually, such paper will mark angles in degree measure. Hence, we shall use that measure frequently in graphing.

EXAMPLE 1.   Graph the equation: $r = 4 \sin \theta.$ (2)

Solution.   The following table of coordinates of points on the graph of (2) was the basis for the graph in Figure 171. Notice that all points of the circle are obtained as θ takes on all values on the interval $\{0° \leq \theta \leq 180°\}$. If $\{180° \leq \theta \leq 360°\}$, the corresponding points $\{r, \theta\}$ duplicate those obtained for $\{0° \leq \theta \leq 180°\}$. Thus, with $\theta = 30°$, from (2) we obtain $r = 4 \sin 30° = 2$, and the point $H:[2, 30°]$ as seen in Figure 171. If $\theta = 210°$, then $r = 4 \sin 210° = -4 \sin 30° = -2$. Notice that the points $[2, 30°]$ and $[-2, 210°]$ are identical.

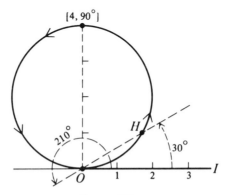

FIGURE 171

$\theta =$	0°	30°	60°	90°	120°	150°	180°	210°	270°	330°	360°
$r =$	0	2	3.5	4	3.5	2	0	−2	−4	−2	0

The following fact will be proved later.

> *For any constant $k \neq 0$, the graph of $r = k \cos \theta$ is a circle tangent to the line $\theta = 90°$, or $\theta = \frac{1}{2}\pi$, at the pole with radius $|k|/2$. The circle is to the right of the line $\theta = 90°$ if $k > 0$, and to the left if $k < 0$.*   (3)

ILLUSTRATION 3.   The student will verify later that the upper semi-circle of the circle $r = 2 \cos \theta$ is obtained with $\{0 \leq \theta \leq 90°\}$, and the lower semicircle with $\{90° \leq \theta \leq 180°\}$, or with $\{-90° \leq \theta \leq 0°\}$.

ILLUSTRATION 4.   The equation $\cos \theta = \frac{1}{2}$ is satisfied if $\theta = \frac{1}{3}\pi$ or if $\theta = -\frac{1}{3}\pi$. Hence, the graph of $\cos \theta = \frac{1}{2}$ consists of the two lines $\theta = \frac{1}{3}\pi$ and $\theta = -\frac{1}{3}\pi$.

EXERCISE 64

Plot the point whose polar coordinates are given. Also, give one other set of coordinates $[r, \theta]$ for the point with $0° \leq \theta < 360°$.
1. $[3, 150°]$.   **2.** $[2, 240°]$.   **3.** $[3, 400°]$.   **4.** $[-1, 90°]$.

5. $[2, \frac{3}{4}\pi]$. **6.** $[1, \frac{3}{2}\pi]$. **7.** $[-1, \frac{1}{2}\pi]$. **8.** $[2, \frac{5}{4}\pi]$.

Graph the equation in polar coordinates $[r, \theta]$.

9. $r = 2$. **10.** $\theta = \frac{1}{3}\pi$. **11.** $\theta = -\frac{3}{4}\pi$.

12. $r = 3\sin\theta$. **13.** $r = -2\sin\theta$. **14.** $r = 2\cos\theta$.

15. $r = -2\cos\theta$. **16.** $\tan\theta = 1$. **17.** $\cos\theta = \frac{1}{2}$.

18. $\sin\theta = -\frac{1}{2}$. **19.** $\cos\theta = \frac{1}{2}\sqrt{2}$. **20.** $\tan\theta = -\sqrt{3}$.

96. RELATIONS CONNECTING POLAR AND RECTANGULAR COORDINATES

Consider a plane supplied with a *pole O* and *polar axis OI* for an $r\theta$-system of polar coordinates. In this plane, create an xy-system of rectangular coordinates with the origin at the pole; the polar axis as the x-axis, with its positive direction as the direction of the ray OI of the polar axis; the y-axis oriented 90° from the x-axis as usual. The resulting coordinates $[r, \theta]$ and (x, y) for a point P in the plane are shown in Figure 172. We apply the distance formula of page 57 to $\overline{OP}$, and recall the definitions of the trigonometric functions on page 158. Then, with the understanding that* $r > 0$, we have $\sin\theta = y/r$ and $\cos\theta = x/r$;

$$x = r\cos\theta; \qquad y = r\sin\theta, \tag{1}$$

$$\frac{y}{x} = \tan\theta; \qquad x^2 + y^2 = r^2. \tag{2}$$

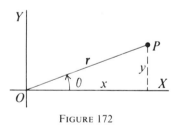

FIGURE 172

Relations (1) and the equation $r^2 = x^2 + y^2$ are true for all (x, y) and the corresponding $[r, \theta]$. The formula $\tan\theta = y/x$ applies when $x \neq 0$. If coordinates (x, y) are given, θ can be found from (1) and (2) by use of a trigonometric table.

ILLUSTRATION 1. To find (x, y) for $P:[r = 2, \theta = \frac{2}{3}\pi]$: from (1),

$$x = 2\cos\tfrac{2}{3}\pi = 2(-\tfrac{1}{2}) = -1; \qquad y = 2\sin\tfrac{2}{3}\pi = 2(\tfrac{1}{2}\sqrt{3}) = \sqrt{3}.$$

*With moderate discussion we could prove (1) and (2) if $r < 0$.

ILLUSTRATION 2.   To obtain $[r, \theta]$ for $P:(3, -4)$, we use (1) and (2):

$$r^2 = 9 + 16 = 25, \text{ or } r = 5; \qquad \sin \theta = \frac{y}{r} = -\frac{4}{5} = -.8.$$

Also, θ is in quadrant IV. With $\sin \alpha = .8$, Table IV gives $\alpha = 53.1°$. Hence $\theta = 360° - 53.1° = 306.9°$. Thus, polar coordinates for P are $[5, 306.9°]$. Another pair would be $[5, -53.1°]$.

Note 1.   Whenever an xy-system and an $r\theta$-system are said to be associated, or are considered together in a problem, we shall assume that Figure 172 and relations (1) and (2) apply.

If a curve in an xy-plane has an equation $f(x, y) = 0$, we can obtain a corresponding $r\theta$-equation by substituting for (x, y) from (1) into $f(x, y) = 0$.

ILLUSTRATION 3.   To obtain an $r\theta$-equation for the circle whose equation is $x^2 + y^2 = h^2$, we use (1) and find

$$r^2 \sin^2 \theta + r^2 \cos^2 \theta = h^2, \text{ or}$$
$$r^2 (\sin^2 \theta + \cos^2 \theta) = h^2, \qquad \text{or} \qquad r^2 = h^2.$$

Hence, $r = +h$ or $r = -h$. We have seen previously that either one of these equations is an equation for the given circle.

ILLUSTRATION 4.   By use of (1), the lines of the form $x = a$ and $y = b$ become, respectively,

$$r \cos \theta = a \qquad and \qquad r \sin \theta = b.$$

It is much more convenient as a rule to change an equation $f(x, y) = 0$ to a form $G(r, \theta) = 0$ by use of (1), than to change a given form $G(r, \theta) = 0$ to a corresponding form $f(x, y) = 0$. In our limited treatment of polar coordinates, we shall emphasize the change from $f(x, y) = 0$ to $G(r, \theta) = 0$.

ILLUSTRATION 5.   To transform the equation $r = 2a \cos \theta$ into rectangular coordinates, first multiply both sides by r to obtain the form $r^2 = 2ar \cos \theta$. Then, from (1) and (2),

$$x^2 + y^2 = 2ax, \text{ or} \tag{3}$$
$$(x^2 - 2ax + a^2) + y^2 = a^2, \qquad \text{or} \qquad (x - a)^2 + y^2 = a^2. \tag{4}$$

Hence the graph of $r = 2a \cos \theta$ and of (4) is a circle with center $(x = a, y = 0)$ and radius $|a|$. Or, if $a > 0$, the graph of $r = 2a \cos \theta$ is a circle whose center is the point $[r = a, \theta = 0]$ and radius is a. This fact was accepted previously without proof in (3) on page 282. In multiplying both sides of $r = 2a \cos \theta$ by r in the preceding details, we added *just the pole,* where $r = 0$, to the set of solutions as originally present.

Since the pole already is a point on the graph of $r = 2a \cos \theta$ (with $\theta = \frac{1}{2}\pi$), the graphs of $r = 2a \cos \theta$ and $r^2 = 2ar \cos \theta$ are the same. Hence our preceding remarks were justified.

EXERCISE 65

Plot the point with the given polar coordinates and find its rectangular coordinates. Use Table IV if necessary.

1. $[2, 30°]$.   **2.** $[3, 315°]$.   **3.** $[1, \frac{1}{2}\pi]$.   **4.** $[-1, \pi]$.

5. $[3, 120°]$.   **6.** $[2, 200°]$.   **7.** $[-2, 210°]$.   **8.** $[3, \frac{4}{3}\pi]$.

Plot the point with the given rectangular coordinates. Then find two sets of polar coordinates for the point. Use Table IV if necessary.

9. $(1, 1)$.   **10.** $(-2, 2)$.   **11.** $(1, -\sqrt{3})$.

12. $(2\sqrt{3}, -2)$.   **13.** $(-3, -3\sqrt{3})$.   **14.** $(-3, -\sqrt{3})$.

15. $(-\sqrt{3}, 1)$.   **16.** $(8, 15)$.   **17.** $(-4, 3)$.

First transform the equation into rectangular coordinates. Then graph the equation. Use the $r\theta$ form in Problems 22–25.

18. $r^2 = 16$.   **19.** $\tan \theta = 2$.   **20.** $r \cos \theta = 2$.

21. $r \sin \theta = -3$.   **22.** $r = 4 \cos \theta$.   **23.** $r = 6 \sin \theta$.

24. $r = -6 \cos \theta$.   **25.** $r = -2 \sin \theta$.   **26.** $\cot \theta = -3$.

Graph the equation. Also, find an equation for the graph in polar coordinates.

27. $y = 3x$.   **28.** $x = 3$.   **29.** $x^2 + y^2 = -4y$.

30. $x^2 + y^2 = 6x$.   **31.** $x^2 + 4y^2 = 4$.   **32.** $y^2 = 9 + 6x$.

33. Prove the fact specified by (1) on page 281.

★97. CERTAIN CURVES IN POLAR COORDINATES

Suppose that the function $f(\theta)$ is periodic with the period 2π, where θ represents an angle in radian measure (which can be replaced by the measure in degrees if desired). That is, for all values of θ,

$$f(\theta + 2\pi) = f(\theta).$$

Then, if the equation $r = f(\theta)$ is graphed in polar coordinates for* $\{0 \leq \theta \leq 2\pi\}$, all points of the graph of the equation will be obtained, with the point for $\theta = 0$ the same as the point for $\theta = 2\pi$. It may happen that part of the graph will be given more than once with θ on the specified interval.

*Or, for $\{-\pi \leq \theta \leq \pi\}$, or for any interval of length 2π.

EXAMPLE 1.   Graph: $r = 3 \sin 2\theta.$ (1)

Solution.   1. Recall that the function $\sin 2\theta$ has the period π:

$$\sin 2\theta = \sin 2(\theta + \pi) = \sin (2\theta + 2\pi).$$

Hence, the function $\sin 2\theta$ also has the period 2π. Thus, all of the graph will be obtained if θ has the domain $\{0 \leq \theta \leq 2\pi\}$.

2. From (1), the maximum value of r occurs when $\sin 2\theta = 1$. The solutions of this equation, with $0 \leq 2\theta \leq 4\pi$:

$$for\ 2\theta:\quad \tfrac{1}{2}\pi\ and\ \tfrac{5}{2}\pi.$$
$$for\ \theta:\quad \tfrac{1}{4}\pi\ and\ \tfrac{5}{4}\pi.$$

The minimum value $r = -3$ occurs when $\sin 2\theta = -1$. The solutions of this equation for 2θ on the interval $\{0 \leq 2\theta \leq 4\pi\}$ are $\{\tfrac{3}{2}\pi, \tfrac{7}{2}\pi\}$; for θ on $\{0 \leq \theta \leq 2\pi\}$, the solutions are $\{\tfrac{3}{4}\pi, \tfrac{7}{4}\pi\}$.

3. The graph passes through the pole when $r = 0$, or $\sin 2\theta = 0$, which has the following solutions for $\theta: \{0, \tfrac{1}{2}\pi, \pi, \tfrac{3}{2}\pi, 2\pi\}$. If $\theta \rightarrow \tfrac{1}{2}\pi$, then the point $P:(r, \theta)$ on the graph, C, of (1) approaches the pole because $r \rightarrow 0$. The secant OP, as in Figure 173, has the line $\theta = \tfrac{1}{2}\pi$ as a limiting position when $\theta \rightarrow \tfrac{1}{2}\pi$. Hence, we agree that the line $\theta = \tfrac{1}{2}\pi$ should be *defined as a tangent* to the graph at the pole. Similarly, for each value of θ which gives $r = 0$ in (1), we obtain a tangent to C at the pole. These tangents are the lines $\theta = 0$, $\theta = \tfrac{1}{2}\pi$, $\theta = \pi$, $\theta = \tfrac{3}{2}\pi$, and $\theta = 2\pi$, which is the same as $\theta = 0$. Also the line $\theta = \tfrac{3}{2}\pi$ is the same as $\theta = \tfrac{1}{2}\pi$.

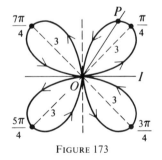

FIGURE 173

4. The following table of coordinates, and previous remarks were the basis for the graph in Figure 173. The arrowheads on the curve, and the values of θ at various points, show how the leaves of the "*four-leaved rose*" are created as θ increases from $\theta = 0$ to $\theta = 2\pi$ in (1). On account of the periodicity of $\sin 2\theta$, coordinates for $\{\pi \leq \theta \leq 2\pi\}$ could be obtained by duplicating values of r from the table. For instance, since

$r = 1.5$ when $\theta = 15°$, we have $r = 1.5$ also when $\theta = 15° + 180° = 195°$. The table gives the leaves to the *right* of the line $\theta = \frac{1}{2}\pi$. For convenience in using a trigonometric table, degree measure as well as radian measure is used in the table. Although $\sin 2\theta$ is periodic with the period π, this does *not* mean that *points* on the graph are duplicated at intervals of π in the values of θ.

$r =$	0	1.5	3	1.5	0	-1.5	-3	-1.5	0
$\theta = w°$	$0°$	$15°$	$45°$	$75°$	$90°$	$105°$	$135°$	$165°$	$180°$
$\theta =$	0		$\frac{1}{4}\pi$		$\frac{1}{2}\pi$		$\frac{3}{4}\pi$		π

For any positive integer n, the graph of $r = a \sin n\theta$ or $r = a \cos n\theta$ is called a **rose curve.** It is found that the rose has $2n$ leaves (as in Figure 173) when n is *even,* and just n leaves when n is *odd.*

The following useful procedure was illustrated in Example 1.

> *If an equation* $G(r, \theta) = 0$ has a solution $(r = 0,$ $\theta = \alpha)$, then as a rule the line $\theta = \alpha$ is a tangent to the graph of $G(r, \theta) = 0$ at the pole.*   (2)

EXAMPLE 2.   Graph: $r - 2(1 - \cos \theta)$.   (3)

Solution.   1. Since the function $\cos \theta$ has the period 2π, all of the graph will be obtained if we consider all values of θ on $\{0 \leq \theta < 2\pi\}$.

2. *Tangents at the pole.* From (3), $r = 0$ when

$$1 - \cos \theta = 0, \quad or \quad \cos \theta = 1.   (4)$$

With $0 \leq \theta < 2\pi$, the only solution of (4) is $\theta = 0$, which then is the equation of the only tangent to the graph at the pole. Or, the polar axis is tangent to the graph.

3. In (3), if we replace θ by $-\theta$, then $\cos (-\theta) = \cos \theta$, by (IV) on page 188, and thus (3) is unaltered. Hence, if $P:[r, \theta]$ is a point on the graph of (3), the point $Q:[r, -\theta]$ also is on the graph. We observe that P and Q are symmetric to the polar axis. Therefore, as a check on later work, we state that the graph should be *symmetric to the polar axis.*

4. The graph in Figure 174 was obtained by use of the coordinates in the following table, which were obtained by substituting values for θ in (3). The graph is called a **cardioid.** In Figure 174, the pole is called a *cusp* of the curve, because, first, there is no tangent to the complete curve at the pole and, second, arcs of the curve from either side are tangent to some line at the point (in this case, the tangent is the polar axis). Arrow-

---

*With G possessing usual desirable properties which we shall not enumerate, but shall assume to exist.

heads on the cardioid show how $P:(r, \theta)$ traces the curve as θ increases from $\theta = 0$ at O to $\theta = 2\pi$ at O.

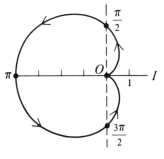

FIGURE 174

$r =$	0	.3	2	3.7	4	3.7	2	.3	0
$\theta = w°$	0°	30°	90°	150°	180°	210°	270°	330°	360°
$\theta =$	0	$\frac{1}{6}\pi$	$\frac{1}{2}\pi$	$\frac{5}{6}\pi$	π	$\frac{7}{6}\pi$	$\frac{3}{2}\pi$	$\frac{11}{6}\pi$	2π

The graph of each of the following equations, where $b \neq 0$, is a cardioid with its cusp at the pole, and either the polar axis or the line $\theta = \frac{1}{2}\pi$ as the tangent at the pole.

$$r = b(1 - \cos \theta); \qquad r = b(1 + \cos \theta); \qquad (5)$$

$$r = b(1 - \sin \theta); \qquad r = b(1 + \sin \theta). \qquad (6)$$

EXAMPLE 3.   Graph: $r^2 = 16 \cos 2\theta.$ (7)

Solution.   1. Since $\cos 2(\theta + \pi) = \cos (2\theta + 2\pi) = \cos 2\theta$, the function $\cos 2\theta$ has the period π, and hence also the period 2π. Therefore, all of the graph is obtained if θ has the domain $\{0 \leq \theta < 2\pi\}$, except for excluded values as discussed later.

2. In (7), if $r = 0$ then $\cos 2\theta = 0$, whose solutions for θ are $\{\frac{1}{4}\pi, \frac{3}{4}\pi, \frac{5}{4}\pi, \frac{7}{4}\pi\}$. Hence, the lines $\theta = \frac{1}{4}\pi$ and $\theta = \frac{3}{4}\pi$ are tangents to the graph at the pole. (The lines $\theta = \frac{1}{4}\pi$ and $\theta = \frac{5}{4}\pi$ are identical; the lines $\theta = \frac{3}{4}\pi$ and $\theta = \frac{7}{4}\pi$ are identical.)

3. If we change r to $-r$ in (7), the equation is unaltered. Thus, if a point $P:[r, \theta]$ is on the graph, then the point $Q:[-r, \theta]$ also is on the graph. Since P and Q are symmetric to the pole, it follows that the graph will be *symmetric to the pole.*

4. In (7), we must have $r^2 \geq 0$. Hence, we exclude any value of θ for which $\cos 2\theta < 0$. The student may verify that this excludes values of θ as follows:

$$\text{when } \tfrac{1}{2}\pi < 2\theta < \tfrac{3}{2}\pi, \qquad or \qquad \tfrac{1}{4}\pi < \theta < \tfrac{3}{4}\pi; \qquad (8)$$

$$\text{when } \tfrac{5}{2}\pi < 2\theta < \tfrac{7}{2}\pi, \qquad or \qquad \tfrac{5}{4}\pi < \theta < \tfrac{7}{4}\pi. \qquad (9)$$

5. The graph in Figure 175 was based on the following table of co-ordinates obtained from (7). The curve is called a **lemniscate of Bernoulli.** Points are obtained for $\{0 \le \theta \le \tfrac{1}{4}\pi\}$ and then for $\{\tfrac{3}{4}\pi \le \theta \le \pi\}$. The same points would be obtained if we should use the intervals $\{\pi \le \theta \le \tfrac{5}{4}\pi\}$ and $\{\tfrac{7}{4}\pi \le \theta \le 2\pi\}$. To obtain the graph, first the tangent lines $\theta = \tfrac{1}{4}\pi$ and $\theta = \tfrac{3}{4}\pi$ were drawn, as broken lines in Figure 175. Then the curve was drawn through the points from the table in such a manner as to be tangent to the tangent lines at the pole.

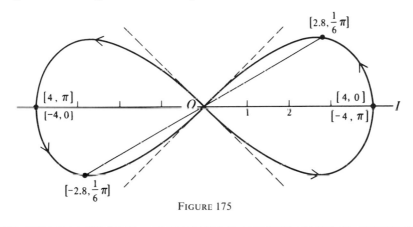

FIGURE 175

$r =$	± 4	$\pm 2\sqrt{2} = \pm 2.8$	0	no value	0	± 2.8	± 4
$\theta =$	0	$\tfrac{1}{6}\pi$	$\tfrac{1}{4}\pi$	$\tfrac{1}{4}\pi < \theta < \tfrac{3}{4}\pi$	$\tfrac{3}{4}\pi$	$\tfrac{5}{6}\pi$	π

★EXERCISE 66

First find the equation of each tangent to the graph of the equation at the pole. Then graph the equation.

1. $r = 2(1 + \cos\theta)$.   2. $r = 3(1 - \sin\theta)$.
3. $r = 2(1 + \sin\theta)$.   4. $r = -2(1 - \cos\theta)$.
5. $r = 2\cos 3\theta$.   6. $r = 4\sin 3\theta$.
7. $r = -\sin 3\theta$.   8. $r = 2\cos\theta$.
9. $r = 4\cos 2\theta$.   10. $r = -4\sin 2\theta$.
11. $r^2 = 4\cos 2\theta$.   12. $r^2 = 9\sin 2\theta$.

chapter 11

Introduction to Solid Analytic Geometry

98. RECTANGULAR COORDINATES IN SPACE

In three-dimensional space (hereafter called simply *three-space,* or *space*), suppose that a unit has been assigned for measuring distance in any direction, and for use on any number scale to be mentioned. Consider three mutually perpendicular lines intersecting at a point *O,* to be called the **origin** of coordinates. We specify these lines as *coordinate axes,* and set up a number scale on each axis with *O* representing zero. With the positive directions on the axes indicated by arrowheads, labeled $\{X, Y, Z\}$ in Figure 176, we refer to *OX* as the *x*-axis, *OY* as the *y*-axis, and *OZ* as the *z*-axis. These axes determine three mutually perpendicular *coordinate planes,* the *xy*-plane *OXY,* the *xz*-plane *OXZ,* and the *yz*-plane, *OYZ.* In Figure 176, think of the *yz*-plane as *vertical* in the plane of the paper facing us, with the positive direction on *OX* directed at us, so that the *xy*-plane is horizontal. We agree that any *line segment* or *distance* parallel to a coordinate axis will be *directed,* with the same positive direction as on that axis.

Let *P* be any point of space. Then the (x, y, z) coordinates of *P* are defined as the *directed perpendicular distances of P from the yz-plane, the xz-plane, and the xy-plane,* respectively. We write $P:(x, y, z)$ to mean "*P with coordinates* (x, y, z)," and say that we have defined an *xyz*-system of rectangular coordinates in space.

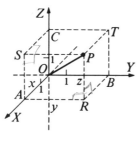

FIGURE 176

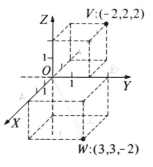

FIGURE 177

ILLUSTRATION 1.   For any point $P:(x,y,z)$, pass planes through P perpendicular to $\{OX, OY, OZ\}$ and intersecting these axes at $\{A, B, C\}$, respectively, as shown in a perspective fashion in Figure 176.   Then $\overline{OA} = x$; $\overline{OB} = y$; $\overline{OC} = z$. The planes drawn through P, together with the coordinate planes, intersect to block out a rectangular parallelepiped, to be called the *coordinate box* for P, with vertices $\{O, A, R, B\}$ in the xy-plane and $\{C, S, P, T\}$ for the upper face in Figure 176. In the box, the diagonal OP is called the *radius vector* of P. We may go to P from O on the path $\overline{OA} = x$, then $\overline{AR} = y$, and then $\overline{RP} = z$. We shall call $OARP$ a *coordinate path* for P.   Points $W:(3,3,-2)$ and $V:(-2,2,2)$ are shown in Figure 177.

The agreements for Figure 176 represent each point P of space by a point on the undistorted yz-plane. We chose angle XOY arbitrarily as 135°. The actual unit for distance in space is the unit used on OY, OZ, and for measuring any other distance in the yz-plane. Arbitrarily, the distorted unit on OX in Figure 176 is taken as approximately 2/3 of the actual unit in space. Then, any point $P:(x,y,z)$ of space is located on the figure as the endpoint of the corresponding coordinate path $OARP$, as in Figure 176. It can be proved* that the agreements for plotting points in Figure 176 are equivalent to projecting each point of space onto the yz-plane by a line with a certain fixed direction. This method is called **parallel projection.** Configurations in any plane parallel to the yz-plane are undistorted by this projection. Any line in space becomes a line or a point in the figure. Lines† parallel in space become parallel lines in the figure.

The coordinate planes divide space into eight parts called **octants.** We shall refer to the octant where all coordinates (x,y,z) are positive as the *first octant,* or the *visible octant,* in the typical figure. With the (*first* axis, *second* axis, *third* axis), or (OX, OY, OZ), oriented as in Figure 176, the xyz-system is referred to as a **right-handed‡** system of rectangular coordinates in space.

Note 1.   We may visualize the xy-plane in Figure 176 as the floor of a room with two walls V and W, which are perpendicular to each

---

*Suppose that the unit on OX in Figure 176 is $\frac{1}{2}\sqrt{2}$ or $.707\cdots$ times the actual space unit for distance as seen on OY and OZ. Then, it can be proved that the perspective system used in Figure 176 locates on the plane a point P' corresponding to $P:(x,y,z)$ in space as follows: P' is the intersection of the yz-plane and a line L through P, with L parallel to a vector with O as its initial point and the point $(2,1,1)$ as the endpoint.

†Any line parallel to the vector mentioned in the preceding footnote becomes simply a *point* in the figure.

‡To obtain a **left-handed** xyz-system, OX in Figure 166 would become OY and OY would become OX. For various reasons which become important in calculus and its applications, right-handed systems deserve the major emphasis. We shall use right-handed systems exclusively.

other and to the floor. To simulate the first octant, think of standing on the floor of the room facing the intersection of V and W as the z-axis, with wall V (the xz-plane) at the left and wall W (the yz-plane) at the right.

The *projection* (perpendicular) of a point P onto a line L is defined as the point P' where a plane (or a line) through P perpendicular to L will meet L. The projection of a line segment PQ onto L is the segment $P'Q'$ joining the projections of P and Q on L, as in Figure 178. In Figure 176 on page 290, $\{A, B, C\}$ are the projections of $P:(x, y, z)$ on the coordinate axes. Then, we have seen that

$$\overline{OA} = x; \qquad \overline{OB} = y; \qquad \overline{OC} = z. \qquad (1)$$

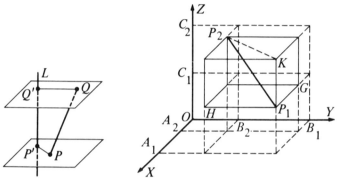

FIGURE 178   FIGURE 179

Through each of the points $P_1:(x_1, y_1, z_1)$ and $P_2:(x_2, y_2, z_2)$, as in Figure 179, pass a plane parallel to each coordinate plane. The resulting six planes intersect to form a rectangular parallelepiped, or box, with $P_1 P_2$ as a diagonal and each edge parallel to a coordinate axis. Figure 179 shows the projections $A_1 A_2$, $B_1 B_2$, and $C_1 C_2$ of the segment $P_1 P_2$ on the axes. Since $\overline{OA_1} = x_1$, $\overline{OA_2} = x_2$, $\overline{OB_1} = y_1$, etc., we obtain

$$|\overline{A_1 A_2}| = |x_2 - x_1|; |\overline{B_1 B_2}| = |y_2 - y_1|; |\overline{C_1 C_2}| = |z_2 - z_1|. \quad (2)$$

In (2), absolute values were used because we desire the lengths of $\{A_1 A_2, B_1 B_2, C_1 C_2\}$. On the box, observe that the vertical dimension $|\overline{P_1 K}| = |\overline{C_1 C_2}| = |z_2 - z_1|$. Thus the dimensions of the box are the absolute values in (2). For any rectangular parallelepiped, with dimensions a, b, and c units, by the Pythagorean theorem it is seen that, first, the length of a diagonal of a face is $\sqrt{a^2 + b^2}$, and then the length of a diagonal of the box is $\sqrt{a^2 + b^2 + c^2}$. Let d be the length of $P_1 P_2$, or $|\overline{P_1 P_2}|$, in Figure 179. Then, by use of (2),

$$d = \sqrt{(x_2 - x_1)^2 + (y_2 - y_1)^2 + (z_2 - z_1)^2}. \qquad (3)$$

ILLUSTRATION 2.   The distance between $(1,2,-3)$ and $(2,-1,4)$ is

$$d = \sqrt{(2-1)^2 + (-1-2)^2 + (4+3)^2} = \sqrt{59}.$$

EXERCISE 67

Plot the point. Show the coordinate box for each point and label a coordinate path from the origin to the point.

1. $(2, 3, 2)$.   **2.** $(-2, 2, 3)$.   **3.** $(3, -2, 2)$.
4. $(3, 2, -3)$.   **5.** $(0, 2, 0)$.   **6.** $(0, 0, 3)$.
7. $(3, 0, 0)$.   **8.** $(-2, -3, 2)$.   **9.** $(-2, 3, -2)$.

Compute the distance between the points.

10. $(0, 0, 0); (2, 4, 3)$.   **11.** $(0, 0, 0); (3, 5, -1)$.
12. $(-1, 2, 3); (-3, 4, 4)$.   **13.** $(-1, 2, 0); (2, 4, 2)$.
14. $(1, -2, -1); (3, 1, -2)$.   **15.** $(-1, 2, 1); (-1, -2, 3)$.

99. EQUATION OF A SURFACE

The graph of an equation

$$f(x, y, z) = 0 \qquad\qquad (1)$$

is the set, T, of points $P:(x, y, z)$ such that the coordinates (x, y, z) form a solution of (1). That is, the graph of (1) is the graph of its solution set in the xyz-system of coordinates. As a rule, T will qualify to be described as a *two-dimensional** set of points, and we shall refer to T as a *surface*. However, examples of (1) will be met when T is just a point, or just a line. An equation for a surface T is an equation $f(x, y, z) = 0$ whose graph is T.

In referring to a *plane section,* or simply a *section* of a surface T, we shall mean the curve of intersection of T and a plane.

ILLUSTRATION 1.   Any section of a plane is a line, because the intersection of any two nonparallel planes is a line. Any section of a sphere is a circle.

ILLUSTRATION 2.   If $P:(x, y, z)$ is any point in the plane perpendicular to the y-axis where $y = 3$, then the equation of the plane is $y = 3$.

Similarly, as in Illustration 2, we obtain the following results.

If a, b, and c are constants, the graphs of the equations $x = a, y = b, and z = c$ are planes which are perpendicular to the x-axis, y-axis, and z-axis, respectively. $\qquad$ (2)

The point $P:(a, b, c)$ is the intersection of the planes described in (2).

*We accept this name as an undefined term.

In a later section, we shall prove that the graph of any linear equation in x, y, and z is a plane and that, conversely, any plane in xyz-space has an equation linear in x, y, and z. We accept these facts for immediate use.

ILLUSTRATION 3.   An equation for the xy-plane is $z = 0$; of the yz-plane is $x = 0$; of the xz-plane is $y = 0$.

The *x-intercepts* of the graph, T, of an equation $f(x, y, z) = 0$ are the values of x where the graph meets the x-axis. If $P:(x, y, z)$ is on the x-axis, from Figure 176 on page 290 it is seen that $y = 0$ and $z = 0$. Hence, we reach the following conclusion concerning x-intercepts, and similar conclusions about the y-intercepts and z-intercepts.

To obtain the intercepts *for the graph of* $f(x, y, z) = 0$:

Place $y = 0$ *and* $z = 0$, *and solve for* x *to find the x-intercepts.*
Place $x = 0$ *and* $z = 0$, *and solve for* y *to find the y-intercepts.*
Place $x = 0$ *and* $y = 0$, *and solve for* z *to find the z-intercepts.*

The intersection of any surface T and a coordinate plane is called the **trace** of T in that plane. We call the trace in the xy-plane the *xy-trace,* and similarly refer to the *xz-trace,* and the *yz-trace.*

EXAMPLE 1.   Graph: $$3x + 2y + 2z = 6. \qquad (3)$$

Solution.   1. *The intercepts.* If $y = 0$ and $z = 0$ in (3), then $x = 2$, the x-intercept; if $x = 0$ and $z = 0$ then $y = 3$, the y-intercept; if $x = 0$ and $y = 0$ then $z = 3$, the z-intercept. This gives the following points on T, as seen in Figure 180: $A:(2, 0, 0)$; $B:(0, 3, 0)$; $C:(0, 0, 3)$.

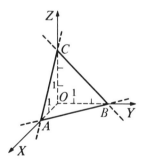

FIGURE 180

2. Since A and B are on the plane T which is the graph of (3), line AB is on T, and AB lies in the xy-plane because both A and B are in that plane. Thus, AB is the trace of T in the xy-plane. AB is continued as a broken line "behind" the xz-plane and the yz-plane, where we act as if these planes are opaque. In Figure 180, BC is the trace of T in the

yz-plane; *AC* is the trace of *T* in the *xz*-plane. The extensions of *AC* and *BC* below the *xy*-plane are shown as broken lines, where we act as if the *xy*-plane is opaque. The triangle *ABC* on *T*, together with the broken line extensions of the traces of *T* is considered as a satisfactory graph of (3).

100. EQUATIONS OF A CURVE

Consider the system of equations

$$f(x, y, z) = 0 \quad and \quad g(x, y, z) = 0. \tag{1}$$

We define the graph of (1) as the set of points $P:(x, y, z)$ where (x, y, z) is a solution of (1). That is, the graph, *T*, of (1) is the graph of its solution set in *xyz*-space. Thus, *T* is the *intersection of the surfaces* whose equations are in (1). We shall call the intersection of any two surfaces a *curve*. In view of preceding remarks, we must refer to the EQUATIONS (plural) of any curve *C*, meaning the equations of two surfaces, as in (1), whose intersection is *C*.

ILLUSTRATION 1.   The *x*-axis is the intersection of the *xz*-plane and the *xy*-plane, whose equations are $y = 0$ and $z = 0$, respectively. Hence, the EQUATIONS of the *x*-axis are $\{y = 0 \text{ and } z = 0\}$.

ILLUSTRATION 2.   In Figure 180, the *xy*-trace *AB* is the line with the equations

$$3x + 2y + 2z = 6 \quad and \quad z = 0. \tag{2}$$

On using $z = 0$ in the first equation we obtain $3x + 2y = 6$, as an equation whose graph in the *xy*-plane is *AB*.

To obtain an equation whose graph in a coordinate plane is the trace of a given surface $f(x, y, z) = 0$ in that plane, proceed as follows:

For the xy-trace: *place* $z = 0$ *in* $f(x, y, z) = 0$.
For the xz-trace: *place* $y = 0$ *in* $f(x, y, z) = 0$.
For the yz-trace: *place* $x = 0$ *in* $f(x, y, z) = 0$.

EXAMPLE 1.   Graph: $3y + 2z = 6.$ (3)

Solution.   1. If $y = z = 0$ in (3), we obtain $0 = 6$, which is a contradiction. Hence, the plane, *T*, which is the graph of (3) has no *x*-intercept, and thus *T* is perpendicular to the *yz*-plane.

2. The *equations* of the *yz*-trace are

$$3y + 2z = 6 \quad and \quad x = 0. \tag{4}$$

Since "$x = 0$" does not affect the equation at the left in (4), we have $3y + 2z = 6$ as an equation whose graph in the *yz*-plane is the *yz*-trace, which is the line AB in Figure 181. The equations of the xz-trace of T are $\{3y + 2z = 6$ *and* $y = 0\}$; or, the xz-trace is the graph of $2z = 6$, or $z = 3$ in the xz-plane, shown as DB in Figure 181. The xy-trace has the equations $\{3y + 2z = 6$ *and* $z = 0\}$; or, the xy-trace is the graph of $3y = 6$, or $y = 2$, in the xy-plance, shown as AC in Figure 181.

3. In Figure 181, CD is the intersection of the plane T with a plane CDE through E perpendicular to the x-axis. Thus the quadrilateral $ABDC$ is a piece of the graph of (3), and is accepted as our result.

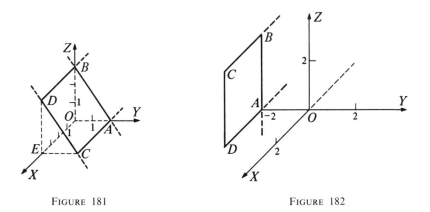

FIGURE 181   FIGURE 182

In Example 1, we observe a special case of the following fact: *If one variable is missing in a linear equation* $ax + by + cz = d$, *the graph of the equation is a plane perpendicular to the plane of the other two variables.*

ILLUSTRATION 3. Figure 182 shows a piece $ABCD$ of the plane $y + 2 = 0$. The yz-trace is the line AB which is the graph of $y = -2$ in the yz-plane. The xy-trace is the line DA which is the graph of $y = -2$ in the xy-plane. Parallelogram $ABCD$ represents the piece of the plane $y + 2 = 0$ cut off by a plane parallel to the xy-plane, giving the intersection BC, and a plane parallel to the yz-plane, giving the intersection CD, in Figure 182.

ILLUSTRATION 4. The student may verify that the quadrilateral $ABCD$ in Figure 183 is a piece of the plane $4x + 3y - 3z = 12$. CD is the intersection of this plane with the plane CED parallel to the xy-plane. Thus CD is parallel to AB; DE is parallel to OX; EC is parallel to OY.

ILLUSTRATION 5. The graph of $x^2 + y^2 + z^2 = 9$ is a sphere whose center is the origin and radius is 3 because, by (3) on page 292, the equation states that the square of the length of the radius vector $\overline{OP}$ from O to $P:(x, y, z)$ is 3^2 or 9. A graph, T, of the sphere is in Figure 184. Each trace of the sphere in a coordinate plane is a circle. In Figure 184, the *yz*-trace is shown without distortion as a circle with radius 3 and center at 0. The *xz*-trace and the *xy*-trace are distorted* from circles into ellipses in the figure. The *yz*-trace has the equations

$$x^2 + y^2 + z^2 = 9 \quad and \quad x = 0.$$

Or, the *yz*-trace is the graph of $y^2 + z^2 = 9$ in the *yz*-plane.

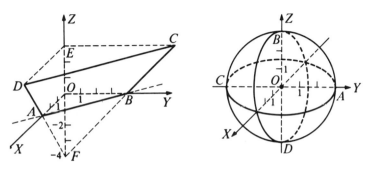

FIGURE 183   FIGURE 184

EXERCISE 68

Describe the plane whose equation is given.
1. $y = 0$.   **2.** $z = 0$.   **3.** $x = 0$.   **4.** $x = 2$.
5. $y = 3$.   **6.** $z = -2$.   **7.** $z = 4$.   **8.** $x = -5$.

Draw a rectangle which is on the plane.
9. $z = 2$.   **10.** $y = 4$.   **11.** $x = -2$.   **12.** $z = -3$.

Graph the equation, that is, obtain a triangle or quadrilateral which is on the plane involved. Show all traces of the plane in the coordinate planes.
13. $2x + y + z = 4$.   **14.** $3x + 2y - 2z = 6$.
15. $2y - 3x - 3z = -6$.   **16.** $4x - 3y - 6z + 12 = 0$.
17. $3x + 4y = 12$.   **18.** $3x + 4y = 4z - 12$.
19. $2z + y = 4$.   **20.** $x + 2z = 6$.   **21.** $2x - y = 4$.
22. $3x = y$.   **23.** $2x = z$.   **24.** $2y = 3z$.
25. $3x - 2y + 2z = 6$.   **26.** $-x + 2y + 3z = 6$.
27. Write a pair of equations for the *z*-axis; the *y*-axis; the *x*-axis.
28. Prove that the graph of $x^2 + y^2 + z^2 = 0$ is a point. (What point?)
29. Prove that the graph of $x^2 + y^2 = 0$ is a line. (What line?)
30. Give a single equation whose graph is the *x*-axis; the *y*-axis.

*The traces in Figure 184 were drawn with "artistic license" to make the figure look somewhat like we feel a sphere should appear. Actually, the ellipses would "bulge" outside the *yz*-trace if parallel projection were adhered to strictly. Such license will be taken also in later figures.

101. DIRECTION ANGLES AND DIRECTION COSINES

In Figure 185 we refer to the *direction* $\overrightarrow{PS}$, where the location of the initial point P in space and length of PS are immaterial. On any line L in space, there are two opposite directions, as shown by $\overrightarrow{PS}$ and $\overrightarrow{PS_1}$ in Figure 185.

To define the angle, θ, *between,* or *made by* two directions, first we represent them by direction arrows $\overrightarrow{PR}$ and $\overrightarrow{PS}$ with the same initial point. Then, θ is specified as the smallest angle, positive or zero, having $\overrightarrow{PR}$ and $\overrightarrow{PS}$ as sides. Thus, in radian measure, $0 \leqq \theta \leqq \pi$, where $\theta = 0$ if $\overrightarrow{PR}$ and $\overrightarrow{PS}$ are the same direction, and $\theta = \pi$ if $\overrightarrow{PR}$ and $\overrightarrow{PS}$ are opposite directions. If $\overrightarrow{PS_1}$ is the direction opposite to $\overrightarrow{PS}$, as in Figure 185, and ϕ is the angle made by $\overrightarrow{PR}$ and $\overrightarrow{PS_1}$, then $\phi = \pi - \theta$. That is,

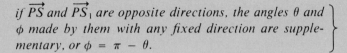

$$\text{if } \overrightarrow{PS} \text{ and } \overrightarrow{PS_1} \text{ are opposite directions, the angles } \theta \text{ and } \phi \text{ made by them with any fixed direction are supplementary, or } \phi = \pi - \theta. \tag{1}$$

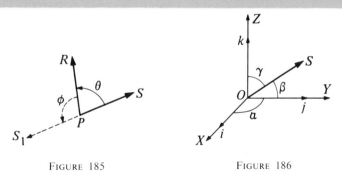

| FIGURE 185 | FIGURE 186 |

A line L in space will be called a *directed line* if one of its directions is specified as the *positive* direction on L, and the opposite direction is called the *negative* direction on L. Then the positive direction is called THE direction of L. Let L_1 and L_2 be two lines in space which have been assigned directions (the lines may not be intersecting lines). Then, the angle θ made by L_1 and L_2 is defined as the angle between their directions. To exhibit θ, we would draw direction arrows for the lines with the same initial point, as in Figure 185.

In an *xyz*-system of coordinates, let the positive direction on each axis be represented by a direction arrow with the origin as the initial point and length 1. Let $\{i, j, k\}$ be the corresponding direction arrows for $\{OX, OY, OZ\}$. Let any direction in space be represented by $\overrightarrow{OS}$, as in Figure 186.

DEFINITION I. *The* **direction angles** $\{\alpha,\beta,\gamma\}$ *of any direction* $\overrightarrow{OS}$ *are the angles made by* $\overrightarrow{OS}$ *with the directions of* $\{OX, OY, OZ\}$, *respectively. The* **direction cosines** *of* $\overrightarrow{OS}$ *are the cosines of its direction angles, or* $\{\cos \alpha, \cos \beta, \cos \gamma\}$.

Let the direction angles of a direction $\overrightarrow{RP}$ be $\{\alpha, \beta, \gamma\}$, and of the opposite direction $\overrightarrow{PR}$ be $\{\alpha', \beta', \gamma'\}$. By (1),

$$\alpha' = \pi - \alpha; \qquad \beta' = \pi - \beta; \qquad \gamma' = \pi - \gamma. \tag{2}$$

Recall the trigonometric identity $\cos (\pi - \theta) = -\cos \theta$. Then, from (2), $\cos \alpha' = \cos (\pi - \alpha) = -\cos \alpha$; $\cos \beta' = -\cos \beta$; $\cos \gamma' = -\cos \gamma$. Thus

$$\left. \begin{array}{l} \textit{if a direction is reversed, its direction cosines are} \\ \textit{multiplied by} -1. \end{array} \right\} \tag{3}$$

THEOREM I. *Let* $P_1:(x_1, y_1, z_1)$ *and* $P_2:(x_2, y_2, z_2)$ *be distinct points. Then, the direction cosines of* $\overrightarrow{P_1 P_2}$ *are as follows, where*

$$d = | \overline{P_1 P_2} | = \sqrt{(x_2 - x_1)^2 + (y_2 - y_1)^2 + (z_2 - z_1)^2}: \tag{4}$$

$$\cos \alpha = \frac{x_2 - x_1}{d}; \qquad \cos \beta = \frac{y_2 - y_1}{d}; \qquad \cos \gamma = \frac{z_2 - z_1}{d}. \tag{5}$$

Proof. 1. Through each of P_1 and P_2, in Figure 187, pass planes parallel to the coordinate planes, and thus block out a box with $P_1 P_2$ as a diagonal. In Figure 187,

$$\overline{OC_1} = z_1; \qquad \overline{OC_2} = z_2; \qquad z_2 - z_1 = \overline{C_1 C_2} = \overline{P_1 K}.$$

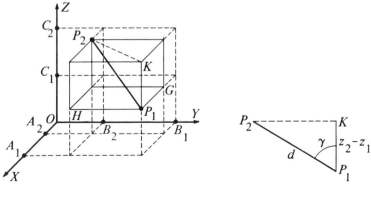

FIGURE 187   FIGURE 188

2. Triangle P_1KP_2 is taken from Figure 187 and drawn without distortion in Figure 188. The direction $\overrightarrow{P_1K}$ is the direction of the z-axis. Hence $\gamma = \angle P_2P_1K$. By trigonometry,* from triangle P_1KP_2 we obtain

$$\cos \gamma = \frac{\overline{P_1K}}{d} = \frac{z_2 - z_1}{d}. \tag{6}$$

Similarly, the other results in (5) are obtained.

Let any direction in space be represented by an arrow $\overrightarrow{P_1P_2}$ with P_1 and P_2 chosen properly. Then, from (5),

$$\cos^2 \alpha + \cos^2 \beta + \cos^2 \gamma = \frac{(x_2 - x_1)^2 + (y_2 - y_1)^2 + (z_2 - z_1)^2}{d^2} = 1.$$

That is, for any direction, the direction cosines satisfy

$$\cos^2 \alpha + \cos^2 \beta + \cos^2 \gamma = 1. \tag{7}$$

ILLUSTRATION 1. To find the direction cosines of a direction $\overrightarrow{P_1P_2}$ with $P_1:(3, 2, -6)$ and $P_2:(1, -2, 4)$, we use (5) with $d = \sqrt{2^2 + 4^2 + 10^2} = 2\sqrt{30}$. Then we obtain

$$\cos \alpha = -\frac{1}{\sqrt{30}}; \qquad \cos \beta = -\frac{2}{\sqrt{30}}; \qquad \cos \gamma = \frac{5}{\sqrt{30}}.$$

If (5) is applied with P_1 as $(0,0,0)$ and P_2 as $P:(x, y, z)$, we obtain the direction cosines of OP, the radius vector from O to P, as follows, where $\rho = \sqrt{x^2 + y^2 + z^2}$ (read ρ as *rho*):

$$\cos \alpha = \frac{x}{\rho}; \qquad \cos \beta = \frac{y}{\rho}; \qquad \cos \gamma = \frac{z}{\rho}. \tag{8}$$

If two of the direction cosines in (7) are assigned arbitrarily, from (7) we obtain just one value for the third cosine, or no value, or two values, depending on the data.

ILLUSTRATION 2. If $\cos \alpha = \frac{1}{2}$ and $\cos \gamma = \frac{1}{2}$, from (7) we obtain

$$\tfrac{1}{4} + \cos^2 \beta + \tfrac{1}{4} = 1, \qquad or \qquad \cos^2 \beta = \tfrac{1}{2}.$$

Hence, $\cos \beta = \pm\sqrt{\tfrac{1}{2}} = \pm\tfrac{1}{2}\sqrt{2}$.

Hereafter, as a standard notation, for any direction with the direction angles $\{\alpha, \beta, \gamma\}$, we shall let† $\lambda = \cos \alpha$, $\mu = \cos \beta$, and $\nu = \cos \gamma$. Then, we may refer to the direction $\{\lambda, \mu, \nu\}$. The direction $\{-\lambda, -\mu, -\nu\}$ is the opposite of $\{\lambda, \mu, \nu\}$, because of (3).

---

*The figure applies only when $0 \leq \gamma \leq \frac{1}{2}\pi$. A more elaborate argument, which we omit, would prove (6) if $\frac{1}{2}\pi < \gamma$.

†Read $\{\lambda, \mu, \nu\}$ as $\{lambda, mu, nu\}$.

From (7), for any set of direction cosines $\{\lambda, \mu, \nu\}$,

$$\lambda^2 + \mu^2 + \nu^2 = 1. \tag{9}$$

Note 1. If we refer to the *direction angles or cosines of a line L,* this will mean the angles or cosines for a particular direction (of the two available) which has been assigned to L.

EXERCISE 69

Let P be the given point. In Problems 1–6 draw OP. In each problem find the direction cosines of $\overrightarrow{OP}$.

1. $(2, -2, 1)$.　　**2.** $(-2, 1, 2)$.　　**3.** $(-3, -2, 6)$.
4. $(-3, 4, 0)$.　　**5.** $(0, 0, 2)$.　　**6.** $(0, 4, -3)$.
7. $(2, 0, 0)$.　　**8.** $(9, 6, -2)$.　　**9.** $(14, -2, 5)$.

Obtain direction cosines for the direction from the first point to the second point.

10. $(-1, 2, 3)$; $(1, 4, 2)$.　　　**11.** $(3, -2, 5)$; $(4, -4, 7)$.
12. $(-2, 3, 0)$; $(1, 1, -6)$.　　　**13.** $(2, -4, -7)$; $(-4, -2, 2)$.

14. Find the direction angles and the direction cosines of the y-axis; x-axis; z-axis.

Obtain the missing direction angle and all direction cosines, if this is possible.

15. $\alpha = 45°$; $\gamma = 45°$.　　　**16.** $\beta = \frac{1}{3}\pi$; $\gamma = \frac{3}{4}\pi$.

17. Find the missing direction cosine, if $\lambda = \frac{2}{3}$ and $\mu = \frac{1}{3}$.

18. Find the direction angles and direction cosines of the line $y = x$ in the xy-plane with this line directed into quadrant III in that plane.

19. Repeat Problem 18 for the line $y = -x$ directed into the second quadrant of the xy-plane.

20. Find the direction cosines and direction angles of a direction making equal angles with the three axes of the xyz-system. Use a trigonometric table.

102. DIRECTION NUMBERS FOR A DIRECTION

Recall the following terminology. To state that the numbers of an ordered triple $\{a, b, c\}$ are *proportional* to the numbers of another ordered triple $\{r, s, t\}$, means that there exists a constant of proportionality $k \neq 0$ such that

$$a = kr; \quad b = ks; \quad c = kt. \tag{1}$$

When (1) is true, we write

$$a:b:c = r:s:t, \tag{2}$$

which is read "*a is to b is to c as r is to s is to T*." It is easy to show that, if (2) is true and if $r:s:t = u:v:w$, then also $a:b:c = u:v:w$. If $a:b: c = r:s:t$ and h is any number not zero, then $a:b:c = hr:hs:ht$, because the presence of h would merely change the constant of proportionality in (1). Thus, if $a:b:c = \frac{1}{2}:\frac{1}{3}:\frac{1}{4}$, we may multiply each number on the right by 12 and obtain $a:b:c = 6:4:3$, where only *integers* are involved.

If three numbers $\{a,b,c\}$ are proportional to the direction cosines of a direction ω (read *omega*), then

$$a:b:c = \cos \alpha : \cos \beta : \cos \gamma, \tag{3}$$

and there exists $k \neq 0$ such that

$$\cos \alpha = ka; \qquad \cos \beta = kb; \qquad \cos \gamma = kc. \tag{4}$$

In case (3) is true, we say that "$a:b:c$" are **direction numbers** for the direction ω. We read "$a:b:c$," written alone, simply as "a, b, c," and place colons between the letters in writing them merely to emphasize that they are proportional to the cosines. One set of direction numbers for a given direction is its set of direction cosines; in this case, $k = 1$ in (4).

If $a:b:c$ are any direction numbers, then $\{a,b,c\}$ are *not all zero* because, from (4),

$$k^2(a^2 + b^2 + c^2) = \cos^2 \alpha + \cos^2\beta + \cos^2 \gamma = 1,$$

where $k \neq 0$, and hence $a^2 + b^2 + c^2 \neq 0$.

THEOREM II.   *If $a:b:c$ are direction numbers for a direction with direction angles $\{\alpha, \beta, \gamma\}$, then*

$$\left.\begin{array}{c}
\cos \alpha = \dfrac{a}{\pm\sqrt{a^2 + b^2 + c^2}}; \qquad \cos \beta = \dfrac{b}{\pm\sqrt{a^2 + b^2 + c^2}}; \\[4mm]
\cos \gamma = \dfrac{c}{\pm\sqrt{a^2 + b^2 + c^2}},
\end{array}\right\} \tag{5}$$

where "+" is used throughout, or "−" throughout.

Proof.   From (4),

$$\cos^2 \alpha + \cos^2 \beta + \cos^2 \gamma = 1 = k^2(a^2 + b^2 + c^2).$$

Hence, $k = 1/\pm\sqrt{a^2 + b^2 + c^2}$. If we use k with the plus sign, one set of direction cosines is obtained from (4). Another set is obtained by use of the minus sign, as shown in (5). Two opposite directions are given in (5).

ILLUSTRATION 1.   If the direction numbers are $3:-2:1$, from (5) we obtain

$$\cos \alpha = \frac{3}{\sqrt{14}}, \qquad \cos \beta = -\frac{2}{\sqrt{14}}, \qquad \cos \gamma = \frac{1}{\sqrt{14}}; \ or$$

$$\cos \alpha = -\frac{3}{\sqrt{14}}, \qquad \cos \beta = \frac{2}{\sqrt{14}}, \qquad \cos \gamma = -\frac{1}{\sqrt{14}}.$$

Because of (5), a set of direction numbers for a given line serves as direction numbers for both of the directions on the line.

For brevity, we may refer to "*a direction $a:b:c$*," meaning a direction having the given direction numbers. Since the direction cosines $\{\lambda, \mu, \nu\}$ are one choice for direction numbers, we may refer to "*the direction $\lambda:\mu:\nu$.*"

Because of Theorem II, any three numbers $\{a,b,c\}$ which are not all zero can be assigned arbitrarily as direction numbers, and then yield two corresponding sets of direction cosines. This is one feature which makes it more convenient to deal with direction numbers than direction cosines. It is not possible to assign a set of direction cosines arbitrarily because the sum of their squares must be 1.

In (5) on page 299, the direction cosines for $\overrightarrow{P_1 P_2}$ are seen to be proportional to $(x_2 - x_1):(y_2 - y_1):(z_2 - z_1)$, with $1/d$ as the constant of proportionality. For the *opposite* direction $\overrightarrow{P_2 P_1}$, the proportionality factor would be $-1/d$. Hence, we have the following result.

> *A set of direction numbers for either direction on the line through the distinct points $P_1:(x_1,y_1,z_1)$ and $P_2:(x_2,y_2,z_2)$ is*
> $$(x_2 - x_1):(y_2 - y_1):(z_2 - z_1).$$
(6)

ILLUSTRATION 2.   From (6), direction numbers for the line through $P_1:(2,-3,4)$ and $P_2:(-1,2,-2)$ are $(-1-2):(2+3):(-2-4)$, or $-3:5:-6$. From these, if desired, we may find direction cosines for the line by use of (5).

THEOREM III.   *If $a_1:b_1:c_1$ and $a_2:b_2:c_2$ are direction numbers for lines L_1 and L_2, respectively, then L_1 and L_2 are parallel when and only when*

$$a_1:b_1:c_1 = a_2:b_2:c_2.$$
(7)

Proof.   1. If (7) is true, then $a_1:b_1:c_1$ can be used as direction numbers for *both* L_1 and L_2 in (5). Hence, L_1 and L_2 have the same direction cosines, or those of L_1 are the *negatives* of the cosines for L_2. Thus, the directions of L_1 and L_2 are the *same* or are *opposites,* and hence L_1 and L_2 are parallel.

2. Suppose that L_1 and L_2 are parallel, with direction cosines $\lambda_1 : \mu_1 : \nu_1$ and $\lambda_2 : \mu_2 : \nu_2$, respectively. Then, the directions of L_1 and L_2 either are the same, and thus have the same direction cosines, or are opposites, and then $\{\lambda_2, \mu_2, \nu_2\}$ are the negatives of $\{\lambda_1, \mu_1, \nu_1\}$. In either case, $\lambda_1 : \mu_1 : \nu_1 = \lambda_2 : \mu_2 : \nu_2$. Since $\lambda_1 : \mu_1 : \nu_1 = a_1 : b_1 : c_1$ and $\lambda_2 : \mu_2 : \nu_2 = a_2 : b_2 : c_2$, it is seen that $a_1 : b_1 : c_1 = a_2 : b_2 : c_2$. Hence, Theorem III has been proved.

ILLUSTRATION 3.   The directions $2 : -3 : 5$ and $-4 : 6 : -10$ are parallel because $2 : -3 : 5 = -4 : 6 : -10$, with a constant of proportionality -2 involved in going from left to right: $-4 = -2(2)$, $6 = -2\,(-3)$, $-10 = -2(5)$.

ILLUSTRATION 4.   From (6) with P_1 as $O:(0,0,0)$, the direction $\overrightarrow{OP}$ has the direction numbers $a : b : c$ if P is the point $P : (a, b, c)$.

103. THE ANGLE BETWEEN TWO DIRECTIONS

THEOREM IV.   Let θ be the angle made by two directions with the direction angles $\{\alpha_1, \beta_1, \gamma_1\}$ and $\{\alpha_2, \beta_2, \gamma_2\}$, respectively. Then

$$\cos\theta = \cos\alpha_1 \cos\alpha_2 + \cos\beta_1 \cos\beta_2 + \cos\gamma_1 \cos\gamma_2. \tag{1}$$

Proof.   1. Let $\{\lambda_1, \mu_1, \nu_1\}$ and $\{\lambda_2, \mu_2, \nu_2\}$ be the direction cosines for the directions. In the xyz-system of coordinates, locate $P_1 : (\lambda_1, \mu_1, \nu_1)$ and $P_2 : (\lambda_2, \mu_2, \nu_2)$, and construct OP_1, OP_2, and P_1P_2, as in Figure 189.

Then
$$| \overrightarrow{OP_1} | = \sqrt{\lambda_1^2 + \mu_1^2 + \nu_1^2} = 1,$$

and similarly $| \overrightarrow{OP_2} | = 1$. Also, by (8) on page 300, the direction cosines of $\overrightarrow{OP_1}$ are $\lambda_1 : \mu_1 : \nu_1$. Similarly, the direction $\overrightarrow{OP_2}$ has the direction cosines $\lambda_2 : \mu_2 : \nu_2$. Hence, θ is the angle at $O : (0,0,0)$ in ΔOP_1P_2.

2. Let $d = | P_1P_2 |$. By use of the distance formula on page 292,

$$d^2 = (\lambda_1 - \lambda_2)^2 + (\mu_1 - \mu_2)^2 + (\nu_1 - \nu_2)^2$$
$$= (\lambda_1^2 + \mu_1^2 + \nu_1^2) + (\lambda_2^2 + \mu_2^2 + \nu_2^2) - 2(\lambda_1\lambda_2 + \mu_1\mu_2 + \nu_1\nu_2), \text{ or}$$
$$d^2 = 2 - 2(\lambda_1\lambda_2 + \mu_1\mu_2 + \nu_1\nu_2). \tag{2}$$

3. By use of the law of cosines for triangles from page 226, as applied to ΔOP_1P_2,

$$d^2 = \overline{P_1P_2^2} = \overline{OP_1^2} + \overline{OP_2^2} - 2\,| \overline{OP_1} | \cdot | \overline{OP_2} | \cos\theta, \text{ or}$$
$$d^2 = 1 + 1 - 2\cos\theta, \quad or \quad d^2 = 2 - 2\cos\theta. \tag{3}$$

From (2) and (3) we obtain

$$\cos \theta = \lambda_1\lambda_2 + \mu_1\mu_2 + \nu_1\nu_2, \tag{4}$$

which is (1) in different notation.

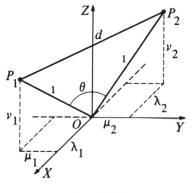

FIGURE 189

Two lines L_1 and L_2 in space may be parallel, in which case the two possible directions on L_1 are the same as the two directions on L_2. In this case L_1 and L_2 may coincide.* Or, L_1 and L_2 may not be parallel. In such a case they may be intersecting lines. If two nonparallel lines L_1 and L_2 do *not* intersect, they are called *skew lines.* In any one of the preceding situations, we define the angle *made by L_1 and L_2* as that angle which is made by selected directions on L_1 and L_2, respectively. Thus, to obtain a particular angle made by the lines, first we must select a direction on each line, and then find the angle made by these directions.

EXAMPLE 1.   Obtain the angles between two lines with the direction numbers $2:3:-6$ and $2:-6:-9$.

Solution.   1. By use of (5) on page 302, since $\sqrt{4 + 9 + 36} = 7$ and $\sqrt{4 + 36 + 81} = 11$, one choice of direction cosines for the directions of the lines is as follows:

$$\frac{2}{7} : \frac{3}{7} : -\frac{6}{7} \qquad and \qquad \frac{2}{11} : -\frac{6}{11} : -\frac{9}{11}. \tag{5}$$

Then, from (1),

$$\cos \theta = \frac{2}{7} \cdot \frac{2}{11} + \frac{3}{7}\left(-\frac{6}{11}\right) + \left(-\frac{6}{7}\right)\left(-\frac{9}{11}\right) = \frac{40}{77} = .519.$$

From Table IV, by interpolation, $\theta = 58.7°$.

---

*Thus, we agree to refer to *coincident* parallel lines. This agreement is convenient because initially two lines of this nature may be defined by distinct algebraic means.

2. The cosines in (5) were obtained by taking the plus sign with each radical as used in (5) on page 302. If a plus sign is taken with one radical and a minus sign with the other radical, we obtain $\cos \theta = -.519$. Then $\theta = 180° - 58.7° = 121.3°$. Thus, with various choices for directions on the two lines, the angle between their directions is either $58.7°$ or $121.3°$.

Since $\cos 90° = 0$, from (1) we obtain the following result.

THEOREM V.   Lines L_1 and L_2 are perpendicular when and only when the sum of the products of corresponding direction cosines for directions on the lines is zero:

$$\cos \alpha_1 \cos \alpha_2 + \cos \beta_1 \cos \beta_2 + \cos \gamma_1 \cos \gamma_2 = 0. \qquad (6)$$

ILLUSTRATION 1.   The directions with the direction cosines $\frac{1}{3}$: $-\frac{2}{3} : \frac{2}{3}$ and $\frac{2}{3} : \frac{2}{3} : \frac{1}{3}$ are perpendicular because $\frac{1}{3}(\frac{2}{3}) - \frac{2}{3}(\frac{2}{3}) + \frac{2}{3}(\frac{1}{3}) = 0$.

THEOREM VI.   Two directions ω_1 and ω_2 having the direction numbers $a_1 : b_1 : c_1$ and $a_2 : b_2 : c_2$, respectively, are perpendicular when and only when

$$a_1 a_2 + b_1 b_2 + c_1 c_2 = 0. \qquad (7)$$

Proof.   With $\cos \alpha_1 = \dfrac{\pm a_1}{\sqrt{a_1^2 + b_1^2 + c_1^2}}$, $\cos \alpha_2 = \dfrac{\pm a_2}{\sqrt{a_2^2 + b_2^2 + c_2^2}}$, etc., as obtained from page 302, (6) becomes

$$\frac{a_1 a_2 + b_1 b_2 + c_1 c_2}{\sqrt{a_1^2 + b_1^2 + c_1^2} \sqrt{a_2^2 + b_2^2 + c_2^2}} = 0. \qquad (8)$$

Hence, (6) is true when and only when (7) is true, which proves Theorem VI.

ILLUSTRATION 2.   If $9 : 6 : 2$ and $-2 : 6 : -9$ are direction numbers for two directions, they are perpendicular because $9(-2) + 6(6) + 2(-9) = 0$.

In applications, (7) will be found more convenient than (6) when perpendicularity is involved.

EXERCISE 70

Find the direction cosines of each direction having the given direction numbers. First change to integral direction numbers if necessary.

1. $3 : 2 : -6$.   **2.** $6 : -9 : -2$.   **3.** $0 : 0 : -2$.

4. $-\frac{2}{5} : \frac{2}{5} : -\frac{1}{5}$.   **5.** $-\frac{5}{2} : \frac{11}{4} : -\frac{1}{2}$.   **6.** $3 : -4 : 0$.

7. $14 : -5 : -2$.   **8.** $\frac{8}{7} : -\frac{12}{7} : \frac{9}{7}$.   **9.** $12 : 0 : -5$.

Plot a point $P : (x, y, z)$ so that $\overrightarrow{OP}$ will have the given direction numbers.
10. $3 : 2 : 4$.   **11.** $-2 : 3 : 2$.   **12.** $0 : 3 : -2$.

Obtain direction numbers for the line through the points.
13. $(2, -3, 2); (-1, 2, -4)$.   **14.** $(-1, 2, -3); (-4, 2, -5)$.
15. $(0, 0, 0); (3, -2, 4)$.   **16.** $(-3, 0, -2); (0, 0, 0)$.

Are the two directions parallel or are they perpendicular?
17. $-3 : 2 : -1$ and $6 : -4 : 2$.   **18.** $2 : -3 : 4$ and $-4 : 6 : -8$.
19. $7 : 4 : -5$ and $1 : 2 : 3$.   **20.** $\frac{1}{2} : \frac{3}{4} : 2$ and $2 : -3 : 8$.
21. $2 : -1 : 2$ and $7 : 4 : -5$.   **22.** $2 : -3 : \frac{3}{2}$ and $12 : -18 : 9$.
23. $\frac{1}{3} : \frac{3}{2} : -2$ and $-6 : -16 : -13$.

Are the given points collinear?
24. $(2, -1, 3); (5, 1, 2); (-4, -5, 5)$.

Hint. They are collinear if and only if the line through the first two points is parallel to the line through the last two.
25. $(-1, 2, -3); (-3, 3, -7); (1, 1, 1)$.

Find the angle made by the lines with the given direction cosines.
26. $\frac{2}{3} : \frac{2}{3} : -\frac{1}{3}$ and $\frac{2}{7} : -\frac{3}{7} : \frac{6}{7}$.
27. $\frac{9}{11} : -\frac{6}{11} : \frac{2}{11}$ and $-\frac{3}{7} : \frac{2}{7} : \frac{6}{7}$.
28. $\frac{3}{5} : 0 : \frac{4}{5}$ and $-\frac{5}{13} : \frac{12}{13} : 0$.

Find the angles made by two lines with the given direction numbers.
29. $6 : -9 : 2$ and $3 : 6 : 2$.   **30.** $2 : 3 : -6$ and $4 : 0 : -3$.

104. ONE-POINT, NORMAL DIRECTION FORM FOR A PLANE

Hereafter, any line perpendicular to a plane will be called a *normal* to the plane. For any plane W, there is an unique normal line perpendicular to the plane from (or at) the origin. Let $A : B : C$ be direction numbers for this normal line. We shall say that W has the *normal direction* $A : B : C$.

THEOREM VII.   *Every plane W has an equation linear in x, y, and z. If the normal direction for W has the direction numbers $A : B : C$, and if the point $M : (x_0, y_0, z_0)$ is on W, then an equation for W is*

$$A(x - x_0) + B(y - y_0) + C(z - z_0) = 0. \tag{1}$$

Proof.   1. At M on W, erect the normal line MN, as in Figure 190 on page 308, with the direction $\overrightarrow{MN}$. It has the direction numbers $A : B : C$.

2. Let $P : (x, y, z)$ be any point in space, and draw MP. By (6) on page 303, direction numbers for MP are $(x - x_0) : (y - y_0) : (z - z_0)$.

Then, P is on the plane W if and only if MP is perpendicular to MN or, by (7) on page 306, if and only if the sum of the products of corresponding direction numbers for MN and MP is zero. This gives (1) as an equation of the plane.

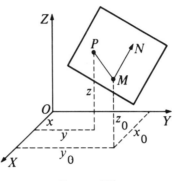

FIGURE 190

ILLUSTRATION 1.   An equation for the plane through $(2, -3, 4)$ with the normal direction $4 : -1 : 2$ is

$$4(x - 2) - (y + 3) + 2(z - 4) = 0, \qquad or \qquad 4x - y + 2z = 19.$$

THEOREM VIII.   *If A, B, and C are not all zero, the graph of*

$$Ax + By + Cz = D \tag{2}$$

is a plane with the normal direction $A : B : C$. That is, the graph of any linear equation in x, y, and z is a plane.

Proof.   1. By assigning values to two of the variables $\{x, y, z\}$ in (2), and then computing the value of the third variable to satisfy (2), we can obtain at least one solution (x_0, y_0, z_0) of (2). Then

$$Ax_0 + By_0 + Cz_0 = D. \tag{3}$$

2. On subtracting the sides of (3) from corresponding sides of (2), we obtain

$$A(x - x_0) + B(y - y_0) + C(z - z_0) = 0, \tag{4}$$

which is equivalent to (2). Notice that (4) is of the form (1). Thus (4), and hence (2), is the equation of a plane through $M : (x_0, y_0, z_0)$ with the normal direction $A : B : C$.

ILLUSTRATION 2.   The plane $3x + 5y - 7z = 15$ has the normal direction $3 : 5 : -7$.

EXAMPLE 1.   Obtain an equation of the plane W through $(2, -3, 1)$ which is parallel to the plane $T : \{4x - 2y + z = 2\}$.

Solution.   Since W and T are parallel, they have the same normal direction. Hence, the normal direction for W is $4 : -2 : 1$, and an equation for W, through $(2, -3, 1)$, is

$$4(x - 2) - 2(y + 3) + (z - 1) = 0, \quad or \quad 4x - 2y + z = 15.$$

Consider two planes:

$$W_1 : \{A_1 x + B_1 y + C_1 z + D_1 = 0\}; \tag{5}$$

$$W_2 : \{A_2 x + B_2 y + C_2 z + D_2 = 0\}. \tag{6}$$

Their normals have the direction numbers $A_1 : B_1 : C_1$ and $A_2 : B_2 : C_2$, respectively. The planes are parallel* when and only when their normals are parallel. The planes are perpendicular when and only when their normals are perpendicular. Hence, by (7) on page 303 concerning parallelism, and (7) on page 306 concerning perpendicularity, we have the following results.

Planes W_1 and W_2 parallel: $A_1 : B_1 : C_1 = A_2 : B_2 : C_2.$ (7)

Planes W_1 and W_2 perpendicular: $A_1 A_2 + B_1 B_2 + C_1 C_2 = 0.$ (8)

ILLUSTRATION 3.   The planes

$$3x + 4y - 2z + 5 = 0 \quad and \quad 6x + 8y - 4z - 7 = 0$$

are parallel because $3 : 4 : -2 = 6 : 8 : -4$. The planes

$$6x - 9y + 2z = 3 \quad and \quad 6x + 2y - 9z = 8$$

are perpendicular because $6(6) + 2(-9) + (-9)(2) = 0$.

EXERCISE 71

Obtain the equation of the plane through the point P with the given normal direction.

1. $P : (2, 1, -2)$; normal direction $3 : -1 : 4$.
2. $P : (-1, 2, -3)$; normal direction $-1 : 2 : -1$.
3. $P : (2, 3, 0)$; normal direction $4 : -1 : 0$.
4. $P : (-3, 1, 0)$; normal direction $3 : 0 : -1$.

Write an equation for the specified plane.

5. Through $(-1, 2, -3)$ parallel to $2x - 3y + z = 8$.
6. Through $(0, -1, 2)$ parallel to $-x + 2y - z = 3$.

*If two linear equations in $\{x, y, z\}$ are equivalent, so that the graphs of the equations are the same plane, we refer to the graphs as coincident parallel planes.

7. Through $(3,-2,-1)$ parallel to $x - 2y - 3z = 4$.

8. Through $(2,-1,3)$ parallel to the xz-plane.

9. Through $(-1,0,2)$ parallel to the yz-plane.

10. Through $(-4,1,5)$ parallel to the xy-plane.

11. Through $(1,2,-1)$ perpendicular to a line with the direction cosines $-\frac{9}{11} : \frac{2}{11} : \frac{6}{11}$.

12. Through $(4,-2,1)$ perpendicular to the direction $3:5:-2$.

13. Prove that the equation of the plane with x-intercept $a \neq 0$, y-intercept $b \neq 0$, and z-intercept $c \neq 0$ is

$$\frac{x}{a} + \frac{y}{b} + \frac{z}{c} = 1,$$

which is called the **intercept form** of a plane.

Change the equation to the intercept form and find the intercepts.

14. $3x - 2y - 5z = 30$.   **15.** $x + 3y - z = 4$.

16. Find an equation for the plane through $(2,-3,1)$ perpendicular to the line through $(1,4,2)$ and $(-2,3,1)$.

17. Find an equation for the plane through $(-1,5,0)$ perpendicular to the line through $(-1,1,2)$ and $(2,-3,1)$.

Are the two planes parallel or perpendicular?

18. $3x + 2y - z = 5;$ $-6x - 4y + 2z = 1$.

19. $2x + y - 2z = 1;$ $7x - 4y + 5z = 0$.

20. $\frac{1}{3}x + 2y + \frac{3}{2}z = 5;$ $-6x + 13y - 16z = 0$.

21. $2x - \frac{3}{2}y - 3z = 2;$ $12x - 9y - 18z = 5$.

22. Prove that W_1 and W_2 of (5) and (6) on page 309 are the same plane if $A_2 = hA_1$, $B_2 = hB_1$, $C_2 = hC_1$, and $D_2 = hD_1$. *Conversely,* if W_1 and W_2 are the same plane, prove that the preceding conditions are satisfied. Thus, show that W_1 and W_2 are identical if and only if $A_1 : B_1 : C_1 : D_1 = A_2 : B_2 : C_2 : D_2$.

Hint. In the converse case, $\{A_2 = kA_1, B_2 = kB_1, C_2 = kC_1\}$. Why? Show that, also, $D_2 = kD_1$ by use of the two equations.

★**23.** Find the measures of the dihedral angles (equal in pairs) formed by the planes $3x - 2y + 6z = 4$ and $-2x - 2y + z = 5$.

Hint. A figure showing the dihedral angles formed by two planes would indicate that the measures of these angles are equal to the measures of the plane angles formed by the normals to the planes.

★**24.** Find an equation for the plane through the given points:

$$U:(1,2,-3); \quad V:(3,-1,1); \quad W:(-2,1,0).$$

Hint. Obtain direction numbers for UV and VW. Then obtain direction numbers for the normal direction, which will be perpendicular to *both* UV and VW. Or, substitute the coordinates of each point in $Ax + By + Cz + D = 0$, and solve the resulting three equations for $\{A, B, C\}$ in terms of D; then divide out D.

★**25.** Suppose that OP is the normal line from the origin to $P:(x_0, y_0, z_0)$ on a plane, W, and that $\overrightarrow{OP}$ has the direction angles $\{\alpha, \beta, \gamma\}$. With $|\overline{OP}| = p$, prove that an equation for the plane is $x \cos \alpha + y \cos \beta + z \cos \gamma = p$. (If $p = 0$, we agree to take $|\overline{OP}| = 1$, and direct $\overrightarrow{OP}$ upward, or to the right if W is vertical.) This form of equation is referred to as the **normal form** for the equation of a plane.

Hint. First obtain $\{x_0, y_0, z_0\}$ in terms of $\{\alpha, \beta, \gamma, p\}$.

105. A LINE DEFINED BY TWO PLANES

If the following linear equations do *not* represent parallel planes, the graph of the system is the line L which is the intersection of the planes W_1 and W_2:

$$W_1:\{A_1x + B_1y + C_1z = D_1\}; \quad W_2:\{A_2x + B_2y + C_2z = D_2\}. \quad (1)$$

An unlimited number of pairs of equations as in (1) can be written for any given line L, because an unlimited number of pairs of noncoincident planes, W_1 and W_2, can be drawn through L. We refer to (1) as the *equations* of L.

EXAMPLE 1.   Sketch the line L with the equations

$$3x + 2y = 6 \quad and \quad y + 2z = 2. \quad (2)$$

Solution.   1. First we draw the planes

$$W_1:\{3x + 2y = 6\} \quad and \quad W_2:\{y + 2z = 2\}, \quad (3)$$

in Figure 191. W_1 is perpendicular to the xy-plane because z is *not present* in $3x + 2y = 6$ (and thus there is no z-intercept: $x = 0 = y$ yields $6 = 0$).

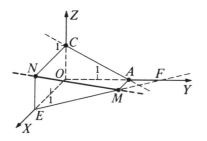

FIGURE 191

W_2 is perpendicular to the yz-plane. The xy-trace of W_1 has the equation $3x + 2y = 6$ in the xy-plane, and is EF in Figure 191; the xz-trace has the equations

$$3x + 2y = 6 \quad and \quad y = 0, or \tag{4}$$

$$3x = 6 \quad and \quad y = 0. \tag{5}$$

The graph of (5) is the line EN with the equation $x = 2$ in the xz-plane. Similarly, W_2 has the yz-trace CA, xz-trace CN, and xy-trace AM in Figure 191.

2. The intersection, M, of the xy-traces of W_1 and W_2 is a point whose coordinates satisfy (2); hence M is on L. Similarly, the intersection, N, of the xz-traces is on L. Hence, MN is the line L. In (3), W_1 *projects L* onto the xy-plane; W_2 *projects L* onto the yz-plane.

In (3), we refer to W_1 and W_2 as *projection planes* for L as defined by (2), because each of W_1 and W_2 is perpendicular to a coordinate plane. The equations of W_1 and W_2 may be called *projection equations* for L.

Suppose that L is defined as the graph of two linear equations where one or both are not the equations of projection planes (that is, not perpendicular to coordinate planes). Then, we may obtain the equations of corresponding projection planes by eliminating one variable at a time by use of the given equations, as in the next example.

EXAMPLE 2.   Obtain two projection planes for the line L which is the graph of the statement

$$4x + y + 2z = 10, and \tag{6}$$

$$x - 2y + 2z = -2. \tag{7}$$

Solution.   Subtract, in the order (7) from (6):

$$3x + 3y = 12, \quad or \quad x + y = 4. \tag{8}$$

Multiply by (2) in (6): $8x + 2y + 4z = 20.$ (9)

Add sides of (7) and (9): $9x + 6z = 18.$ (10)

Hence equations (6) and (7) are equivalent to

$$x + y = 4 \quad and \quad 3x + 2z = 6. \tag{11}$$

Thus L is the graph of (11), where the first plane projects L onto the xy-plane, and the second projects L onto the xz-plane. To graph L, it would be easier to use (11) than (6) and (7).

ILLUSTRATION 1.   The line L which is the intersection of

$$x + 2y = 6 \quad and \quad 3x + y = 6 \tag{12}$$

is shown (in part) as HK in Figure 192. Each plane in (12) is perpendicular to the xy-plane. The intersection of the xy-traces of (12) gives H in Figure 192. The upper sides of the rectangles on the planes from (12) were drawn at the same height and, of course, parallel to the xy-traces. This is done because parallel lines in space become parallel lines in a figure drawn by parallel projection, which we use. HK is vertical.

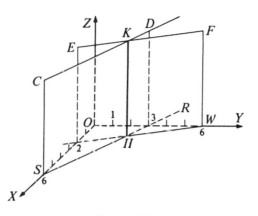

FIGURE 192

EXERCISE 72

Graph the line whose equations are given.

1. $x + z = 4$ and $4x + 3y = 12$.   2. $y + 3z = 3$ and $y + 2x = 4$.
3. $2x + y = 4$ and $2x + 3y = 6$.   4. $x + z = 2$ and $2z - x = 2$.
5. $-3x + 2y = 6$ and $y + 2z = 4$.   6. $y + z = 3$ and $y + 4z = 4$.
7. $y = 2x$ and $x + y = 3$.   8. $y = z$ and $y + z = 2$.

9. Construct the traces in the three coordinate planes for each given plane. By finding the intersections of corresponding traces, obtain three points on the line L whose equations are given. (As a check, the points should fall on a line.) The scale units on the axes must be adhered to accurately. Draw L.

$$3x + 4y + 6z = 12 \quad and \quad 3x - 16y - 6z = -24.$$

Obtain the equations of two projection planes for the line L whose equations are given. Then construct L.

10. $4x + y + 2z = 10$ and $x - 2y + 2z = -2$.
11. $2x + 4z - y = 6$ and $3x + 2y - z = 2$.
12. $x + y + 3z = 5$ and $x - 3y - 3z = -7$.

106. PARAMETRIC AND SYMMETRIC EQUATIONS FOR A LINE

In the preceding section we considered a line L in xyz-space as determined by two nonparallel planes whose intersection is L. In such a case, L is the graph of the solution set of the system of two linear equations in x, y, and z for the planes whose intersection is L.

A line L also may be defined geometrically by specifying a point on L and its direction, or by giving two points on L. In each of these cases, we shall find a system of two linear equations, or other means, for describing L analytically.

THEOREM IX.   A line L through $P_1:(x_1, y_1, z_1)$ with the direction numbers $a:b:c$ has the parametric equations

$$x = x_1 + at, \qquad y = y_1 + bt, \qquad z = z_1 + ct, \tag{1}$$

where t is a parameter whose domain is all real numbers.   Also, L has the equations

$$\frac{x - x_1}{a} = \frac{y - y_1}{b} = \frac{z - z_1}{c}, \tag{2}$$

where we omit any fraction with a zero denominator and use the corresponding equation from (1) *in addition to* (2).

Proof.   1. From (6) on page 303, if $P:(x,y,z)$ is any point in space other than P_1, as in Figure 193, direction numbers for P_1P are $(x - x_1):(y - y_1):(z - z_1)$. Then, P is on L if and only if

$$(x - x_1):(y - y_1):(z - z_1) = a:b:c. \tag{3}$$

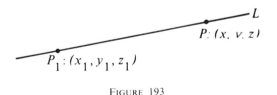

$$L$$
$$P:(x, v, z)$$
$$P_1:(x_1, y_1, z_1)$$

FIGURE 193

2. By the definition of a proportion as on page 301, (3) means that, for each point $P:(x,y,z)$, not P_1, on L, there exists a corresponding number $t \neq 0$ such that

$$x - x_1 = at, \qquad y - y_1 = bt, \qquad z - z_1 = ct, \tag{4}$$

which is equivalent to (1). Also, for every value of $t \neq 0$, (4) gives (x,y,z) satisfying (3), so that $P:(x,y,z)$ is on L. Moreover, if $t = 0$ then

(4) gives $\{x = x_1, y = y_1, z = z_1\}$. Hence, with no exceptions, the coordinates of each point P on L are given by (4) for some value of t; and, for each value of t, (4) gives just one point on L. Thus, there is a "*one-to-one*" correspondence between points on L and values of t, in (4). For this reason we call (4), or (1), *parametric equations of L with t as the parameter.*

3. If $a \neq 0$, from $x - x_1 = at$ we obtain $t = (x - x_1)/a$, etc. for $t = (y - y_1)/b$ and $t = (z - z_1)/c$. Hence, if $\{a, b, c\}$ are all different from zero, the continued equality (2) is obtained.

Notice that (2) abbreviates

$$\frac{x - x_1}{a} = \frac{y - y_1}{b}, \quad and \quad \frac{y - y_1}{b} = \frac{z - z_1}{c}, \quad and \quad \frac{x - x_1}{a} = \frac{z - z_1}{c},$$

where any two of the equations implies the third equation. Thus, (2) is equivalent to

$$\frac{x - x_1}{a} = \frac{y - y_1}{b} \quad and \quad \frac{y - y_1}{b} = \frac{z - z_1}{c}. \tag{5}$$

We refer to (2) as *symmetric equations* for the line L. From (2), as above, we obtain (5) as the equations of two planes whose intersection is L. Notice that any equation derived from (2), as in (5), is a *projection equation* for L, because one variable is missing.

EXAMPLE 1.   Obtain two projection equations for the line L through the point $(-2, 6, -2)$ with the direction $4 : -6 : 3$.

Solution.   From (2), equations for the line are

$$\frac{x + 2}{4} = \frac{y - 6}{-6} = \frac{z + 2}{3}, \ or$$

$$\frac{x + 2}{4} = \frac{y - 6}{-6} \quad and \quad \frac{y - 6}{-6} = \frac{z + 2}{3}. \tag{6}$$

In (6) we obtain $\{3x + 2y = 6 \text{ and } y + 2z = 2\}$ as equations for L.

THEOREM X.   If $P_1 : (x_1, y_1, z_1)$ and $P_2 : (x_2, y_2, z_2)$ are distinct points, then symmetric equations for the line $P_1 P_2$ are*

$$\frac{x - x_1}{x_2 - x_1} = \frac{y - y_1}{y_2 - y_1} = \frac{z - z_1}{z_2 - z_1}. \tag{7}$$

We obtain (7) from (2) by use of (6) on page 303, which states that $P_1 P_2$ has the direction numbers $(x_2 - x_1) : (y_2 - y_1) : (z_2 - z_1)$.

*With the understanding that no zero denominator is used, and that a zero denominator implies use of the corresponding numerator placed equal to zero, in addition to (7).

ILLUSTRATION 1.   From (7), symmetric equations for the line through $(3,-2,1)$ and $(1,-1,-2)$ are as follows with $(3,-2,1)$ used as P_1:

$$\frac{x-3}{-2} = \frac{y+2}{1} = \frac{z-1}{-3}, or$$

$$x + 2y = -1 \quad and \quad 3y + z = -5.$$

EXAMPLE 2. ˙ Obtain projection equations for the line L through the point $P:(4,1,6)$ with direction numbers $2:0:3$.

Solution.   We cannot use the fraction with denominator b in (2). The other fractions, and the equation involving b in (1), yield

$$\frac{x-4}{2} = \frac{z-6}{3} \quad and \quad y - 1 = 0, or$$

$$3x - 2z = 0 \quad and \quad y = 1,$$

as the equations of projection planes through L.

Consider the line P_1P_2 where $P_1:(x_1,y_1,z_1)$ and $P_2:(x_2,y_2,z_2)$ are distinct points; P_1P_2 has the direction numbers $(x_2 - x_1):(y_2 - y_1):(z_2 - z_1)$. By use of (1), parametric equations for P_1P_2 are

$$x = x_1 + t(x_2 - x_1), \quad y = y_1 + t(y_2 - y_1), \quad z = z_1 + t(z_2 - z_1). \quad (8)$$

Suppose that a number scale is imposed on the line P_1P_2, with either direction on the line taken as positive. Then, a proof as in the case of similar equations for a line in a plane, on page 264, would prove the following fact:

For any point $P:(x,y,z)$ on P_1P_2 of (8), the corresponding value of t is defined by

$$\overline{P_1P} = t\overline{P_1P_2}, \quad or \quad t = \frac{\overline{P_1P}}{\overline{P_1P_2}}. \qquad (9)$$

Thus, to obtain P as the midpoint of P_1P_2, we use $t = \frac{1}{2}$ in (8). For the points of trisection of P_1P_2 we use $t = \frac{1}{3}$ and $t = \frac{2}{3}$. To find P so that $\overline{P_1P} = 2\overline{P_1P_2}$, we use $t = 2$.

EXERCISE 73

1. Write two projection equations for the line and then draw it:

$$\frac{x-3}{-3} = \frac{y}{4} = \frac{z-1}{1}.$$

Write equations in symmetric form for the line satisfying the data.
2. Through $(-1,2,3)$ with the direction $2:-3:1$.

3. Through $(2, -1\frac{3}{4})$ with the direction $-1:2:-3$.

4. Through $(0, -1, 3)$ with direction cosines $-\frac{1}{3} : -\frac{2}{3} : \frac{2}{3}$.

5. Through $(2, -1, 3)$ with the direction $2:0:-1$.

6. Through the points $(-1, 2, -1)$ and $(3, 4, -2)$.

7. Through the points $(-3, 1, 0)$ and $(0, 1, -3)$.

8. Through the points $(-2, 3, 1)$ and $(0, -2, 1)$.

9. Through the points $(2, -3, 2)$ and $(2, -5, 0)$.

10. Through $(2, -1, 3)$ perpendicular to the plane $3x + 2y + 5z = 3$.

11. Through $(-1, 3, 4)$ perpendicular to the plane $x - 2y = 2$.

12. Through $(2, -1, 3)$ perpendicular to the plane $3y - 4z = 6$.

13. Through $(2, -1, 3)$ parallel to the line $\dfrac{x - 2}{3} = \dfrac{y}{2} = \dfrac{z + 1}{-1}$.

14. Through $(2, 3, -5)$ parallel to the z-axis.

15. Through $(2, -1, 2)$ parallel to the line through $(1, 3, -1)$ and $(-2, 1, 4)$.

Find the coordinates of the midpoint and the trisection points of the line segment joining the points.

16. $(3, 5, -4); (-3, -1, 8)$.   **17.** $(-2, 1, 3)$ and $(2, 4, -3)$.

A line L goes through the point $(2, -3, 2)$ and has the direction $3 : -2 : 1$. By use of the parametric form for L, find the coordinates of the point where L pierces the specified plane.

18. The xy-plane.   **19.** $z = 3$.   **20.** $3x - 2y + 5z = -14$.

Change the equations of the line defined by the equations to the symmetric form. Then, state direction numbers for L and a point on it.

21. $\dfrac{2x - 1}{3} = \dfrac{2 - y}{2} = \dfrac{3z + 5}{1}$.   **22.** $\dfrac{3x - 2}{4} = \dfrac{5 + y}{3} = \dfrac{1 - 2z}{5}$.

Hint. To change $\dfrac{2x - 1}{3}$ to the form $\dfrac{x - x_1}{a}$, divide both numerator and denominator by 2; etc., for the other fractions.

23. $9x - 5y + 3z = 8$ and $9x - 2y + 12z = 17$.

Hint. A point on L and its direction are needed. Place $y = 0$ and solve for x and z to find a point. The direction $a:b:c$ is perpendicular to $9 : -5 : 3$ and to $9 : -2 : 12$; hence

$$9a - 5b + 3c = 0 \quad and \quad 9a - 2b + 12c = 0.$$

Solve for a and b in terms of c, and then obtain $a:b:c$ in a form not involving c.

24. $x + y - z = 1$ and $4x - y + 6z = 9$.

107. EQUATIONS FOR A SPHERE

Let r be the radius and let $C:(g, h, k)$ be the center of a sphere. Then, it has the equation

$$(x - g)^2 + (y - h)^2 + (z - k)^2 = r^2, \tag{1}$$

because (1) states that the square of the distance between $C:(g,h,k)$ and $P:(x,y,z)$ is equal to r^2. By completing squares with the terms in x, y, and z, respectively, we find that any equation of the form

$$x^2 + y^2 + z^2 + Rx + Sy + Tz + U = 0$$

represents a sphere, real (including a point-sphere) or imaginary.

108. CYLINDERS

If a line L with a fixed direction moves through all points of a curve C in a plane, the surface T thus swept out by L is called a **cylinder** with C as a **directrix curve.** Each position of L on T is called a *ruling* of T. For any cylinder, we may always choose a directrix curve which is a plane section of the cylinder perpendicular to the rulings. Hereafter it will be assumed that the directrix has been chosen in the preceding fashion. In any instance which we shall consider, the rulings will be perpendicular to a coordinate plane, so that the directrix C may be taken as the trace of the cylinder in that plane. We shall say that the cylinder is *perpendicular to this plane,* and *parallel to the rulings.* Frequently, a cylinder is named after its directrix. Thus, the cylinder is called a circular, an elliptic, a hyperbolic, or a parabolic cylinder according as the directrix is a circle, an ellipse, a hyperbola, or a parabola.

EXAMPLE 1.   Investigate the graph of $x^2 + y^2 = 4.$ (1)

Solution.   1. Let T be the graph of (1). The xy-trace of (1) is the circle with center at the origin $O:(0,0,0)$ and radius 2.

2. If $P:(x,y,z)$ is any point in space, the perpendicular projection of P on the xy-plane is a point $Q:(x,y,0)$, as in Figure 194. Then, since (1) does not involve z, P is on the graph of (1) when and only when the projection $Q:(x,y,0)$ has coordinates satisfying (1). This means that Q is on the circle C which is the graph of (1) in the xy-plane. Hence, T consists of those points which are located on the lines perpendicular to the xy-plane at points on the circle C. That is, T is a circular cylinder perpendicular to the xy-plane with the directrix C.

Reasoning similar to that employed in the preceding solution proves the following result.

THEOREM XI.   If an equation $f(x,y,z) = 0$ does not involve one of the variables $\{x,y,z\}$, the graph of the equation is a cylinder perpendicular to the plane of the other variables, with a directrix which is the graph of the equation in that plane.

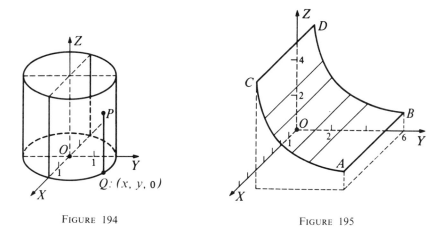

FIGURE 194   FIGURE 195

Any plane section of a cylinder parallel to the rulings intersects the cylinder in one or more of the rulings, as seen for the trace in the xz-plane in Figure 194. Any plane section of the cylinder perpendicular to the rulings is a curve which is congruent to the directrix curve, and may be moved into coincidence with it by translation parallel to the rulings.

ILLUSTRATION 1.   The graph of $yz = 6$ in xyz-space is a hyperbolic cylinder, perpendicular to the yz-plane. A piece of this cylinder is shown in Figure 195.

ILLUSTRATION 2.   We have seen that the graph of $x + 3y = 8$ is a plane perpendicular to the xy-plane. We may refer to the graph as a *linear* cylinder.

To write the equation of a cylinder with its directrix specified as a certain curve in a given coordinate plane, we write the equation in two variables whose graph in that plane is the given curve.

ILLUSTRATION 3.   Let the directrix of a cylinder be the ellipse in the xz-plane whose x-intercepts are ± 5 and z-intercepts are ± 3, with the x-axis and z-axis as axes of symmetry. An equation for this ellipse in the xz-plane is

$$\frac{x^2}{25} + \frac{z^2}{9} = 1.$$

Hence this is the xyz-equation for the cylinder which is perpendicular to xz-plane.

EXERCISE 74

Find the center and radius of the sphere.
1. $x^2 + 2x + y^2 - 6y + z^2 - 5 = 0.$
2. $x^2 + y^2 + z^2 - 4x + 2y + 6z + 10 = 0.$

Show a few of the rulings on each cylinder which is constructed.

Construct the piece of the graph of the equation between the xy-plane and the plane z = 4.

3. $x^2 + y^2 = 25$.   **4.** $4x^2 + 9y^2 = 36$.   **5.** $y = 2x^2$, with $y \leq 5$.

6. Construct the piece of the cylinder $9y^2 - 4x^2 = 36$ between the xy-plane and the plane $z = 5$, with $2 \leq y \leq 10$.

7. Construct a piece of the cylinder $xz = 4$ between the xz-plane and the plane $y = 5$ in the first octant, with $x \leq 4$ and $z \leq 4$.

Graph the equation in xyz-space.

8. $y^2 + z^2 = 4$, with $0 \leq x \leq 6$.

9. $z = y^2$, with $0 \leq x \leq 8$ and $z \leq 4$.

10. $z = x^2$, with $0 \leq y \leq 5$ and $z \leq 4$.

11. $xy = 4$, with $0 \leq x \leq 4, 0 \leq y \leq 4$, and $0 \leq z \leq 5$.

12. $x^2 + z^2 = 25$, with $0 \leq y \leq 4$.

109. SURFACES OF REVOLUTION

If a plane curve C is revolved about a line L in its plane, the surface T which thus is generated is called a *surface of revolution,* with L as the *axis* of revolution and C as the *generatrix.* Any plane section of T by a plane through L is called a *meridian section,* and consists of one of the positions assumed by C in its revolution about L, together with the symmetric reflection of C with respect to L. If the given curve C is symmetric to L, each meridian section of T is simply a position assumed by C in its revolution about L. Any plane section of T by a plane perpendicular to L is a circle. If an ellipse, a hyperbola, or a parabola is revolved about an axis of symmetry of the curve, the surface obtained is called an *ellipsoid, a hyperboloid,* or a *paraboloid of revolution.* An ellipsoid of revolution is called an *oblate* or a *prolate spheroid* according as the ellipse is revolved about its minor or its major axis. We obtain a sphere when the ellipse is a circle. Figure 196 shows a prolate spheroid. The earth is known to be approximately an oblate spheroid.

Illustration 1.   Let the generatrix of a spheroid be the graph in the yz-plane of

$$9y^2 + 4z^2 = 36. \tag{1}$$

A graph of (1) in the yz-plane is the ellipse with y-intercepts ±2 and z-intercepts ±3 shown in Figure 196. If this ellipse is revolved about the z-axis, the prolate spheroid W of Figure 196 is obtained. Any meridian section of W, for instance the xz-trace, is identical with the yz-trace. Any section of W parallel to the xy-plane is a circle. Thus, the circle TUV on W in Figure 196 results from revolution of either U or V about OZ.

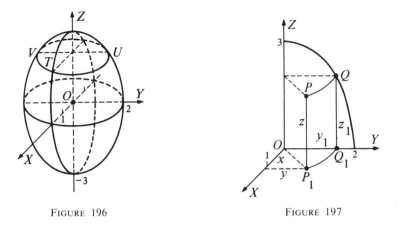

FIGURE 196   FIGURE 197

EXAMPLE 1.   Obtain the equation of the spheroid W obtained by revolving (1) about OZ.

Solution.   1. In Figure 197, let $Q:(0,y_1,z_1)$ be any point on the ellipse C which is the graph of (1) in the yz-plane. Let $P:(x,y,z)$ be any point of W into which Q revolves as the spheroid is generated. Let $P_1:(x,y,0)$ and $Q_1:(0,y_1,0)$ be the projections of P and Q, respectively, on the xy-plane.

2. Since Q is on the ellipse (1), we have

$$9y_1^2 + 4z_1^2 = 36. \tag{2}$$

In Figure 197, $\overline{P_1P} = \overline{Q_1Q}$; $\overline{OQ_1} = y_1$; $\overline{OP_1^2} = \overline{OQ_1^2}$; $\overline{OP_1^2} = x^2 + y^2$. Hence

$$z = z_1; \qquad x^2 + y^2 = y_1^2. \tag{3}$$

When (3) is used in (2), we obtain

$$9(x^2 + y^2) + 4z^2 = 36, \tag{4}$$

or $9x^2 + 9y^2 + 4z^2 = 36$ is an equation for W. Notice that (4) is obtained from (1) on replacing y^2 by $(x^2 + y^2)$.

Let C be the graph of $f(y,z) = 0$, or $f(x,y) = 0$, or $f(x,z) = 0$, in the coordinate plane of the two variables involved. Then, as in Example 1, the equation of the surface generated by revolving C about a coordinate axis in its plane is obtained as follows:

I. **Revolution about OZ:**

If C has the equation $f(y,z) = 0$, *replace* y^2 by $(x^2 + y^2)$.*

If C has the equation $f(x,z) = 0$, *replace x^2 by $(x^2 + y^2)$.*

*That is, replace $|y|$ by $\sqrt{x^2 + y^2}$, and similarly in other cases.

II. Revolution about OY:

If C has the equation $f(x,y) = 0$, replace x^2 by $(x^2 + z^2)$.

If C has the equation $f(y,z) = 0$, replace z^2 by $(x^2 + z^2)$.

III. Revolution about OX:

If C has the equation $f(x,y) = 0$, replace y^2 by $(y^2 + z^2)$.

If C has the equation $f(x,z) = 0$, replace z^2 by $(y^2 + z^2)$.

ILLUSTRATION 2.   If the hyperbola $y^2 - 4z^2 = 4$ (5)

in the yz-plane is revolved about OY, a *hyperboloid of revolution, T,* also called a *hyperboloid of two sheets,* is generated, as shown in Figure 198. An equation for T is obtained by substituting $(x^2 + z^2)$ for z^2 in (5):

$$y^2 - 4(x^2 + z^2) = 4.$$

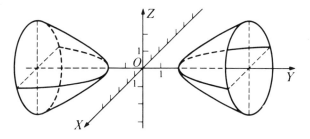

FIGURE 198

From I, II, and III, we observe that the graph of an xyz-equation is a surface of revolution, with OX, OY, or OZ as the axis of revolution, if the equation can be arranged to exhibit the following characteristics, respectively:

I. **Axis OZ:** *x and y occur only in the combination $(x^2 + y^2)$.*

II. **Axis OY:** *x and z occur only in the combination $(x^2 + z^2)$.*

III. **Axis OX:** *y and z occur only in the combination $(y^2 + z^2)$.*

110. CONES OF REVOLUTION

Let C be a fixed plane curve and let V be a fixed point not in the plane of C. Then, the surface generated, or swept out, by a line L drawn through V to a point P on C, as P moves through all positions on C, is called a **cone**, T, whose vertex is V and directrix curve is C. Each position of L on T is called a *ruling* of T. A special case of this definition was met in the introduction to conic sections on page 86, which should be reread at this time to recall the description of a cone.

ILLUSTRATION 1.   Suppose that the directrix C for a cone T is a circle, with the vertex O of the cone on a line L perpendicular to the plane of C at its center, as in Figure 199, where L is the z-axis and C is either one of the circles in the figure. Then T is a surface of revolution about L as the axis, and is called a *right circular cone*. Any circular section of this cone perpendicular to its axis may be considered as a directrix for the cone.

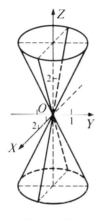

FIGURE 199

EXAMPLE 1.   Obtain an equation for the right circular cone generated by revolving about OZ the line L which is the graph of $z = 2y$ in the yz-plane.

Solution.   By I on page 321, we replace $|y|$ by $\sqrt{x^2 + y^2}$ in $z = 2y$ and obtain

$$\pm 2\sqrt{x^2 + y^2} = z, \tag{1}$$

where "+" applies where $z \geqq 0$ and "−" applies where $z < 0$. On squaring both sides in (1), we obtain $4x^2 + 4y^2 = z^2$ as the equation of the cone, as shown in Figure 199.

EXERCISE 75

Sketch a limited part of the surface* obtained by revolving the graph of the given equation, in the plane of the variables involved, about the specified

*Sketch all of any ellipsoid.

coordinate axis. Show the traces of the surface in the coordinate planes, and perhaps a few circular sections cut by planes perpendicular to the axis of revolution. Also write an equation for the surface.

1. $9y^2 + 25z^2 = 225$:   about OY; about OZ.
2. $9x^2 + 4z^2 = 36$:   about OX; about OZ.
3. $y^2 - z^2 = 16$:   about OY; about OZ.

Note 1.   In Problem 3, the surface of revolution about OZ is called a hyperboloid of revolution of *one sheet.*

4. $z = y^2$:   about OZ.   **5.** $y = z^2$:   about OY.
6. $x = 3z$:   about OZ.   **7.** $2y = z$:   about OY.
8. $4z^2 - x^2 = 4$: about OZ; about OX.

Describe the graph of the equation as a surface of revolution, stating the axis and a generatrix. Then construct the surface.

9. $4(x^2 + y^2) + z^2 = 4$.   **10.** $(x^2 + z^2) - 4y = 0$.
11. $4x^2 + 4y^2 = z^2$.   **12.** $16x^2 + 9y^2 + 9z^2 = 144$.
13. $x^2 - y^2 + z^2 = 9$.   **14.** $9x^2 - y^2 + 9z^2 = 0$.

111. SECTIONS OF A SURFACE, AND SYMMETRY

Recall the discussion of symmetry for graphs in an xy-plane on page 95. Similar remarks can be made about symmetry for graphs in xyz-space. Thus a surface T with an equation $f(x,y,z) = 0$ is symmetric to the xy-plane if the equation is essentially unaltered when z is replaced by $-z$. T is symmetric to the origin if the equation is essentially unaltered when (x,y,z) are replaced by $(-x,-y,-z)$. We shall refer to symmetry only with respect to a coordinate plane or the origin. Cylinders and surfaces of revolution, as considered previously, have obvious types of symmetry which we have not emphasized. To investigate any surface T with a given equation $f(x,y,z) = 0$, test it for intercepts of T on the axes, traces of T in the coordinate planes, domains for the variables, symmetry, and the nature of plane sections of T by planes parallel to each of the coordinate planes.

EXAMPLE 1.   Discuss the surface: $\dfrac{x^2}{4} + \dfrac{y^2}{9} + \dfrac{z^2}{25} = 1$.   (1)

Solution.   1. Observe that (1) is unaltered if (x,y,z) are replaced by $(-x,-y,-z)$. Hence the graph T of (1) is symmetric to the origin. Since (1) is unaltered if z is replaced by $-z$, T is symmetric to the xy-plane, and similarly is symmetric to the xz-plane, and the yz-plane. T has the x-intercepts ± 2; y-intercepts ± 3; z-intercepts ± 5.

2. The xy-trace of T is the curve C with the equations

$$\frac{x^2}{4} + \frac{y^2}{9} + \frac{z^2}{25} = 1 \qquad and \qquad z = 0. \tag{2}$$

Or, C is the graph in the xy-plane of

$$\frac{x^2}{4} + \frac{y^2}{9} = 1, \tag{3}$$

which is an ellipse, in Figure 200. Similarly, the yz-trace of T is the ellipse $25y^2 + 9z^2 = 225$ in the yz-coordinate system of the yz-plane. The xz-trace also is an ellipse.

3. From (1), $$\frac{x^2}{4} + \frac{y^2}{9} = 1 - \frac{z^2}{25}. \tag{4}$$

Hence, if $|z| > 5$, the right-hand side of (4) is negative and (4) has no real solutions for x and y. Therefore the domain for z in (1) is $\{|z| \leqq 5\}$, and similarly is $\{|x| \leqq 2\}$ and $\{|y| \leqq 3\}$ for the other variables.

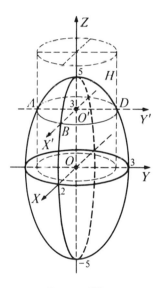

FIGURE 200

4. The plane section, C, of T by a plane $z = 3$ has the equations

$$\frac{x^2}{4} + \frac{y^2}{9} + \frac{z^2}{25} = 1 \qquad and \qquad z = 3, \tag{5}$$

which is equivalent to

$$\frac{x^2}{4} + \frac{y^2}{9} = 1 - \frac{9}{25} \qquad and \qquad z = 3. \tag{6}$$

The equation at the left in (6) represents an elliptic cylinder, H, perpendicular to the xy-plane, as indicated by broken lines in Figure 200. We shall call this cylinder the **projecting cylinder** for the plane section (5). The section of this cylinder by the plane $z = 3$ is the curve C, shown as AD in Figure 200, which is the graph of (6). Hence, the plane section of T by the plane $z = 3$ is an *ellipse*. Or, we may think of a new xy-system of coordinates in the plane $z = 3$, as in Figure 200, and then recognize the equation at the left in (6) as the equation of an ellipse. Similarly, if k is any number such that $|k| \leq 5$, the section of T by the plane $z = k$ is an ellipse. Also, we could show that plane sections of T by planes $x = h$, with $|h| \leq 2$, and by planes $y = g$, with $|g| \leq 3$, are ellipses. The graph of (1) is called an **ellipsoid.**

112. VARIOUS QUADRIC SURFACES

In the next exercise the student will have an opportunity to investigate and graph special cases of the types of quadratic equations in the following illustrations. In the equations, $\{a,b,c,p\}$ are constants, where $\{a,b,c\}$ are taken as positive.

ILLUSTRATION 1.   **General ellipsoid:**

$$\frac{x^2}{a^2} + \frac{y^2}{b^2} + \frac{z^2}{c^2} = 1. \tag{1}$$

A special case of (1) was considered in Example 1 on page 324.

ILLUSTRATION 2.   **Hyperboloid of one sheet,** in Figure 201:

$$\frac{x^2}{a^2} + \frac{y^2}{b^2} - \frac{z^2}{c^2} = 1. \tag{2}$$

Sections parallel to the xy-plane are ellipses. Sections parallel to the xz-plane or yz-plane are hyperbolas, except that sections by planes $x = a$, $x = -a, y = b$, or $y = -b$ in each case consist of two intersecting lines.

ILLUSTRATION 3.   **Elliptic paraboloid,** with $p > 0$ in Figure 202:

$$\frac{x^2}{a^2} + \frac{y^2}{b^2} = pz. \tag{3}$$

Sections parallel to the xy-plane are ellipses. Sections by planes of the form $x = g$ or $y = h$ are parabolas. The domain for z is $\{z \geq 0\}$.

ILLUSTRATION 4.   **Hyperboloid of two sheets,** illustrated in Figure 198 on page 322:

$$-\frac{x^2}{a^2} + \frac{y^2}{b^2} - \frac{z^2}{c^2} = 1. \tag{4}$$

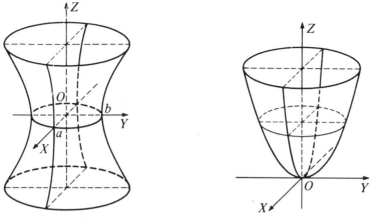

FIGURE 201   FIGURE 202

The domain for y is $\{\,|\,y\,| \geqq b\}$. Any section of (4) by a plane $y = h$ with $|\,h\,| \geqq b$ is an ellipse. Sections by planes of the form $x = g$ or $z = k$ arc hyperbolas.

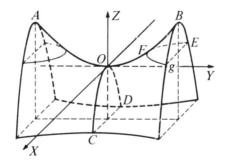

FIGURE 203

ILLUSTRATION 5.   **Hyperbolic paraboloid,** with $p > 0$ in Figure 203:

$$\frac{y^2}{b^2} - \frac{x^2}{a^2} = pz. \tag{5}$$

The surface is said to be *saddle shaped*. The xy-trace consists of two lines. Otherwise, any section by a plane $z = k$, with $k \neq 0$, is a hyperbola having its transverse axis parallel to OY when $k > 0$, and parallel to OX when $k < 0$. Other characteristics of the surface are discussed in the next example.

EXAMPLE 1.   Discuss the hyperbolic paraboloid

$$\frac{y^2}{4} - \frac{x^2}{9} = 2z, \tag{6}$$

Solution.   1. *Intercepts on the axes.* If $x = y = 0$ then $z = 0$, or the surface, W, defined by (6) goes through the origin. Thus, the x-intercept, y-intercept, and z-intercept are all zero.

2. *The xy-trace.* If $z = 0$ in (6) then

$$\frac{y^2}{4} - \frac{x^2}{9} = 0, \quad or \quad \left(\frac{y}{2} - \frac{x}{3}\right)\left(\frac{y}{2} + \frac{x}{3}\right) = 0. \tag{7}$$

From (7), $\dfrac{y}{2} - \dfrac{x}{3} = 0 \quad or \quad \dfrac{y}{2} + \dfrac{x}{3} = 0,$ \hfill (8)

and the xy-trace consists of two lines through the origin; the equations of these lines are in (8). To avoid complicating* Figure 203 for (6), these lines are not shown on the surface.

3. *The yz-trace.* If $x = 0$ in (6), then $y^2 = 8z$, which is the equation in the yz plane of the parabola AOB in Figure 203.

4. *The xz-trace.* If $y = 0$ in (6) then $x^2 = -18z$, which is the equation in the yz plane of the parabola COD in Figure 203.

5. Suppose that $z = 4$ in (6). Then

$$\frac{y^2}{4} - \frac{x^2}{9} = 8, \quad or \quad \frac{y^2}{32} - \frac{x^2}{72} = 1. \tag{9}$$

The curve of intersection of $z = 4$ and W has the equations

$$z = 4 \quad and \quad \frac{y^2}{32} - \frac{x^2}{72} = 1. \tag{10}$$

Equation (9) represents a hyperbolic cylinder perpendicular to the xy-plane. The trace of the cylinder in that plane is a hyperbola with y-intercepts $y = \pm\sqrt{32}$, transverse axis along the y-axis, and hence no x-intercepts. In Figure 203 the branches of the plane section (10) are shown as EFG and its symmetrical image with respect to the xz-plane. This section is typical of any plane section of W made by a plane $z = k$ with $k > 0$.

6. Suppose that $z = -4$ in (6). Then

$$\frac{y^2}{4} - \frac{x^2}{9} = -8, \quad or \quad \frac{x^2}{72} - \frac{y^2}{32} = 1. \tag{11}$$

---

*Scale marks are omitted on the axes in Figure 203, which thus represents *any* surface of type (5), rather than the particular surface (6) being discussed in Example 1.

The curve of intersection of $z = -4$ and W has the equations

$$z = -4 \quad and \quad \frac{x^2}{72} - \frac{y^2}{32} = 1. \tag{12}$$

Equation (11) represents a hyperbolic cylinder perpendicular to the xy-plane. The trace of the cylinder in that plane is a hyperbola with x-intercepts $x = \pm\sqrt{72}$, transverse axis along the x-axis, and hence no y-intercepts. The plane section (12) is not shown in Figure 203, to avoid complicating it. The section (12) is typical of the section of W by any plane $z = k$ where $k < 0$.

7. Equation (6) involves only *even* powers of x and y. Hence, the equation is unaltered when x is replaced by $-x$, or the surface W is symmetric to the yz-plane. Similarly, W is symmetric to the xz-plane.

ILLUSTRATION 6.　The **quadric cone,** as illustrated for a case where $a^2 = b^2$ in Figure 199 on page 323:

$$\frac{x^2}{a^2} + \frac{y^2}{b^2} - \frac{z^2}{c^2} = 0.$$

Suggestions for proving that the graph of this equation satisfies the definition of a cone, as given on page 322, are found in a problem of the next exercise.

The standard forms for surfaces which have been met in this section could be altered by interchanging the roles of (x,y,z) in the equations. Also, new standard forms could be given where each surface is oriented with respect to the point $P:(x_0,y_0,z_0)$ as a new origin, by a transformation of coordinates which translates the origin, as was done in the study of conic sections in a plane. The graph of any equation of the second degree in $\{x,y,z\}$ is called a *quadric surface,* or a *conicoid,* if the graph is not the empty set. It can be proved that a quadric surface is either a surface of one of the types we have considered, or is of a corresponding degenerate type, with planes of symmetry not necessarily parallel to coordinate planes.

EXERCISE 76

Investigate the graph of the equation as was done in Example 1 on page 328. Then draw a substantial part of the graph as illustrated by figures in Section 112.

1. $\dfrac{x^2}{9} + \dfrac{y^2}{4} + \dfrac{z^2}{16} = 1.$　　2. $\dfrac{x^2}{9} + \dfrac{y^2}{4} - \dfrac{z^2}{16} = 1.$

3. $-\dfrac{x^2}{9} + \dfrac{y^2}{4} - \dfrac{z^2}{16} = 1.$ **4.** $\dfrac{x^2}{9} + \dfrac{y^2}{4} - \dfrac{z^2}{16} = 0.$

5. $\dfrac{x^2}{4} + \dfrac{z^2}{9} = 2y.$ **6.** $\dfrac{y^2}{4} - \dfrac{x^2}{9} = 4z.$

7. $\dfrac{x^2}{9} + \dfrac{y^2}{4} = 2z.$ **8.** $\dfrac{x^2}{9} - \dfrac{y^2}{4} + \dfrac{z^2}{16} = 0.$

9. $x^2 = 16y^2.$ **10.** $z^2 = 6y.$ **11.** $x^2 - 4z^2 = 16.$

12. Prove that the graph of the equation in Illustration 6 on page 329 is a cone as defined in Section 110 on page 322.

Hint. Let C be any plane section of the surface made by a plane $z = k$. Let $P:(x_0, y_0, z_0)$ be on C. Prove that every point Q on OP lies on the surface. Use a parametric form for line OP.

chapter **12**

Mathematical Induction

113. THE AXIOM OF INDUCTION

In any logical foundation for the system of real numbers, the following axiom or an equivalent set of postulates is found to be essential.

Axiom of induction. *If W is a set of positive integers with the following properties, then W consists of all positive integers:*

A. *The integer* 1 *belongs to W.*

B. *If the integer h > 0 belongs to W, then (h + 1) also belongs to W.*

Suppose that hypothesis A of the axiom were changed to read "*the integer k > 0 belongs to W.*" Then the axiom would state that every integer $p \geqq k$ belongs to *W*. The axiom is the foundation for a powerful method of proof which we shall illustrate. The method sometimes applies in proving a theorem for which there is a special case corresponding to each value of the positive integer *n*.

From an intuitional standpoint, the axiom states a very elementary fact. Thus, let each positive integer be represented as a rung of a ladder reaching upward to unlimited height. The axiom then makes the following assertion: *if we are able to climb to the first rung of the ladder* (A of the Axiom), *and if we know that we can proceed from any rung to the next higher rung* (B of the Axiom), *then we can continue climbing upward forever.*

114. PROOF BY MATHEMATICAL INDUCTION

EXAMPLE 1.   By use of the axiom of induction, prove that, for every positive integer *n*, the sum of the first *n* even positive integers is equal to *n(n + 1)*. Or, for every *n > 0*,

$$2 + 4 + 6 + \cdots + 2n = n(n + 1). \tag{1}$$

Proof.   Let T_n represent the special case shown in (1). We desire to prove that all of the special cases $\{T_n, n = 1, 2, 3, \cdots\}$, are true.

Part I.   Verification of the first few special cases. In (1), we obtain:

when *n* = 1, $2 = 1(2) = 2,$ *or* $2 = 2;$

when $n = 2$, $2 + 4 = 2(3) = 6$,   *or* $6 = 6$;

when $n = 3$, $2 + 4 + 6 = 3(4) = 12$,   *or* $12 = 12$;

when $n = 4$, $2 + 4 + 6 + 8 = 4(5) = 20$,   *or* $20 = 20$.

Hence, equation (1) is true when $n = 1,2,3$, and 4, or $\{T_1, T_2, T_3, T_4\}$ are true. (Logically, below, only verification of T_1 would be necessary for the proof.)

Part II.   *Auxiliary Lemma.   If h is any positive integer such that equation* (1) *is true when n = h, then* (1) *is true also when n = h + 1.*

Proof of the Lemma.   1. The hypothesis is that (1) is true when $n = h$.

Hypothesis: $2 + 4 + 6 + \cdots + 2h = h(h + 1).$ (2)

With assumption (2), we desire to prove the conclusion that (1) is true also when $n = h + 1$, which is as follows:

Conclusion: $2 + 4 + \cdots + 2h + 2(h + 1) = (h + 1)(h + 2).$ (3)

We are entitled to use (2), and work forward to (3), but SHOULD NOT use (3) in the details of the proof.

2. Observe that the left-hand side of (2) consists of h terms.   The left-hand side of (3) consists of those h terms plus the added $(h + 1)$th term, or $2(h + 1)$. Hence, to prove (3), add $2(h + 1)$ to both sides of (2), and obtain

$$\left. \begin{array}{l} 2 + 4 + \cdots + 2h + 2(h + 1) \\ = h(h + 1) + 2(h + 1) = (h + 1)(h + 2). \end{array} \right\} \qquad (4)$$

In (4), both the left-hand and the right-hand sides are the same as in (3). Hence, (3) is true if (2) is true, or the Auxiliary Lemma has been proved.

Part III.   *Summary of the induction.*   Let W be the set of all values of n such that (1) is true.   We have verified that T_1 is true, or that (1) is true when $n = 1$.   Hence, W contains the integer 1.   Also, in Part II we proved that, if W contains the integer h, that is, if T_h is true, then W also contains the integer $(h + 1)$, or T_{h+1} is true.   Therefore, by the axiom of induction, W contains *all* positive integers, or (1) is true for all values of n. That is, all of the cases $\{T_1, T_2, T_3, \cdots\}$ of (1) are true.

The method of proof used in Example 1 is called **mathematical induction.**   Recall the ladder analogy concerning the axiom of induction in Section 113.   The verification in Part I of the proof shows that we can reach the first rung (or, we may feel more secure if we verify that we can ascend the first few rungs).   In Part II of the proof, we show that, if we can ascend to the hth rung, then we can go to the next higher rung.   Parts I and II thus show that similar parts of the axiom of induction are true for the set W of all values of n for which the theorem $\{T_n\}$ is true. Then, the axiom proves that W contains *all* $n > 0$, and hence all cases of the theorem $\{T_n\}$ are true.

Method of mathematical induction. *Proof of a theorem* $\{T_n\}$, *for all values of the positive integer n.*

   I. *Let W be the set of all values of n for which T_n is true. Verify that T_n is correct when $n = 1$ (and possibly also when $n = 2$ and $n = 3$ for added appreciation of the nature of T_n).*

   II. *Proof of an auxiliary lemma with the hypothesis that T_n is true for any particular value $n = h$, and with the conclusion that, then, T_n is true when $n = h + 1$.*

   III. *A summary of the induction, stating clearly why Parts I and II, and the axiom of induction show that W contains all integers, or that T_n is true for all values of n.*

Both Parts I and II of a proof by mathematical induction are necessary as a basis for Part III. For instance, Part I alone would be insufficient, regardless of how many special cases might be verified, because such verification cannot prove a general theorem.

   ILLUSTRATION 1.   The number $T_n = n^2 - n + 41$ can be verified to be a prime number when $n = 1, 2, 3, \cdots, 40$. Thus, it might be inferred that T_n is a prime number for *all* values of n. Such a conjecture would be incorrect because $T_{41} = (41)^2 - 41 + 41$ is NOT prime.

Also, Part II alone of a proof by mathematical induction would give no basis for a proof. Thus, if we should forget the necessity of Part I of such a proof, we could prove the false statement that, if n is any positive integer,

$$1 + 2 + 3 + \cdots n = 50 + \tfrac{1}{2}n(n + 1). \tag{5}$$

The interested student may prove easily, as in Example 1, that (5) is true when $n = h + 1$ *if* (5) is true when $n = h$.

   It must not be inferred that proof by mathematical induction applies only when the theorem involved can be stated in the form of an equation, as in Example 1. It merely happens that theorems of this variety are useful illustrations. In calculus, for instance, proofs by mathematical induction occur often where no equation need be involved.

   Note 1.   Recall that an **arithmetic progression** is defined as a sequence of terms such that each term, after the first term, is obtained from the preceding term by adding to it a fixed number called the **common**

difference. Thus, the sequence $\{-5, -3, -1, \cdots\}$ is an arithmetic progression (abbreviated by "A.P.") where the first term is -5 and the common difference is 2.

EXAMPLE 2.   Let the first term of an A.P. be b and the common difference be d. By mathematical induction, prove that the nth term, l_n, of the progression is $[b + (n - 1)d]$.

Proof.   *Part* I. *Verification of special cases.* By the data and the definition in Note 1:

1st *term is* b;   (6)

2d *term is* $(b + d)$;   (7)

3d *term is* $(b + d) + d, or (b + 2d)$.   (8)

We observe that, if $l_n = b + (n - 1)d$ and $n = 1, 2,$ and 3, the terms in (6), (7), and (8), respectively, are obtained.   Hence, the result $l_n = [b + (n - 1)d]$ has been verified when $n = 1, 2,$ and 3.

Part II. *Auxiliary Lemma. If the formula* $l_n = b + (n - 1)d$ *is correct when* $n = h$, *then the formula is correct when* $n = h + 1$.

Proof.   By hypothesis,

$$h\text{th } term\ is\ [b + (h - 1)d].   (9)$$

By the definition of an A.P. in Note 1, and (9),

$$(h + 1)\text{th } term\ is\ \ \ \ b + (h - 1)d + d = b + hd.   (10)$$

The result in (10) is seen to be that which is obtained if we replace n by $(h + 1)$ in $[b + (n - 1)d]$. Hence, the auxiliary lemma has been proved.

Part III.   *Summary of the induction.* Let W be the set of all values of n for which $l_n = b + (n - 1)d$, and let T_n represent the special case of this result for any specified integer n. In Part I it was verified that T_n is true when* $n = 1$, etc., as in Part III of the solution of Example 1.

For use later, recall the following results from Theorem I on page 43 and Theorem II on page 44.

$$If\ \ 0 < A\ and\ C \leqq D, then\ AC \leqq AD.   (11)$$

$$If\ A < B\ and\ B < D, then\ A < D.   (12)$$

EXAMPLE 3.   By mathematical induction, prove the following result. *If* $0 < x < 1$ *and* n *is any positive integer greater than* 1, *then*

$$x^n < x.   (13)$$

Solution.   *Part* I.   *Verification for* $n = 2$. Since $0 < x$ and $x < 1$.

*Also T_2 and T_3 were verified, but there is no need for mentioning them above.

because of (11) we obtain

$$x \cdot x < x \cdot 1 \qquad or \qquad x^2 < x. \tag{14}$$

Hence (13) is true when $n = 2$.

Part II.   *Auxiliary Lemma. If* (13) *is true when* $n = h$, *where* h *is any positive integer greater than* 1, *then* (13) *is true when* $n = h + 1$.

Proof.   By hypothesis, $x^h < x$. Then, by use of (11),

$$x \cdot x^h < x \cdot x \qquad or \qquad x^{h+1} < x^2. \tag{15}$$

In (14) and (15) we have $x^{h+1} < x^2$ and $x^2 < x$. Therefore, by use of (12), we obtain $x^{h+1} < x$, which proves the auxiliary theorem.

Part III.   *Summary of the induction.* Let W be the set of values of the positive integer n for which $x^n < x$, and let T_n be the case stating that the result is true for any specified integer n. In Part I, it was verified that T_n is true when $n = 2$. In Part II it was proved that, if T_n is true when $n = h > 1$ then T_n is true when $n = h + 1$. Or, W includes the integer $(h + 1)$ if W includes the integer h. Hence, by the axiom of induction, W includes all positive integers greater than 1, or (13) is true for all values of the positive integer $n \geq 2$.

Note 2.   In the natural sciences, a general conclusion often is reached, although not demonstrated in the mathematical sense, by a consideration of what occurs in a succession of special cases. Such reasoning is referred to as *incomplete induction*. In contrast, mathematical induction can be called *complete induction*.

EXERCISE 77

Prove that the stated result is true for all values of the positive integer n by use of mathematical induction.

1. $3 + 6 + \cdots + 3n = \dfrac{3n(n + 1)}{2}$.

2. $6 + 12 + \cdots + 6n = 3n(n + 1)$.

3. $2 + 2^2 + \cdots + 2^n = 2^{n+1} - 2$.

4. $3 + 3^2 + \cdots + 3^n = \frac{1}{2}(3^{n+1} - 3)$.

5. $1 \cdot 2 + 2 \cdot 3 + \cdots + n(n + 1) = \frac{1}{3}n(n + 1)(n + 2)$.

6. Let S_n represent the sum of the first n terms of the A.P. of Example 2 on page 334, or $S_n = b + (b + d) + (b + 2d) + \cdots + [b + (n - 1)d]$. Prove that $S_n = \frac{1}{2}n[2b + (n - 1)d]$.

Note 1. Recall that a *geometric progression* (abbreviated "**G.P.**") is defined as a sequence of terms where each term after the first term is obtained from the preceding term by multiplying it by a fixed number called the *common ratio* of the G.P. Problem 7 refers to a G.P.

7. In a G.P. with first term b and common ratio r, let l_n be the nth term, and let S_n be the sum of the first n terms. Prove that

$$l_n = br^{n-1} \qquad and \qquad S_n = \frac{b - br^n}{1 - r}.$$

8. If $1 < u$, prove that $u < u^n$ for all $n > 1$.
9. If $0 < R, 0 < S$, and $R < S$, then $R^n < S^n$.
10. If n is any positive integer, then $\frac{1}{3}(n^3 + 2n)$ is an integer.

★11. If $n > 1$, then
$$\frac{1}{\sqrt{1}} + \frac{1}{\sqrt{2}} + \cdots + \frac{1}{\sqrt{n}} > \sqrt{n}.$$

★12. *Prove that:* for every positive integer $n, (x + y)^n$ is given by the expansion (5) on page 31. That is, *prove the binomial theorem by mathematical induction.*

Comment. The resulting demonstration is one of the classical proofs of mathematics. However, the complexity of the proof compares unfavorably with another method which is available in finite mathematics. In Part II of the requested proof, the expansion of $(x + y)^h$ must be written with both the rth and the $(r - 1)$th terms given explicitly. Then, after multiplication to obtain $(x + y)^{h+1}$, it must be shown that the term involving $x^{h+1-r}y^r$ has the correct coefficient.

Tables

TABLE I
SQUARES AND SQUARE ROOTS: 1–200

N	N²	√N	N	N²	√N	N	N²	√N	N	N²	√N
1	1	1.000	51	2,601	7.141	101	10,201	10.050	151	22,801	12.288
2	4	1.414	52	2,704	7.211	102	10,404	10.100	152	23,104	12.329
3	9	1.732	53	2,809	7.280	103	10,609	10.149	153	23,409	12.369
4	16	2.000	54	2,916	7.348	104	10,816	10.198	154	23,716	12.410
5	25	2.236	55	3,025	7.416	105	11,025	10.247	155	24,025	12.450
6	36	2.449	56	3,136	7.483	106	11,236	10.296	156	24,336	12.490
7	49	2.646	57	3,249	7.550	107	11,449	10.344	157	24,649	12.530
8	64	2.828	58	3,364	7.616	108	11,664	10.392	158	24,964	12.570
9	81	3.000	59	3,481	7.681	109	11,881	10.440	159	25,281	12.610
10	100	3.162	60	3,600	7.746	110	12,100	10.488	160	25,600	12.649
11	121	3.317	61	3,721	7.810	111	12,321	10.536	161	25,921	12.689
12	144	3.464	62	3,844	7.874	112	12,544	10.583	162	26,244	12.728
13	169	3.606	63	3,969	7.937	113	12,769	10.630	163	26,569	12.767
14	196	3.742	64	4,096	8.000	114	12,996	10.677	164	26,896	12.806
15	225	3.873	65	4,225	8.062	115	13,225	10.724	165	27,225	12.845
16	256	4.000	66	4,356	8.124	116	13,456	10.770	166	27,556	12.884
17	289	4.123	67	4,489	8.185	117	13,689	10.817	167	27,889	12.923
18	324	4.243	68	4,624	8.246	118	13,924	10.863	168	28,224	12.962
19	361	4.359	69	4,761	8.307	119	14,161	10.909	169	28,561	13.000
20	400	4.472	70	4,900	8.367	120	14,400	10.954	170	28,900	13.038
21	441	4.583	71	5,041	8.426	121	14,641	11.000	171	29,241	13.077
22	484	4.690	72	5,184	8.485	122	14,884	11.045	172	29,584	13.115
23	529	4.796	73	5,329	8.544	123	15,129	11.091	173	29,929	13.153
24	576	4.899	74	5,476	8.602	124	15,376	11.136	174	30,276	13.191
25	625	5.000	75	5,625	8.660	125	15,625	11.180	175	30,625	13.229
26	676	5.099	76	5,776	8.718	126	15,876	11.225	176	30,976	13.266
27	729	5.196	77	5,929	8.775	127	16,129	11.269	177	31,329	13.304
28	784	5.292	78	6,084	8.832	128	16,384	11.314	178	31,684	13.342
29	841	5.385	79	6,241	8.888	129	16,641	11.358	179	32,041	13.379
30	900	5.477	80	6,400	8.944	130	16,900	11.402	180	32,400	13.416
31	961	5.568	81	6,561	9.000	131	17,161	11.446	181	32,761	13.454
32	1,024	5.657	82	6,724	9.055	132	17,424	11.489	182	33,124	13.491
33	1,089	5.745	83	6,889	9.110	133	17,689	11.533	183	33,489	13.528
34	1,156	5.831	84	7,056	9.165	134	17,956	11.576	184	33,856	13.565
35	1,225	5.916	85	7,225	9.220	135	18,225	11.619	185	34,225	13.601
36	1,296	6.000	86	7,396	9.274	136	18,496	11.662	186	34,596	13.638
37	1,369	6.083	87	7,569	9.327	137	18,769	11.705	187	34,969	13.675
38	1,444	6.164	88	7,744	9.381	138	19,044	11.747	188	35,344	13.711
39	1,521	6.245	89	7,921	9.434	139	19,321	11.790	189	35,721	13.748
40	1,600	6.325	90	8,100	9.487	140	19,600	11.832	190	36,100	13.784
41	1,681	6.403	91	8,281	9.539	141	19,881	11.874	191	36,481	13.820
42	1,764	6.481	92	8,464	9.592	142	20,164	11.916	192	36,864	13.856
43	1,849	6.557	93	8,649	9.644	143	20,449	11.958	193	37,249	13.892
44	1,936	6.633	94	8,836	9.695	144	20,736	12.000	194	37,636	13.928
45	2,025	6.708	95	9,025	9.747	145	21,025	12.042	195	38,025	13.964
46	2,116	6.782	96	9,216	9.798	146	21,316	12.083	196	38,416	14.000
47	2,209	6.856	97	9,409	9.849	147	21,609	12.124	197	38,809	14.036
48	2,304	6.928	98	9,604	9.899	148	21,904	12.166	198	39,204	14.071
49	2,401	7.000	99	9,801	9.950	149	22,201	12.207	199	39,601	14.107
50	2,500	7.071	100	10,000	10.000	150	22,500	12.247	200	40,000	14.142
N	N²	√N	N	N²	√N	N	N²	√N	N	N²	√N

TABLE II
THREE-PLACE LOGARITHMS
OF NUMBERS

N	Log N	N	Log N
1.0	.000	5.5	.740
1.1	.041	5.6	.748
1.2	.079	5.7	.756
1.3	.114	5.8	.763
1.4	.146	5.9	.771
1.5	.176	6.0	.778
1.6	.204	6.1	.785
1.7	.230	6.2	.792
1.8	.255	6.3	.799
1.9	.279	6.4	.806
2.0	.301	6.5	.813
2.1	.322	6.6	.820
2.2	.342	6.7	.826
2.3	.362	6.8	.833
2.4	.380	6.9	.839
2.5	.398	7.0	.845
2.6	.415	7.1	.851
2.7	.431	7.2	.857
2.8	.447	7.3	.863
2.9	.462	7.4	.869
3.0	.477	7.5	.875
3.1	.491	7.6	.881
3.2	.505	7.7	.886
3.3	.519	7.8	.892
3.4	.531	7.9	.898
3.5	.544	8.0	.903
3.6	.556	8.1	.908
3.7	.568	8.2	.914
3.8	.580	8.3	.919
3.9	.591	8.4	.924
4.0	.602	8.5	.929
4.1	.613	8.6	.935
4.2	.623	8.7	.940
4.3	.633	8.8	.944
4.4	.643	8.9	.949
4.5	.653	9.0	.954
4.6	.663	9.1	.959
4.7	.672	9.2	.964
4.8	.681	9.3	.968
4.9	.690	9.4	.973
5.0	.699	9.5	.978
5.1	.708	9.6	.982
5.2	.716	9.7	.987
5.3	.724	9.8	.991
5.4	.732	9.9	.996
5.5	.740	1.00	1.000
N	Log N	N	Log N

TABLE III
THREE-PLACE LOGARITHMS
OF FUNCTIONS

→	L Sin *	L Tan *	L Cot	L Cos *	
0°	—	—	—	10.000	90°
1°	8.242	8.242	1.758	10.000	89°
2°	.543	.543	.457	10.000	88°
3°	.719	.719	.281	9.999	87°
4°	.844	.845	.155	.999	86°
5°	8.940	8.942	1.058	9.998	85°
6°	9.019	9.022	0.978	9.998	84°
7°	.086	.089	.911	.997	83°
8°	.144	.148	.852	.996	82°
9°	.194	.200	.800	.995	81°
10°	9.240	9.246	0.754	9.993	80°
11°	9.281	9.289	0.711	9.992	79°
12°	.318	.327	.673	.990	78°
13°	.352	.363	.637	.989	77°
14°	.384	.397	.603	.987	76°
15°	9.413	9.428	0.572	9.985	75°
16°	9.440	9.458	0.543	9.983	74°
17°	.466	.485	.515	.981	73°
18°	.490	.512	.488	.978	72°
19°	.513	.537	.463	.976	71°
20°	9.534	9.561	0.439	9.973	70°
21°	9.554	9.584	0.416	9.970	69°
22°	.574	.606	.394	.967	68°
23°	.592	.628	.372	.964	67°
24°	.609	.649	.351	.961	66°
25°	9.626	9.669	0.331	9.957	65°
26°	9.642	9.688	0.312	9.954	64°
27°	.657	.707	.293	.950	63°
28°	.672	.726	.274	.946	62°
29°	.686	.744	.256	.942	61°
30°	9.699	9.761	0.239	9.938	60°
31°	9.712	9.779	0.221	9.933	59°
32°	.724	.796	.204	.928	58°
33°	.736	.813	.187	.924	57°
34°	.748	.829	.171	.919	56°
35°	9.759	9.845	0.155	9.913	55°
36°	9.769	9.861	0.139	9.908	54°
37°	.779	.877	.123	.902	53°
38°	.789	.893	.107	.897	52°
39°	.799	.908	.092	.891	51°
40°	9.808	9.924	0.076	9.884	50°
41°	9.817	9.939	0.061	9.878	49°
42°	.826	.954	.046	.871	48°
43°	.834	.970	.030	.864	47°
44°	.842	.985	.015	.857	46°
45°	9.849	10.000	0.000	9.849	45°
	L Cos *	L Cot *	L Tan	L Sin *	←

* Subtract 10 from each entry in this column.

TABLE IV
THREE-PLACE VALUES OF TRIGONOMETRIC FUNCTIONS
AND
DEGREES IN RADIAN MEASURE

Rad.	Deg.	Sin	Tan	Sec	Csc	Cot	Cos	Deg.	Rad.
.000	0°	.000	.000	1.000	——	——	1.000	90°	1.571
.017	1°	.017	.017	1.000	57.30	57.29	1.000	89°	1.553
.035	2°	.035	.035	1.001	28.65	28.64	0.999	88°	1.536
.052	3°	.052	.052	1.001	19.11	19.08	.999	87°	1.518
.070	4°	.070	.070	1.002	14.34	14.30	.998	86°	1.501
.087	5°	.087	.087	1.004	11.47	11.43	.996	85°	1.484
.105	6°	.105	.105	1.006	9.567	9.514	.995	84°	1.466
.122	7°	.122	.123	1.008	8.206	8.144	.993	83°	1.449
.140	8°	.139	.141	1.010	7.185	7.115	.990	82°	1.431
.157	9°	.156	.158	1.012	6.392	6.314	.988	81°	1.414
.175	10°	.174	.176	1.015	5.759	5.671	.985	80°	1.396
.192	11°	.191	.194	1.019	5.241	5.145	.082	79°	1.379
.209	12°	.208	.213	1.022	4.810	4.705	.978	78°	1.361
.227	13°	.225	.231	1.026	4.445	4.331	.974	77°	1.344
.244	14°	.242	.249	1.031	4.134	4.011	.970	76°	1.326
.262	15°	.259	.268	1.035	3.864	3.732	.966	75°	1.309
.279	16°	.276	.287	1.040	3.628	3.487	.901	74°	1.292
.297	17°	.292	.300	1.046	3.420	3.271	.956	73°	1.274
.314	18°	.309	.325	1.051	3.236	3.078	.951	72°	1.257
.332	19°	.326	.344	1.058	3.072	2.904	.946	71°	1.239
.349	20°	.342	.364	1.064	2.924	2.747	.940	70°	1.222
.367	21°	.358	.384	1.071	2.790	2.605	.934	69°	1.204
.384	22°	.375	.404	1.079	2.669	2.475	.927	68°	1.187
.401	23°	.391	.424	1.086	2.559	2.356	.921	67°	1.169
.419	24°	.407	.445	1.095	2.459	2.246	.914	66°	1.152
.436	25°	.423	.466	1.103	2.366	2.145	.906	65°	1.134
.454	26°	.438	.488	1.113	2.281	2.050	.899	64°	1.117
.471	27°	.454	.510	1.122	2.203	1.963	.891	63°	1.100
.489	28°	.469	.532	1.133	2.130	1.881	.883	62°	1.082
.506	29°	.485	.554	1.143	2.063	1.804	.875	61°	1.065
.524	30°	.500	.577	1.155	2.000	1.732	.866	60°	1.047
.541	31°	.515	.601	1.167	1.942	1.664	.857	59°	1.030
.559	32°	.530	.625	1.179	1.887	1.600	.848	58°	1.012
.576	33°	.545	.649	1.192	1.836	1.540	.839	57°	0.995
.593	34°	.559	.675	1.206	1.788	1.483	.829	56°	0.977
.611	35°	.574	.700	1.221	1.743	1.428	.819	55°	0.960
.628	36°	.588	.727	1.236	1.701	1.376	.809	54°	0.942
.646	37°	.602	.754	1.252	1.662	1.327	.799	53°	0.925
.663	38°	.616	.781	1.269	1.624	1.280	.788	52°	0.908
.681	39°	.629	.810	1.287	1.589	1.235	.777	51°	0.890
.698	40°	.643	.839	1.305	1.556	1.192	.766	50°	0.873
.716	41°	.656	.869	1.325	1.524	1.150	.755	49°	0.855
.733	42°	.669	.900	1.346	1.494	1.111	.743	48°	0.838
.750	43°	.682	.933	1.367	1.466	1.072	.731	47°	0.820
.768	44°	.695	0.966	1.390	1.440	1.036	.719	46°	0.803
.785	45°	.707	1.000	1.414	1.414	1.000	.707	45°	0.785
Rad.	Deg.	Cos	Cot	Csc	Sec	Tan	Sin	Deg.	Rad.

TABLE V
Four-Place Logarithms of Numbers

N	0	1	2	3	4	5	6	7	8	9	Prop. Parts
10	.0000	0043	0086	0128	0170	0212	0253	0294	0334	0374	
11	.0414	0453	0492	0531	0569	0607	0645	0682	0719	0755	
12	.0792	0828	0864	0899	0934	0969	1004	1038	1072	1106	
13	.1139	1173	1206	1239	1271	1303	1335	1367	1399	1430	
14	.1461	1492	1523	1553	1584	1614	1644	1673	1703	1732	
15	.1761	1790	1818	1847	1875	1903	1931	1959	1987	2014	
16	.2041	2068	2095	2122	2148	2175	2201	2227	2253	2279	
17	.2304	2330	2355	2380	2405	2430	2455	2480	2504	2529	
18	.2553	2577	2601	2625	2648	2672	2695	2718	2742	2765	
19	.2788	2810	2833	2856	2878	2900	2923	2945	2967	2989	
20	.3010	3032	3054	3075	3096	3118	3139	3160	3181	3201	
21	.3222	3243	3263	3284	3304	3324	3345	3365	3385	3404	
22	.3424	3444	3464	3483	3502	3522	3541	3560	3579	3598	
23	.3617	3636	3655	3674	3692	3711	3729	3747	3766	3784	
24	.3802	3820	3838	3856	3874	3892	3909	3927	3945	3962	
25	.3979	3997	4014	4031	4048	4065	4082	4099	4116	4133	
26	.4150	4166	4183	4200	4216	4232	4249	4265	4281	4298	
27	.4314	4330	4346	4362	4378	4393	4409	4425	4440	4456	
28	.4472	4487	4502	4518	4533	4548	4564	4579	4594	4609	
29	.4624	4639	4654	4669	4683	4698	4713	4728	4742	4757	
30	.4771	4786	4800	4814	4829	4843	4857	4871	4886	4900	
31	.4914	4928	4942	4955	4969	4983	4997	5011	5024	5038	
32	.5051	5065	5079	5092	5105	5119	5132	5145	5159	5172	
33	.5185	5198	5211	5224	5237	5250	5263	5276	5289	5302	
34	.5315	5328	5340	5353	5366	5378	5391	5403	5416	5428	
35	.5441	5453	5465	5478	5490	5502	5514	5527	5539	5551	
36	.5563	5575	5587	5599	5611	5623	5635	5647	5658	5670	
37	.5682	5694	5705	5717	5729	5740	5752	5763	5775	5786	
38	.5798	5809	5821	5832	5843	5855	5866	5877	5888	5899	
39	.5911	5922	5933	5944	5955	5966	5977	5988	5999	6010	
40	.6021	6031	6042	6053	6064	6075	6085	6096	6107	6117	
41	.6128	6138	6149	6160	6170	6180	6191	6201	6212	6222	
42	.6232	6243	6253	6263	6274	6284	6294	6304	6314	6325	
43	.6335	6345	6355	6365	6375	6385	6395	6405	6415	6425	
44	.6435	6444	6454	6464	6474	6484	6493	6503	6513	6522	
45	.6532	6542	6551	6561	6571	6580	6590	6599	6609	6618	
46	.6628	6637	6646	6656	6665	6675	6684	6693	6702	6712	
47	.6721	6730	6739	6749	6758	6767	6776	6785	6794	6803	
48	.6812	6821	6830	6839	6848	6857	6866	6875	6884	6893	
49	.6902	6911	6920	6928	6937	6946	6955	6964	6972	6981	
50	.6990	6998	7007	7016	7024	7033	7042	7050	7059	7067	
N	0	1	2	3	4	5	6	7	8	9	

Prop. Parts

	28	27	26
1	2.8	2.7	2.6
2	5.6	5.4	5.2
3	8.4	8.1	7.8
4	11.2	10.8	10.4
5	14.0	13.5	13.0
6	16.8	16.2	15.6
7	19.6	18.9	18.2
8	22.4	21.6	20.8
9	25.2	24.3	23.4

	22	21	20
1	2.2	2.1	2.0
2	4.4	4.2	4.0
3	6.6	6.3	6.0
4	8.8	8.4	8.0
5	11.0	10.5	10.0
6	13.2	12.6	12.0
7	15.4	14.7	14.0
8	17.6	16.8	16.0
9	19.8	18.9	18.0

	16	15	14
1	1.6	1.5	1.4
2	3.2	3.0	2.8
3	4.8	4.5	4.2
4	6.4	6.0	5.6
5	8.0	7.5	7.0
6	9.6	9.0	8.4
7	11.2	10.5	9.8
8	12.8	12.0	11.2
9	14.4	13.5	12.6

	13	12	11
1	1.3	1.2	1.1
2	2.6	2.4	2.2
3	3.9	3.6	3.3
4	5.2	4.8	4.4
5	6.5	6.0	5.5
6	7.8	7.2	6.6
7	9.1	8.4	7.7
8	10.4	9.6	8.8
9	11.7	10.8	9.9

	43	42	41	40	39		38	37	36	35	34		33	32	31	30	29	
1	4.3	4.2	4.1	4.0	3.9	1	3.8	3.7	3.6	3.5	3.4	1	3.3	3.2	3.1	3.0	2.9	1
2	8.6	8.4	8.2	8.0	7.8	2	7.6	7.4	7.2	7.0	6.8	2	6.6	6.4	6.2	6.0	5.8	2
3	12.9	12.6	12.3	12.0	11.7	3	11.4	11.1	10.8	10.5	10.2	3	9.9	9.6	9.3	9.0	8.7	3
4	17.2	16.8	16.4	16.0	15.6	4	15.2	14.8	14.4	14.0	13.6	4	13.2	12.8	12.4	12.0	11.6	4
5	21.5	21.0	20.5	20.0	19.5	5	19.0	18.5	18.0	17.5	17.0	5	16.5	16.0	15.5	15.0	14.5	5
6	25.8	25.2	24.6	24.0	23.4	6	22.8	22.2	21.6	21.0	20.4	6	19.8	19.2	18.6	18.0	17.4	6
7	30.1	29.4	28.7	28.0	27.3	7	26.6	25.9	25.2	24.5	23.8	7	23.1	22.4	21.7	21.0	20.3	7
8	34.4	33.6	32.8	32.0	31.2	8	30.4	29.6	28.8	28.0	27.2	8	26.4	25.6	24.8	24.0	23.2	8
9	38.7	37.8	36.9	36.0	35.1	9	34.2	33.3	32.4	31.5	30.6	9	29.7	28.8	27.9	27.0	26.1	9

TABLE V (*continued*)
FOUR-PLACE LOGARITHMS OF NUMBERS

Prop. Parts			N	0	1	2	3	4	5	6	7	8	9
			50	.6990	6998	7007	7016	7024	7033	7042	7050	7059	7067
			51	.7076	7084	7093	7101	7110	7118	7126	7135	7143	7152
25	**24**	**23**	52	.7160	7168	7177	7185	7193	7202	7210	7218	7226	7235
			53	.7243	7251	7259	7267	7275	7284	7292	7300	7308	7316
1 2.5	2.4	2.3	54	.7324	7332	7340	7348	7356	7364	7372	7380	7388	7396
2 5.0	4.8	4.6	**55**	.7404	7412	7419	7427	7435	7443	7451	7459	7466	7474
3 7.5	7.2	6.9	56	.7482	7490	7497	7505	7513	7520	7528	7536	7543	7551
4 10.0	9.6	9.2	57	.7559	7566	7574	7582	7589	7597	7604	7612	7619	7627
5 12.5	12.0	11.5	58	.7634	7642	7649	7657	7664	7672	7679	7686	7694	7701
6 15.0	14.4	13.8	59	.7709	7716	7723	7731	7738	7745	7752	7760	7767	7774
7 17.5	16.8	16.1	**60**	.7782	7789	7796	7803	7810	7818	7825	7832	7839	7846
8 20.0	19.2	18.4	61	.7853	7860	7868	7875	7882	7889	7896	7903	7910	7917
9 22.5	21.6	20.7	62	.7924	7931	7938	7945	7952	7959	7966	7973	7980	7987
			63	.7993	8000	8007	8014	8021	8028	8035	8041	8048	8055
19	**18**	**17**	64	.8062	8069	8075	8082	8089	8096	8102	8109	8116	8122
1 1.9	1.8	1.7	**65**	.8129	8136	8142	8149	8156	8162	8169	8176	8182	8189
2 3.8	3.6	3.4	66	.8195	8202	8209	8215	8222	8228	8235	8241	8248	8254
3 5.7	5.4	5.1	67	.8261	8267	8274	8280	8287	8293	8299	8306	8312	8319
4 7.6	7.2	6.8	68	.8325	8331	8338	8344	8351	8357	8363	8370	8376	8382
5 9.5	9.0	8.5	69	.8388	8395	8401	8407	8414	8420	8426	8432	8439	8445
6 11.4	10.8	10.2	**70**	.8451	8457	8463	8470	8476	8482	8488	8494	8500	8506
7 13.3	12.6	11.9	71	.8513	8519	8525	8531	8537	8543	8549	8555	8561	8567
8 15.2	14.4	13.6	72	.8573	8579	8585	8591	8597	8603	8609	8615	8621	8627
9 17.1	16.2	15.3	73	.8633	8639	8645	8651	8657	8663	8669	8675	8681	8686
			74	.8692	8698	8704	8710	8716	8722	8727	8733	8739	8745
10	**9**		**75**	.8751	8756	8762	8768	8774	8779	8785	8791	8797	8802
1 1.0	0.9		76	.8808	8814	8820	8825	8831	8837	8842	8848	8854	8859
2 2.0	1.8		77	.8865	8871	8876	8882	8887	8893	8899	8904	8910	8915
3 3.0	2.7		78	.8921	8927	8932	8938	8943	8949	8954	8960	8965	8971
4 4.0	3.6		79	.8976	8982	8987	8993	8998	9004	9009	9015	9020	9025
5 5.0	4.5		**80**	.9031	9036	9042	9047	9053	9058	9063	9069	9074	9079
6 6.0	5.4		81	.9085	9090	9096	9101	9106	9112	9117	9122	9128	9133
7 7.0	6.3		82	.9138	9143	9149	9154	9159	9165	9170	9175	9180	9186
8 8.0	7.2		83	.9191	9196	9201	9206	9212	9217	9222	9227	9232	9238
9 9.0	8.1		84	.9243	9248	9253	9258	9263	9269	9274	9279	9284	9289
			85	.9294	9299	9304	9309	9315	9320	9325	9330	9335	9340
8	**7**		86	.9345	9350	9355	9360	9365	9370	9375	9380	9385	9390
1 0.8	0.7		87	.9395	9400	9405	9410	9415	9420	9425	9430	9435	9440
2 1.6	1.4		88	.9445	9450	9455	9460	9465	9469	9474	9479	9484	9489
3 2.4	2.1		89	.9494	9499	9504	9509	9513	9518	9523	9528	9533	9538
4 3.2	2.8		**90**	.9542	9547	9552	9557	9562	9566	9571	9576	9581	9586
5 4.0	3.5		91	.9590	9595	9600	9605	9609	9614	9619	9624	9628	9633
6 4.8	4.2		92	.9638	9643	9647	9652	9657	9661	9666	9671	9675	9680
7 5.6	4.9		93	.9685	9689	9694	9699	9703	9708	9713	9717	9722	9727
8 6.4	5.6		94	.9731	9736	9741	9745	9750	9754	9759	9763	9768	9773
9 7.2	6.3		**95**	.9777	9782	9786	9791	9795	9800	9805	9809	9814	9818
			96	.9823	9827	9832	9836	9841	9845	9850	9854	9859	9863
6	**5**	**4**	97	.9868	9872	9877	9881	9886	9890	9894	9899	9903	9908
1 0.6	0.5	0.4	98	.9912	9917	9921	9926	9930	9934	9939	9943	9948	9952
2 1.2	1.0	0.8	99	.9956	9961	9965	9969	9974	9978	9983	9987	9991	9996
3 1.8	1.5	1.2	**N**	0	1	2	3	4	5	6	7	8	9
4 2.4	2.0	1.6											
5 3.0	2.5	2.0											
6 3.6	3.0	2.4											
7 4.2	3.5	2.8											
8 4.8	4.0	3.2											
9 5.4	4.5	3.6											

TABLE VI
RADIAN MEASURE: VALUES OF FUNCTIONS

α Rad.	Degrees in α	Sin α	Cos α	Tan α	α Rad.	Degrees in α	Sin α	Cos α	Tan α
.00	0° 00.0′	.00000	1.0000	.00000	.60	34° 22.6′	.56464	.82534	.68414
.01	0° 34.4′	.01000	.99995	.01000	.61	34° 57.0′	.57287	.81965	.69892
.02	1° 08.8′	.02000	.99980	.02000	.62	35° 31.4′	.58104	.81388	.71391
.03	1° 43.1′	.03000	.99955	.03001	.63	36° 05.8′	.58914	.80803	.72911
.04	2° 17.5′	.03999	.99920	.04002	.64	36° 40.2′	.59720	.80210	.74454
.05	2° 51.9′	.04998	.99875	.05004	.65	37° 14.5′	.60519	.79608	.76020
.06	3° 26.3′	.05996	.99820	.06007	.66	37° 48.9′	.61312	.78999	.77610
.07	4° 00.6′	.06994	.99755	.07011	.67	38° 23.3′	.62099	.78382	.79225
.08	4° 35.0′	.07991	.99680	.08017	.68	38° 57.7′	.62879	.77757	.80866
.09	5° 09.4′	.08988	.99595	.09024	.69	39° 32.0′	.63654	.77125	.82534
.10	5° 43.8′	.09983	.99500	.10033	.70	40° 06.4′	.64422	.76484	.84229
.11	6° 18.2′	.10978	.99396	.11045	.71	40° 40.8′	.65183	.75836	.85953
.12	6° 52.5′	.11971	.99281	.12058	.72	41° 15.2′	.65938	.75181	.87707
.13	7° 26.9′	.12963	.99156	.13074	.73	41° 49.6′	.66687	.74517	.89492
.14	8° 01.3′	.13954	.99022	.14092	.74	42° 23.9′	.67429	.73847	.91309
.15	8° 35.7′	.14944	.98877	.15114	.75	42° 58.3′	.68164	.73169	.93160
.16	9° 10.0′	.15932	.98723	.16138	.76	43° 32.7′	.68892	.72484	.95045
.17	9° 44.4′	.16918	.98558	.17166	.77	44° 07.1′	.69614	.71791	.96967
.18	10° 18.8′	.17903	.98384	.18197	.78	44° 41.4′	.70328	.71091	.98926
.19	10° 53.2′	.18886	.98200	.19232	.79	45° 15.8′	.71035	.70385	1.0092
.20	11° 27.5′	.19867	.98007	.20271	.80	45° 50.2′	.71736	.69671	1.0296
.21	12° 01.9′	.20846	.97803	.21314	.81	46° 24.6′	.72429	.68950	1.0505
.22	12° 36.3′	.21823	.97590	.22362	.82	46° 59.0′	.73115	.68222	1.0717
.23	13° 10.7′	.22798	.97367	.23414	.83	47° 33.3′	.73793	.67488	1.0934
.24	13° 45.1′	.23770	.97134	.24472	.84	48° 07.7′	.74464	.66746	1.1156
.25	14° 19.4′	.24740	.96891	.25534	.85	48° 42.1′	.75128	.65998	1.1383
.26	14° 53.8′	.25708	.96639	.26602	.86	49° 16.5′	.75784	.65244	1.1616
.27	15° 28.2′	.26673	.96377	.27676	.87	49° 50.8′	.76433	.64483	1.1853
.28	16° 02.6′	.27636	.96106	.28755	.88	50° 25.2′	.77074	.63715	1.2097
.29	16° 36.9′	.28595	.95824	.29841	.89	50° 59.6′	.77707	.62941	1.2346
.30	17° 11.3′	.29552	.95534	.30934	.90	51° 34.0′	.78333	.62161	1.2602
.31	17° 45.7′	.30506	.95233	.32033	.91	52° 08.3′	.78950	.61375	1.2864
.32	18° 20.1′	.31457	.94924	.33139	.92	52° 42.7′	.79560	.60582	1.3133
.33	18° 54.5′	.32404	.94604	.34252	.93	53° 17.1′	.80162	.59783	1.3409
.34	19° 28.8′	.33349	.94275	.35374	.94	53° 51.5′	.80756	.58979	1.3692
.35	20° 03.2′	.34290	.93937	.36503	.95	54° 25.9′	.81342	.58168	1.3984
.36	20° 37.6′	.35227	.93590	.37640	.96	55° 00.2′	.81919	.57352	1.4284
.37	21° 12.0′	.36162	.93233	.38786	.97	55° 34.6′	.82489	.56530	1.4592
.38	21° 46.3′	.37092	.92866	.39941	.98	56° 09.0′	.83050	.55702	1.4910
.39	22° 20.7′	.38019	.92491	.41105	.99	56° 43.4′	.83603	.54869	1.5237
.40	22° 55.1′	.38942	.92106	.42279	1.00	57° 17.7′	.84147	.54030	1.5574
.41	23° 29.5′	.39861	.91712	.43463	1.01	57° 52.1′	.84683	.53186	1.5922
.42	24° 03.9′	.40776	.91309	.44657	1.02	58° 26.5′	.85211	.52337	1.6281
.43	24° 38.2′	.41687	.90897	.45862	1.03	59° 00.9′	.85730	.51482	1.6652
.44	25° 12.6′	.42594	.90475	.47078	1.04	59° 35.3′	.86240	.50622	1.7036
.45	25° 47.0′	.43497	.90045	.48306	1.05	60° 09.6′	.86742	.49757	1.7433
.46	26° 21.4′	.44395	.89605	.49545	1.06	60° 44.0′	.87236	.48887	1.7844
.47	26° 55.7′	.45289	.89157	.50797	1.07	61° 18.4′	.87720	.48012	1.8270
.48	27° 30.1′	.46178	.88699	.52061	1.08	61° 52.8′	.88196	.47133	1.8712
.49	28° 04.5′	.47063	.88233	.53339	1.09	62° 27.1′	.88663	.46249	1.9171
.50	28° 38.9′	.47943	.87758	.54630	1.10	63° 01.5′	.89121	.45360	1.9648
.51	29° 13.3′	.48818	.87274	.55936	1.11	63° 35.9′	.89570	.44466	2.0143
.52	29° 47.6′	.49688	.86782	.57256	1.12	64° 10.3′	.90010	.43568	2.0660
.53	30° 22.0′	.50553	.86281	.58592	1.13	64° 44.7′	.90441	.42666	2.1198
.54	30° 56.4′	.51414	.85771	.59943	1.14	65° 19.0′	.90863	.41759	2.1759
.55	31° 30.8′	.52269	.85252	.61311	1.15	65° 53.4′	.91276	.40849	2.2345
.56	32° 05.1′	.53119	.84726	.62695	1.16	66° 27.8′	.91680	.39934	2.2958
.57	32° 39.5′	.53963	.84190	.64097	1.17	67° 02.2′	.92075	.39015	2.3600
.58	33° 13.9′	.54802	.83646	.65517	1.18	67° 36.5′	.92461	.38092	2.4273
.59	33° 48.3′	.55636	.83094	.66956	1.19	68° 10.9′	.92837	.37166	2.4979
.60	34° 22.6′	.56464	.82534	.68414	1.20	68° 45.3′	.93204	.36236	2.5722

TABLE VI (*continued*)
RADIAN MEASURE: VALUES OF FUNCTIONS

α Rad.	Degrees in α	Sin α	Cos α	Tan α	α Rad.	Degrees in α	Sin α	Cos α	Tan α
1.20	68° 45.3′	.93204	.36236	2.5722	1.40	80° 12.8′	.98545	.16997	5.7979
1.21	69° 19.7′	.93562	.35302	2.6503	1.41	80° 47.2′	.98710	.16010	6.1654
1.22	69° 54.1′	.93910	.34365	2.7328	1.42	81° 21.6′	.98865	.15023	6.5811
1.23	70° 28.4′	.94249	.33424	2.8198	1.43	81° 56.0′	.99010	.14033	7.0555
1.24	71° 02.8′	.94578	.32480	2.9119	1.44	82° 30.4′	.99146	.13042	7.6018
1.25	71° 37.2′	.94898	.31532	3.0096	1.45	83° 04.7′	.99271	.12050	8.2381
1.26	72° 11.6′	.95209	.30582	3.1133	1.46	83° 39.1′	.99387	.11057	8.9886
1.27	72° 45.9′	.95510	.29628	3.2236	1.47	84° 13.5′	.99492	.10063	9.8874
1.28	73° 20.3′	.95802	.28672	3.3413	1.48	84° 47.9′	.99588	.09067	10.983
1.29	73° 54.7′	.96084	.27712	3.4672	1.49	85° 22.2′	.99674	.08071	12.350
1.30	74° 29.1′	.96356	.26750	3.6021	1.50	85° 56.6′	.99749	.07074	14.101
1.31	75° 03.4′	.96618	.25785	3.7471	1.51	86° 31.0′	.99815	.06076	16.428
1.32	75° 37.8′	.96872	.24818	3.9033	1.52	87° 05.4′	.99871	.05077	19.670
1.33	76° 12.2′	.97115	.23848	4.0723	1.53	87° 39.8′	.99917	.04079	24.498
1.34	76° 46.6′	.97348	.22875	4.2556	1.54	88° 14.1′	.99953	.03079	32.461
1.35	77° 21.0′	.97572	.21901	4.4552	1.55	88° 48.5′	.99978	.02079	48.078
1.36	77° 55.3′	.97786	.20924	4.6734	1.56	89° 22.9′	.99994	.01080	92.620
1.37	78° 29.7′	.97991	.19945	4.9131	1.57	89° 57.3′	1.0000	.00080	1255.8
1.38	79° 04.1′	.98185	.18964	5.1774	1.58	90° 31.6′	.99996	− .00920	− 108.65
1.39	79° 38.5′	.98370	.17981	5.4707	1.59	91° 06.0′	.99982	− .01920	− 52.067
1.40	80° 12.8′	.98545	.16997	5.7979	1.60	91° 40.4′	.99957	− .02920	− 34.233

DEGREES IN RADIANS

1°	0.01745	16°	0.27925	31°	0.54105	46°	0.80285	61°	1.06465	76°	1.32645
2	0.03491	17	0.29671	32	0.55851	47	0.82030	62	1.08210	77	1.34390
3	0.05236	18	0.31416	33	0.57596	48	0.83776	63	1.09956	78	1.36136
4	0.06981	19	0.33161	34	0.59341	49	0.85521	64	1.11701	79	1.37881
5	0.08727	20	0.34907	35	0.61087	50	0.87266	65	1.13446	80	1.39626
6	0.10472	21	0.36652	36	0.62832	51	0.89012	66	1.15192	81	1.41372
7	0.12217	22	0.38397	37	0.64577	52	0.90757	67	1.16937	82	1.43117
8	0.13963	23	0.40143	38	0.66323	53	0.92502	68	1.18682	83	1.44862
9	0.15708	24	0.41888	39	0.68068	54	0.94248	69	1.20428	84	1.46008
10	0.17453	25	0.43633	40	0.69813	55	0.95993	70	1.22173	85	1.48353
11	0.19199	26	0.45379	41	0.71558	56	0.97738	71	1.23918	86	1.50098
12	0.20944	27	0.47124	42	0.73304	57	0.99484	72	1.25664	87	1.51844
13	0.22689	28	0.48869	43	0.75049	58	1.01229	73	1.27409	88	1.53589
14	0.24435	29	0.50615	44	0.76794	59	1.02974	74	1.29154	89	1.55334
15	0.26180	30	0.52360	45	0.78540	60	1.04720	75	1.30900	90	1.57080

1° = .01745329 rad.   log .01745329 = 8.24187737 − 10.
1′ = .0002908882 rad.   log .0002908882 = 6.46372612 − 10.
1″ = .0000048481368 rad. log .0000048481368 = 4.68557487 − 10.

MINUTES IN RADIANS

1′	0.00029	11′	0.00320	21′	0.00611	31′	0.00902	41′	0.01193	51′	0.01484
2	0.00058	12	0.00349	22	0.00640	32	0.00931	42	0.01222	52	0.01513
3	0.00087	13	0.00378	23	0.00669	33	0.00960	43	0.01251	53	0.01542
4	0.00116	14	0.00407	24	0.00698	34	0.00989	44	0.01280	54	0.01571
5	0.00145	15	0.00436	25	0.00727	35	0.01018	45	0.01309	55	0.01600
6	0.00175	16	0.00465	26	0.00756	36	0.01047	46	0.01338	56	0.01629
7	0.00204	17	0.00495	27	0.00785	37	0.01076	47	0.01367	57	0.01658
8	0.00233	18	0.00524	28	0.00814	38	0.01105	48	0.01396	58	0.01687
9	0.00262	19	0.00553	29	0.00844	39	0.01134	49	0.01425	59	0.01716
10	0.00291	20	0.00582	30	0.00873	40	0.01164	50	0.01454	60	0.01745

TABLE VII
Natural Logarithms

N	0	1	2	3	4	5	6	7	8	9
1.0	0.0 0000	0995	1980	2956	3922	4879	5827	6766	7696	8618
1.1	0.0 9531	*0436	*1333	*2222	*3103	*3976	*4842	*5700	*6551	*7395
1.2	0.1 8232	9062	9885	*0701	*1511	*2314	*3111	*3902	*4686	*5464
1.3	0.2 6236	7003	7763	8518	9267	*0010	*0748	*1481	*2208	*2930
1.4	0.3 3647	4359	5066	5767	6464	7156	7844	8526	9204	9878
1.5	0.4 0547	1211	1871	2527	3178	3825	4469	5108	5742	6373
1.6	0.4 7000	7623	8243	8858	9470	*0078	*0682	*1282	*1879	*2473
1.7	0.5 3063	3649	4232	4812	5389	5962	6531	7098	7661	8222
1.8	0.5 8779	9333	9884	*0432	*0977	*1519	*2058	*2594	*3127	*3658
1.9	0.6 4185	4710	5233	5752	6269	6783	7294	7803	8310	8813
2.0	0.6 9315	9813	*0310	*0804	*1295	*1784	*2271	*2755	*3237	*3716
2.1	0.7 4194	4669	5142	5612	6081	6547	7011	7473	7932	8390
2.2	0.7 8846	9299	9751	*0200	*0648	*1093	*1536	*1978	*2418	*2855
2.3	0.8 3291	3725	4157	4587	5015	5442	5866	6289	6710	7129
2.4	0.8 7547	7963	8377	8789	9200	9609	*0016	*0422	*0826	*1228
2.5	0.9 1629	2028	2426	2822	3216	3609	4001	4391	4779	5166
2.6	5551	5935	6317	6698	7078	7456	7833	8208	8582	8954
2.7	0.9 9325	9695	*0063	*0430	*0796	*1160	*1523	*1885	*2245	*2604
2.8	1.0 2962	3318	3674	4028	4380	4732	5082	5431	5779	6126
2.9	6471	6815	7158	7500	7841	8181	8519	8856	9192	9527
3.0	1.0 9861	*0194	*0526	*0856	*1186	*1514	*1841	*2168	*2493	*2817
3.1	1.1 3140	3462	3783	4103	4422	4740	5057	5373	5688	6002
3.2	6315	6627	6938	7248	7557	7865	8173	8479	8784	9089
3.3	1.1 9392	9695	9996	*0297	*0597	*0896	*1194	*1491	*1788	*2083
3.4	1.2 2378	2671	2964	3256	3547	3837	4127	4415	4703	4990
3.5	5276	5562	5846	6130	6413	6695	6976	7257	7536	7815
3.6	1.2 8093	8371	8647	8923	9198	9473	9746	*0019	*0291	*0563
3.7	1.3 0833	1103	1372	1641	1909	2176	2442	2708	2972	3237
3.8	3500	3763	4025	4286	4547	4807	5067	5325	5584	5841
3.9	6098	6354	6609	6864	7118	7372	7624	7877	8128	8379
4.0	1.3 8629	8879	9128	9377	9624	9872	*0118	*0364	*0610	*0854
4.1	1.4 1099	1342	1585	1828	2070	2311	2552	2792	3031	3270
4.2	3508	3746	3984	4220	4456	4692	4927	5161	5395	5629
4.3	5862	6094	6326	6557	6787	7018	7247	7476	7705	7933
4.4	1.4 8160	8387	8614	8840	9065	9290	9515	9739	9962	*0185
4.5	1.5 0408	0630	0851	1072	1293	1513	1732	1951	2170	2388
4.6	2606	2823	3039	3256	3471	3687	3902	4116	4330	4543
4.7	4756	4969	5181	5393	5604	5814	6025	6235	6444	6653
4.8	6862	7070	7277	7485	7691	7898	8104	8309	8515	8719
4.9	1.5 8924	9127	9331	9534	9737	9939	*0141	*0342	*0543	*0744
5.0	1.6 0944	1144	1343	1542	1741	1939	2137	2334	2531	2728
5.1	2924	3120	3315	3511	3705	3900	4094	4287	4481	4673
5.2	4866	5058	5250	5441	5632	5823	6013	6203	6393	6582
5.3	6771	6959	7147	7335	7523	7710	7896	8083	8269	8455
5.4	1.6 8640	8825	9010	9194	9378	9562	9745	9928	*0111	*0293
5.5	1.7 0475	0656	0838	1019	1199	1380	1560	1740	1919	2098
5.6	2277	2455	2633	2811	2988	3166	3342	3519	3695	3871
5.7	4047	4222	4397	4572	4746	4920	5094	5267	5440	5613
5.8	5786	5958	6130	6302	6473	6644	6815	6985	7156	7326
5.9	7495	7665	7834	8002	8171	8339	8507	8675	8842	9009
6.0	1.7 9176	9342	9509	9675	9840	*0006	*0171	*0336	*0500	*0665
N	0	1	2	3	4	5	6	7	8	9

TABLE VII (*continued*)
NATURAL LOGARITHMS

N	0	1	2	3	4	5	6	7	8	9
6.0	1.7 9176	9342	9509	9675	9840	*0006	*0171	*0336	*0500	*0665
6.1	1.8 0829	0993	1156	1319	1482	1645	1808	1970	2132	2294
6.2	2455	2616	2777	2938	3098	3258	3418	3578	3737	3896
6.3	4055	4214	4372	4530	4688	4845	5003	5160	5317	5473
6.4	5630	5786	5942	6097	6253	6408	6563	6718	6872	7026
6.5	7180	7334	7487	7641	7794	7947	8099	8251	8403	8555
6.6	1.8 8707	8858	9010	9160	9311	9462	9612	9762	9912	*0061
6.7	1.9 0211	0360	0509	0658	0806	0954	1102	1250	1398	1545
6.8	1692	1839	1986	2132	2279	2425	2571	2716	2862	3007
6.9	3152	3297	3442	3586	3730	3874	4018	4162	4305	4448
7.0	4591	4734	4876	5019	5161	5303	5445	5586	5727	5869
7.1	6009	6150	6291	6431	6571	6711	6851	6991	7130	7269
7.2	7408	7547	7685	7824	7962	8100	8238	8376	8513	8650
7.3	1.9 8787	8924	9061	9198	9334	9470	9606	9742	9877	*0013
7.4	2.0 0148	0283	0418	0553	0687	0821	0956	1089	1223	1357
7.5	1490	1624	1757	1890	2022	2155	2287	2419	2551	2683
7.6	2815	2946	3078	3209	3340	3471	3601	3732	3862	3992
7.7	4122	4252	4381	4511	4640	4769	4898	5027	5156	5284
7.8	5412	5540	5668	5796	5924	6051	6179	6306	6433	6560
7.9	6686	6813	6939	7065	7191	7317	7443	7568	7694	7819
8.0	7944	8069	8194	8318	8443	8567	8691	8815	8939	9063
8.1	2.0 9186	9310	9433	9556	9679	9802	9924	*0047	*0169	*0291
8.2	2.1 0413	0535	0657	0770	0000	1021	1112	1263	1384	1505
8.3	1626	1746	1866	1986	2106	2226	2346	2465	2585	2704
8.4	2823	2942	3061	3180	3298	3417	3535	3653	3771	3889
8.5	4007	4124	4242	4359	4476	4593	4710	4827	4943	5060
8.6	5176	5292	5409	5524	5640	5756	5871	5987	6102	6217
8.7	6332	6447	6562	6677	6791	6905	7020	7134	7248	7361
8.8	7475	7589	7702	7816	7929	8042	8155	8267	8380	8493
8.9	8605	8717	8830	8942	9054	9165	9277	9389	9500	9611
9.0	2.1 9722	9834	9944	*0055	*0166	*0276	*0387	*0497	*0607	*0717
9.1	2.2 0827	0937	1047	1157	1266	1375	1485	1594	1703	1812
9.2	1920	2029	2138	2246	2354	2462	2570	2678	2786	2894
9.3	3001	3109	3216	3324	3431	3538	3645	3751	3858	3965
9.4	4071	4177	4284	4390	4496	4601	4707	4813	4918	5024
9.5	5129	5234	5339	5444	5549	5654	5759	5863	5968	6072
9.6	6176	6280	6384	6488	6592	6696	6799	6903	7006	7109
9.7	7213	7316	7419	7521	7624	7727	7829	7932	8034	8136
9.8	8238	8340	8442	8544	8646	8747	8849	8950	9051	9152
9.9	2.2 9253	9354	9455	9556	9657	9757	9858	9958	*0058	*0158
10.0	2.3 0259	0358	0458	0558	0658	0757	0857	0956	1055	1154
N	0	1	2	3	4	5	6	7	8	9

NOTE 1. The base for natural logarithms is $e = 2.71828\ 18284\ 59045\ \cdots$:

$$\log_e 10 = 2.3025\ 8509. \qquad \log_{10} e = 0.4342\ 9448. \tag{1}$$

NOTE 2. If $N > 10$ or $N < 1$, then we may write $N = P \cdot 10^k$ where k is an integer and $1 \leqq P < 10$. Then, to find $\log_e N$, use the following relation with $\log_e P$ obtained from the preceding table and $\log_e 10$ obtained from (1):

$$\log_e N = \log_e (P \cdot 10^k) = \log_e P + k \log_e 10.$$

TABLE VIII
e^x AND e^{-x}

x	e^x	e^{-x}	x	e^x	e^{-x}
0.0	1.00	1.00	3.0	20.1	.0498
0.1	1.11	.905	3.1	22.2	.0450
0.2	1.22	.819	3.2	24.5	.0408
0.3	1.35	.741	3.3	27.1	.0369
0.4	1.49	.670	3.4	30.0	.0334
0.5	1.65	.607	3.5	33.1	.0302
0.6	1.82	.549	3.6	36.6	.0273
0.7	2.01	.497	3.7	40.4	.0247
0.8	2.23	.449	3.8	44.7	.0224
0.9	2.46	.407	3.9	49.4	.0202
1.0	2.72	.368	4.0	54.6	.0183
1.1	3.00	.333	4.1	60.3	.0166
1.2	3.32	.301	4.2	66 7	.0150
1.3	3.67	.273	4.3	73.7	.0136
1.4	4.06	.247	4.4	81.5	.0123
1.5	4.48	.223	4.5	90.0	.0111
1.6	4.95	.202	4.6	99.5	.0101
1.7	5.47	.183	4.7	110.	.0091
1.8	6.05	.165	4.8	122.	.0082
1.9	6.69	.150	4.9	134.	.0074
2.0	7.39	.135	5.0	148.	.0067
2.1	8.17	.122	5.1	164.	.0061
2.2	9.02	.111	5.2	181.	.0055
2.3	9.97	.100	5.3	200.	.0050
2.4	11.02	.091	5.4	221.	.0045
2.5	12.18	.082	5.5	245.	.0041
2.6	13.46	.074	5.6	270.	.0037
2.7	14.88	.067	5.7	299.	.0033
2.8	16.44	.061	5.8	330.	.0030
2.9	18.17	.055	5.9	365.	.0027
3.0	20.1	.0498	6.0	403.	.0025

Answers to Exercises

Answers to most of the odd-numbered problems are given here. Answers to even-numbered problems are available in a separate pamphlet when ordered for students by the instructor.

EXERCISE 1. Page 8.

27. $1/25$. **29.** $3/2$. **31.** 25. **33.** $16/9$. **35.** $-3/7$.
37. $3/5$. **39.** $10/3$. **41.** $2/21$. **43.** $25/21$. **45.** $2ad/cx$.
47. $x^2/5z$. **49.** $9x/2y$. **51.** $125x^3$. **53.** $-x^6y^9$.
55. $8x^3/27y^3$. **57.** a^4x^8/b^4y^{12}. **59.** $1/b^4$. **61.** $5/x^5$.
63. $125a^2/81b^5$. **65.** $27b^3/125a^6$. **67.** $1/xy^2z^2$.
69. $81b^4/625a^4$.

EXERCISE 2. PAGE 13.

1. ± 7. **3.** $\pm.8$. **5.** $2/5$. **7.** $.5$. **9.** 15. **11.** $7/15$.
13. $10/3$. **15.** $2u^2$. **17.** $3x^6$. **19.** $3x/b^2$. **21.** $7a^2/bc^3$.
23. $2a^2b/3c^3$. **25.** $15a^2/14c^3$. **27.** 133. **29.** $156ab^3$.
31. $4\sqrt{3}$. **33.** $10\sqrt{5}$. **35.** $\sqrt{5}/5$. **37.** $\sqrt{15}/6$.
39. $5xy\sqrt{2y}$. **41.** $\sqrt{6xy}/2y$. **43.** $3x\sqrt{10xy}/5y^2$. **45.** $5\sqrt{2}/6$.

EXERCISE 3. PAGE 15.

1. $u^2 + 2u - 15$. **3.** $a^2 - 4b^2$. **5.** $6x^2 - 11xy - 10y^2$.
7. $3u^2v^2 - 5uv - 12$. **9.** $u^2 + 4uv + v^2$.
11. $c^3 + 6c^2d + 12cd^2 + 8d^3$. **13.** $u^3 + 8v^3$. **15.** $u(3 + 5v)$.
17. $b(4y - 3x)$. **19.** $3x^2(2a - x)$. **21.** $(u - v)(u + v)$.
23. $(2x - 3y)(2x + 3y)$. **25.** $(3c - d)(3c + d)$. **27.** $(u + 9)^2$.
29. $(3 - 2w)^2$. **31.** $(x + 3y)^2$. **33.** $(x - 7)(x + 4)$.
35. $(x - 9)(x + 3)$. **37.** $(3u - 5v)(u - v)$.
39. $(x - w)(x^2 + wx + w^2)$. **41.** $(a + 3b)(a^2 - 3ab + 9b^2)$.
43. $(u^2 + 4v^2)(u - 2v)(u + 2v)$.

45. $(u - w)(u + w)(u^2 - uw + w^2)(u^2 + uw + w^2)$.
47. $(2u + 3v)(4u^2 - 6uv + 9v^2)$.
49. $(4x^2 + 25u^2)(2x - 5u)(2x + 5u)$.
51. $(u^2 + 4w^2)(u^4 - 4u^2w^2 + 16w^4)$.
53. $(9a^2 + 16)(3a - 4)(3a + 4)$.   **57.** $(11 + 5\sqrt{5})/4$.
59. $4\sqrt{6} - 11$.   **61.** $(9 + 7\sqrt{3})/11$.

EXERCISE 4.   PAGE 19.

1. $560/27$.   **3.** $(2y + 9x)(9y - 6x)$.   **5.** $(a + 4)/(a - 3)$.
7. $(2y + a)/(y - 3a)$.   **9.** $(x^2 - 10x + 25)/(x^2 - 4x)$.
11. $- 2(x^2 + 6)/[(x - 3)^2(x + 3)]$.   **13.** $(x^2y + 2x)/y$.
15. $- 3cd/(3d + 2c)$.   **17.** $a^2b + ab^2$.   **19.** $(4u + 6v)/uv$.
21. $(2v^2 + 3u - 2v)/(u^3 - v^3)$.
23. $(y + 3)(4y^2 + 2xy + x^2)/(3 - y)$.

EXERCISE 5.   PAGE 22.

1. $4i$.   **3.** $3i\sqrt{3}$.   **5.** $3i\sqrt{10}$.   **7.** $5i/3$.   **9.** $2i/7$.   **11.** $.3i$.
13. $i\sqrt{10}/5$.   **15.** $i\sqrt{6}/4$.   **17.** $6ui$.   **19.** $5a^3i\sqrt{2}$.
21. $u^2i\sqrt{3}/5x$.   **23.** $\pm 7i$.   **25.** $\pm 12i$.   **27.** -1.   **29.** i.
31. $-i$.   **33.** $-7 - 17i$.   **35.** $-5 + 12i$.   **37.** $8 - 6i$.
39. $-9 - 46i$.   **41.** $8 + 38i$.   **43.** -25.   **45.** $16 + 4i$.

EXERCISE 6.   PAGE 26.

1. 11.   **3.** 15.   **5.** $11/2$.   **7.** 4.   **9.** $1/2$.   **11.** 2.   **13.** 3.
15. 10.   **17.** 7.   **19.** $.2$.   **21.** -5.   **23.** -2.   **25.** x.
27. xy^2.   **29.** $2aw$.   **31.** $2xy^2/5z$.   **33.** $2z^3/x^2y$.
35. $2ay^2\sqrt{3ay}$.   **37.** $3a\sqrt[3]{3a^2}$.   **39.** $2x\sqrt[4]{2xy^2}$.
41. $3ax\sqrt[4]{a^2}/y^2$.   **43.** $\sqrt[4]{15}$.   **45.** $2\sqrt[3]{9}$.   **47.** $\sqrt[3]{4}$.   **49.** $\sqrt[3]{5}$.
51. $a/2$.   **53.** $2xy\sqrt[3]{2x^2y}$.   **55.** $54x$.   **57.** $\sqrt[4]{36b^3}/2b$.
59. $\sqrt[4]{8a^2c}/2a$.   **61.** $(u + v^2)\sqrt{3}/v^2$.   **63.** $\sqrt{uv^2 + v^4}/u^2v$.

EXERCISE 7.   PAGE 29.

1. 2.   **3.** 2.   **5.** 4.   **7.** $1/4$.   **9.** $1/3$.   **11.** 8.   **13.** 27.
15. 8.   **17.** $3/2$.   **19.** $1/27$.   **21.** $3x^{-2}$.   **23.** $2x^{2/3}$.
25. $\sqrt[5]{y^3}$.   **27.** $a\sqrt[5]{u^3}$.   **29.** $3\sqrt[4]{a}$.   **31.** $a^{3/4}$.   **33.** $x^{7/5}$.

35. $4u^4$.   **37.** $x^{7/6}$.   **39.** $y^{13/2}$.   **41.** $ax^{8/3}$.   **43.** $y^{13/3}$.
45. $\sqrt[3]{y}$.   **47.** $\sqrt{x}$.   **49.** $\sqrt{3x}$.   **51.** $\sqrt{3a}$.   **53.** $\sqrt[4]{3}$.
55. $9a^2$.   **57.** $\sqrt[3]{x}$.   **59.** $\sqrt[6]{x^5}$.   **61.** $\sqrt{15}/5$.
63. $\sqrt[6]{648x^5y^4}/2x$.   **65.** $\sqrt[4]{2000u^5v}/5u$.   **67.** $4y\sqrt[6]{y}$

EXERCISE 8.   PAGE 32.

1. $u^5 - 5u^4v + 10u^3v^2 - 10u^2v^3 + 5uv^4 - v^5$.
3. $729 + 1458x + 1215x^2 + 540x^3 + 135x^4 + 18x^5 + x^6$.
5. $x^4 + 12x^3a + 54x^2a^2 + 108xa^3 + 81a^4$.
7. $u^7 - 14u^6x + 84u^5x^2 - 280u^4x^3 + 560u^3x^4 - 672u^2x^5 + 448\,ux^6$
$- 128x^7$.
9. $x^{5/2} - 5x^2y + 10x^{3/2}y^2 - 10xy^3 + 5x^{1/2}y^4 - y^5$
11. $a^{4/3} + 4ab^{1/2} + 6a^{2/3}b + 4a^{1/3}b^{3/2} + b^2$.
13. $c^{10} + 30c^9 + 405c^8 + 3240c^7$.
15. $a^{50} + 25a^{48}d + 300a^{46}d^2 + 2300a^{44}d^3$.
17. 1.268.   **19.** 2.191.   **21.** $4368a^{10}x^{11}$.   **23.** 15.   **25.** 10.

EXERCISE 9.   PAGE 36.

1. -3.   **3.** ± 2.   **5.** $\pm 5i/3$.   **7.** $\pm\sqrt{21}/7$.   **9.** $7, -4$.
11. $2, -3$.   **13.** $0, 1/2$.   **15.** $0, 9/5$.   **17.** $\pm 1/5$.   **19.** $-2, 3/2$.
21. $1/3, -5/2$.   **23.** $-2, -5/2$.   **25.** $1/3, 1/3$.   **27.** $-2, -2$.
29. $-3/4, 5/2$.   **31.** $\pm 3i, \pm i$.   **33.** $\pm 5/2, \pm 5i/2$.   **35.** $\pm 5, \pm 5i$.
37. $-3, -5/2, 7/3$.

EXERCISE 10.   PAGE 39.

1. $-2, 4$.   **3.** $2/3, 2/3$.   **5.** $-1/2, -1/2$.   **7.** $3, -1/2$.
9. $3, 1/3$.   **11.** $(3 \pm \sqrt{29})/10$.   **13.** $3 \pm 2i$.   **15.** $(2 \pm i)/2$.
17. $\pm 5i/2$.   **19.** $(3 \pm 2i)/2$.   **21.** Disc. 9; real, unequal, rational.
23. Disc. 0; real, equal, rational.
25. Disc. 16; real, unequal, rational.   **27.** $25/24$.   **29.** $0, 20/9$.
31. $2, (-1 + i\sqrt{3}), (-1 - i\sqrt{3})$.
33. $5, 5(-1 + i\sqrt{3})/2, 5(-1 - i\sqrt{3})/2$;
$-5, 5(1 + i\sqrt{3})/2, 5(1 - i\sqrt{3})/2$.

EXERCISE 11.   PAGE 42

1. 10.   **3.** No sol.; $x = -2$ is extraneous.
5. 9/4, and one extraneous.   **7.** 2.   **9.** 1, and one extraneous.
11. 10.   **13.** No sol.; 1 and $-1/2$ are extraneous.   **15.** 0, 12/5.

EXERCISE 12.   PAGE 47

1. $x < 5$.   **3.** $x < 14/3$.   **5.** $x < 2$.   **7.** $x > -11/2$.
9. $x > 5/21$.   **11.** $x \geqq 96/25$.   **13.** $3 < x < 9$; $-4 < x < 3$.
15. $-2 \leqq x \leqq 6$.   **17.** $\frac{5}{7} < x \leqq 2$.   **19.** Inconsistent.
21. $\frac{50}{39} \leqq x < \frac{17}{3}$.   **23.** $1 \leqq x \leqq 3$.   **25.** $x < -\frac{1}{9}$
27. $\{-1 < x + 2 < 1\}$,   or $\{-3 < x < -1\}$.
29. $\{-2 < x - a < 2\}$,   or $\{a - 2 < x < a + 2\}$.
31. $\{x - a < -d \ \ or \ \ d < x - a\}$,   or $\{x < a - d \ \ or \ \ a + d < x\}$.
33. $\{|x| < \frac{5}{2}\}$,   or $\{-\frac{5}{2} < x < \frac{5}{2}\}$.

EXERCISE 13.   PAGE 53.

1. $S' = \{1$ and all integers $n > 8\}$; $R \cap S = \{4, 5, 6, 7, 8\}$; $T \cap R' = \{9, 10, \cdots, 15\}$; *etc.*
3. $\{$John$\}$, $\{$Harry$\}$, $\{$Bill$\}$, $\{$John, Harry$\}$, $\{$John, Bill$\}$, $\{$Bill, Harry$\}$.
5. 55.   **7.** $-4 < x < 0$.   **9.** $T' = \{5 \leqq x\}$.
11. $A = \{-4 < x\}$; $B = \{x \leqq 2\}$.
13. $A = \{-2 < x\}$; $B = \{x < 2\}$.   **15.** $\{|x| > 3\}$.
17. $A = \{x < -1\}$; $B = \{1 < x\}$.   **19.** $S = \{-1 \leqq x\}$; $T = \{x < 4\}$.

EXERCISE 14.   PAGE 59

3. $-2, 2$.   **5.** $-7, 7$.   **7.** $-4, 4$.   **9.** $2, 2$.
11. 13.   **13.** $4\sqrt{2}$.   **15.** $\sqrt{65}$.   **17.** 8   **25.** 9/2.

EXERCISE 16.   PAGE 69

3. $H = \{(-3, 7), (-2, 2), (-1, -1), (0, -2), (1, -1), (2, 2), (3, 7)\}$;
$R = \{-2, -1, 2, 7\}$.

11. -6.   **13.** -2.   **15.** 36.   **17.** -48.
19. $a + b - a^2 - 2ab - b^2$.   **21.** 4.
23. $9 - 3h - h^2$.

EXERCISE 17.   PAGE 73.

1. $y = \dfrac{2x}{3} - \dfrac{8}{3}; \quad x = \dfrac{3y}{2} + 4$.   **3.** $y = 2x - \dfrac{4}{3}; \quad x = \dfrac{y}{2} + \dfrac{2}{3}$.

5. $x = y^2 - 6y + \dfrac{8}{3}$.   **7.** $y = 2x^3 - 3x^2 - 12x + 8$.

9. $y = \dfrac{3x}{5}$ and $y = -\dfrac{3x}{5}$.

EXERCISE 18.   PAGE 77.

1. $-1/5; 5$.   **3.** $1/6; -6$.   **5.** $1; -1$.   **15.** $v = -3/2$.

EXERCISE 19.   PAGE 80.

1. $y = 5 - 2x$.   **3.** $y = 6$.   **5.** $y = 5x - 14$.   **7.** $x + y = 1$.
9. $y = 2x - 10$.   **11.** $x - 2$.   **13.** $2x + 3y = 6$.
15. $x - 3y = -5$.   **17.** $4y + 5x = 0$.   **19.** $3x + 4y = 11$.
21. $5x - 3y = 10$.   **23.** $x + 2y = 4$.   **25.** $x + 5y = 3$.
27. $x - 2y = 5; 2x + y = 0$.   **29.** Slope $-3/2$; y-int. $5/2$.
31. Slope $1/3$; y-int. $-5/3$.   **33.** $x + 4y = -13; 4x - y = -1$.
35. $2x - 5y = 21; 5x + 2y = -20$.   **37.** $x = 0; y = -3$.

EXERCISE 20.   PAGE 84.

1. $x = -1, y = 4$.   **3.** $x = 1.5, y = 2$.   **5.** $x = -5/2, y = -4$.
7. Inconsistent; distinct parallel lines.
9. $x = -7/2, y = 5/3$.   **11.** Not concurrent.
13. Family of non-vertical lines with y-intercept 3.
15. Family of all nonvertical lines through $(2, -3)$.
19. $7x - 7y + 9 = 0$.

EXERCISE 21.   PAGE 91.

1. $(x - 3)^2 + (y - 4)^2 = 4$.   **3.** $(x - 3)^2 + (y + 2)^2 = 16$.
5. $x^2 + y^2 = 16$.   **7.** $(x + 2)^2 + y^2 = 0$.

9. $x^2 + (y - b)^2 = b^2$.   **11.** $C:(3, 2); r = 4$.
13. $C:(-3, 2); r = \sqrt{10}$.   **15.** $C:(3, 0); r = 2\sqrt{5}$.
17. Graph is ϕ.   **19.** $C:(-1, 2); r = \frac{4}{3}$.
21. $x^2 + y^2 + 4y = 0$.   **23.** $x^2 + 10x + y^2 - 8y + 37 = 0$.
25. $x^2 + y^2 = 25$.
27. $x^2 - 12x + y^2 - 8y + 27 = 0$, *or* $x^2 + 4x + y^2 - 8y = 5$.
29. $x^2 - 4x + y^2 - 5y + 4 = 0$.   **33.** $x^2 + y^2 + 4x - 20y + 39 = 0$.

EXERCISE 23.   PAGE 104

3. Asymptotes, $x = 0$ and $y = 0$.
7. Asymptotes, $5y - 2x = 0$ and $5y + 2x = 0$.
13. Asymptotes $y - 3x = 0$ and $y + 3x = 0$.
15. Asymptotes $x = y$ and $x = -y$.
19. $16x^2 + 9y^2 = 144$.   **21.** $y^2 - 4x^2 = 4$.
23. $4x^2 - 9y^2 = 36$, or $4y^2 - 9x^2 = 36$.

EXERCISE 24.   PAGE 110.

15. $\{x = -.4 \text{ and } x = 2.4\}; \{-.4 < x < 2.4\}; \{x \leqq -4\} \cup \{2.4 \leqq x\}$.
17. $\{-3 < x < 2\}; \{x \leqq -3\} \cup \{2 \leqq x\}$.
19. $\{-\frac{5}{2} < x < 3\}; \{x \leqq -\frac{5}{2}\} \cup \{3 \leqq x\}$.
21. $9y = x^2$ or $x = -3y^2$.

EXERCISE 25.   PAGE 113.

1. In (x', y') coordinates: $(1, 3); (-6, 2); (-3, -6); (-2, -4)$.
3. $x'^2 + y'^2 = 4$.   **5.** $x'y' = 6$.
7. $16x'^2 - 9y'^2 = 144$.   **9.** $x'^2 - 4y'^2 = 4$.

EXERCISE 27.   PAGE 119.

1. $4(x - 2)^2 + 9(y - 6)^2 = 36$.   **3.** $(y + 3)^2 - (x + 4)^2 = 9$.
5. $16(x + 4)^2 + 9(y + 6)^2 = 144$.   **7.** $\dfrac{(x - x_0)^2}{a^2} + \dfrac{(y - y_0)^2}{b^2} = 1$.
9. $y - 4 = 2(x - 2)^2$.   **11.** $x - 2 = \frac{1}{2}(y + 2)^2$.

EXERCISE 28.   PAGE 124.

1. $y^2 = 16x$.   **3.** $x^2 = 2py$.   **5.** $y^2 - 4y = 4x - 20$.

EXERCISE 29. PAGE 128.

1. $16x^2 + 25y^2 = 400.$
3. $b^2y^2 + a^2x^2 = a^2b^2$, where $b^2 = a^2 - c^2.$
5. $25x^2 - 150x + 9y^2 - 108y + 324 = 0.$

EXERCISE 30. PAGE 132.

1. $16y^2 - 9x^2 = 144.$ **3.** $b^2y^2 - a^2x^2 = a^2b^2.$
5. $7y^2 - 84y - 9x^2 + 54x + 108 = 0.$

EXERCISE 31. PAGE 136.

1. $4x^2 + 8x + 13 + \dfrac{31}{x - 2}.$ **3.** $5x + 7 + \dfrac{16}{x - 2}.$

5. $4x^2 - 10x + 25 - \dfrac{56}{x + 2}.$ **7.** $2x^2 - 4x + 2 - \dfrac{4}{x - .5}.$

9. $f(3) = 293; f(-2) = 43.$ **11.** $g(.2) = 8.808; g(-.3) = 6.773.$
13. $x + 3.$ **15.** No.   **17.** Yes; $x^4 - 2x^3 + 4x^2 - 8x + 16.$
19. $x^6 - a^6 = (x - a)(x^5 + ax^4 + a^2x^3 + a^3x^2 + a^4x + a^5).$
21. $x^6 - ax^5 + a^2x^4 - a^3x^3 + a^2x^4 - ax^5 + a^6.$
23. $-2/3; 5/3.$

EXERCISE 32. PAGE 140.

1. $\{2, -3, 1\}.$ **3.** $\{5/2, -5/2, 6i - 6i\}.$
9. $x^3 - 7x^2 + 16x - 12 = 0.$ **11.** $3x^3 - 2x^2 + 3x - 2 = 0.$
13. $x^3 - 8x^2 + 22x - 20 = 0.$ **15.** $4x^3 - 15x - 2 = 0.$
17. $x^2 - 6x + 9 = 0.$ **19.** $x^3 - 5x^2 + 8x - 6 = 0.$

EXERCISE 33. PAGE 145.

1. $\{-2, -2, 3\}.$ **3.** $\{4, (-1 + \sqrt{2}), (-1 - \sqrt{2}).$
5. No rational solution.
7. $\{-3, \frac{1}{4}(3 \pm i\sqrt{7})\}.$ **9.** $\{-\frac{2}{3}, \frac{1}{2}(3 \pm \sqrt{5})\}.$
11. One positive and one negative.
13. One positive, one negative, and two imaginary.
15. One positive, one negative, and two imaginary.
17. One negative and two imaginary.
19. One positive, one negative and two imaginary; or, one positive and three negative.

21. One positive and four imaginary; or, one positive, two negative, and two imaginary.

EXERCISE 34. PAGE 149.

1. $P(x) = 0$: $\{-3.6, .25, 3.3\}$.
  $P(x) < 0$: $\{-3.6 < x < 2.5\} \cup \{3.3 < x\}$.
  $P(x) > 0$; $\{x < -3.6\} \cup \{.25 < x < 3.3\}$.
5. $\{-2 < x < -1\} \cup \{4 < x\}$.
13. "< 0": $\{x < -3\} \cup \{2 < x < 4\}$.
  "> 0"; $\{-3 < x < 2\} \cup \{4 < x\}$.
21. $x \geqq -3$.

EXERCISE 35. PAGE 156.

1. $30°$. **3.** $120°$. **5.** $150°$. **7.** $225°$. **9.** $-120°$.
11. $540°$. **13.** $-171°53'$. **15.** $154°42'$. **17.** $\frac{1}{6}\pi$. **19.** $\frac{3}{4}\pi$.
21. $-\pi$. **23.** $\frac{5}{6}\pi$. **25.** $\frac{3}{2}\pi$. **27.** $\frac{11}{6}\pi$.
29. $-\frac{5}{6}\pi$. **31.** $.04073$. **33.** 1.44222.

EXERCISE 36. PAGE 162.

Note. In the answers to exercises, in any problem where all trigonometric functions of an angle are requested, as a rule, only the *sine*, *cosine*, and *tangent*, in that order, will be given. The student should check the other function values by taking reciprocals.
1. $4/5, 3/5, 4/3$. **3.** $-4/5, 3/5, -4/3$. **5.** $5/13, 12/13, 5/12$.
7. $24/25, -7/25, -24/7$. **9.** $-5\sqrt{29}/29, 2\sqrt{29}/29, -5/2$.
11. $3\sqrt{13}/13, -2\sqrt{13}/13, -3/2$.
31. For quadrant II: $12/13, -5/13, -12/5$.
  For quadrant III: $-12/13, -5/13, 12/5$.
33. $\tan\theta = -24/7$, $\csc\theta = -25/24$, $\sin\theta = -24/25$, $\cos\theta = 7/25$,
  $\sec\theta = 25/7$, obtained in this order.

EXERCISE 37. PAGE 167.

Note. All trigonometric functions are given or mentioned for each quadrantal angle.
1. $0, 1, 0$; $\cot 0°$ and $\csc 0°$ do not exist; $\sec 0° = 1$.
3. Same as Problem 1. **5.** $\sqrt{2}/2, -\sqrt{2}/2, -1$.

7. $-1/2, -\sqrt{3}/2, \sqrt{3}/3$.   **9.** $-\sqrt{2}/2, \sqrt{2}/2, -1$.
11. $-\sqrt{3}/2, 1/2, -\sqrt{3}$.   **13.** $\sqrt{2}/2, \sqrt{2}/2, 1$.
15. $1, 0$; $\tan 2.5\pi$ and $\sec 2.5\pi$ do not exist; $\csc 2.5\pi = 1$; $\cot 2.5\pi = 0$.
17. $-\sqrt{3}/2, -1/2, \sqrt{3}$.
19. $1, 0$; $\tan(-270°)$ and $\sec(-270°)$ do not exist; $\csc(-270°) = 1$; $\cot(-270°) = 0$.
21. $-\sqrt{2}/2, \sqrt{2}/2, -1$.   **23.** $-\sqrt{3}/2, 1/2, -\sqrt{3}$.

EXERCISE 38.   PAGE 170.

1. α: $\{3/5, 4/5, 3/4\}$; β: $\{4/5, 3/5, 4/3\}$.
3. α: $\{24/25, 7/25, 24/7\}$; β: $\{7/25, 24/25, 7/24\}$.
5. α: $\{24/25, 7/25, 24/7\}$; β: $\{7/25, 24/25, 7/24\}$.
7. $-1/2$.   **9.** -1.   **11.** $1/2$.   **13.** $-\sqrt{3}/3$.   **15.** $\sqrt{3}/3$.
17. $-\sqrt{2}/2$.   **19.** 1.   **21.** $-\sqrt{3}$.   **23.** $\sqrt{2}/2$.   **25.** -1.
27. $7\pi/6$.   **29.** $\pi/4$ and $5\pi/4$.   **31.** $\pi/6$ and $11\pi/6$.
41. -3.487.   **43.** $-.358$.   **45.** $.643$.   **47.** -57.30.
49. $-.921$.   **51.** $-.532$.   **53.** $.44395$.   **55.** 10.983.
57. $-.16997$.   **59.** -2.2958.   **61.** $33°; 147°$.   **63.** $38°; 218°$.
65. $154°; 206°$.
67. $\sin(-\pi/3) = -\sin(\pi/3) = -\sqrt{3}/2$.
69. $-\sqrt{3}/3$.   **71.** $-\sqrt{2}/2$.   **73.** $-1/2$.

EXERCISE 39.   PAGE 179.

13. $\{-2\pi, 0, 2\pi\}$.   **15.** $\{-2\pi, -\pi, 0, \pi, 2\pi\}$.
17. $\{-11\pi/6, -7\pi/6, \pi/6, 5\pi/6\}$.   **19.** $\{-5\pi/3, -4\pi/3, \pi/3, 2\pi/3\}$.
21. $\{-5\pi/4, -3\pi/4, 3\pi/4, 5\pi/4\}$.   **23.** $\{-4\pi/3, -\pi/3, 2\pi/3, 5\pi/3\}$.
25. $\{-11\pi/6, -\pi/6, \pi/6, 11\pi/6\}$.

EXERCISE 40.   PAGE 186.

1. $\frac{1}{2}\sqrt{3}, \frac{1}{2}, \sqrt{3}$.   **3.** $-\frac{1}{2}\sqrt{2}, \frac{1}{2}\sqrt{2}, -1$.   **5.** $\frac{1}{2}, -\frac{1}{2}\sqrt{3}, -\frac{1}{3}\sqrt{3}$.
7. $0, -1, 0$.   **9.** $\frac{1}{2}\sqrt{3}, -\frac{1}{2}, -\sqrt{3}$.   **11.** $-\frac{1}{2}, \frac{1}{2}\sqrt{3}, -\frac{1}{3}\sqrt{3}$.
13. Tangent and secant undefined: $\{-\frac{3}{2}\pi, -\frac{1}{2}\pi, \frac{1}{2}\pi, \frac{3}{2}\pi\}$.
   Cosecant undefined: $\{-2\pi, -\pi, 0, \pi, 2\pi\}$.
23. $-.53119$.   **25.** $.48887$.   **27.** $.89; 5.39$.
29. $4.15; 5.27$.   **31.** $2.91; 6.05$.

EXERCISE 41.   PAGE 193.

1. $-\sin x.$ **3.** $-\sin x.$ **5.** $(1 + \tan x)/(1 - \tan x).$

7. $(1 + \sqrt{3})/(\sqrt{3} - 1).$ **9.** $\frac{1}{2}\sqrt{2}(\cos x - \sin x).$

11. $-\cos x.$ **13.** $-\tan x.$ **15.** $\frac{1}{2}(\sqrt{3}\sin x - \cos x).$

17. $(\sqrt{3} - \tan x)/(1 + \sqrt{3}\tan x).$ **19.** $\frac{1}{2}(\sqrt{3}\cos x - \sin x).$

21. $(\sqrt{3} - 1)/(1 + \sqrt{3}).$ **25.** $\frac{1}{4}(\sqrt{6} + \sqrt{2}); \frac{1}{4}(\sqrt{2} - \sqrt{6}).$

27. $\frac{1}{4}(\sqrt{6} + \sqrt{2}); \frac{1}{4}(\sqrt{6} - \sqrt{2}); (2 + \sqrt{3}).$

29. $\frac{1}{4}(\sqrt{6} - \sqrt{2}); -\frac{1}{4}(\sqrt{2} + \sqrt{6}); (\sqrt{3} - 2).$

37. $\sin 3x = 2\sin\frac{3}{2}x\cos\frac{3}{2}x; \cos 3x = \cos^2\frac{3}{2}x - \sin^2\frac{3}{2}x.$

$$\tan 3x = \frac{2\tan\frac{3}{2}x}{1 - \tan^2\frac{3}{2}x}.$$

39. $\sin 2x = \pm\sqrt{\dfrac{1 - \cos 4x}{2}}; \cos 2x = \pm\sqrt{\dfrac{1 + \cos 4x}{2}};$

$\tan 2x = \pm\sqrt{\dfrac{1 - \cos 4x}{1 + \cos 4x}}.$ (The sign "+" or "−" depends on the quadrant for angle $2x$.)

EXERCISE 42.   PAGE 196

1. $-\cos x.$ **3.** $-\cos x.$ **5.** $-\cos x.$ **7.** $\sin x.$

9. $\tan x.$ **11.** $-\cos x.$

15. $\sin\left(\frac{3}{2}\pi - x\right) = -\cos x; \cos\left(\frac{3}{2}\pi - x\right) = -\sin x;$
$\tan\left(\frac{3}{2}\pi - x\right) = \cot x.$

17. $-\cot x.$ **19.** $-\csc x.$ **21.** $\tan x.$ **23.** $-\cot x.$

25. $\sec y.$ **27.** $\cos x.$ **29.** $-\cot x.$ **31.** $-\tan w.$

33. One form of the results:
$\sin 4x = 4\sin x\cos x - 8\sin^3 x\cos x;$
$\cos 4x = 1 - 2\sin^2 x - 4\sin^2 x\cos^2 x.$

EXERCISE 43.   PAGE 199.

1. $\dfrac{2\cos^2 x + \sin^2 x}{\sin x\cos x}.$ **3.** $\dfrac{1}{\sin x\cos x}.$

5. $\frac{1}{2}\sqrt{2}(2\sin x\cos x - \cos^2 x + \sin^2 x).$

7. $\dfrac{\sin x + \cos x}{\sin x - \cos x}.$ **9.** $\dfrac{1 + 2\sin x\cos x}{\cos^2 x}.$ **11.** $\dfrac{\sin x - \cos x}{\cos x + \sin x}.$

13. $\dfrac{\sin x\cos x + \cos x}{\sin x}.$ **15.** $\dfrac{\sin x\cos y + \cos x\sin y}{\cos x\cos y - \sin x\sin y}.$

EXERCISE 44.   PAGE 203.

1. $\{\frac{1}{4}\pi, \frac{7}{4}\pi\}$.   3. $\{\frac{7}{6}\pi, \frac{11}{6}\pi\}$.   5. $\{\frac{3}{4}\pi, \frac{7}{4}\pi\}$.   7. $\{\frac{5}{4}\pi, \frac{7}{4}\pi\}$.
9. $\{\frac{1}{2}\pi, \frac{3}{2}\pi\}$.   11. $\{\frac{1}{6}\pi, \frac{7}{6}\pi\}$.   13. $\{0, \pi, 2\pi\}$.
15. $\{\frac{1}{4}\pi, \frac{3}{4}\pi, \frac{5}{4}\pi, \frac{7}{4}\pi\}$.   17. $\{2.112, 4.172\}$.   19. $\{\frac{1}{3}\pi, \frac{2}{3}\pi, \frac{4}{3}\pi, \frac{5}{3}\pi\}$.
21. $\{\frac{1}{6}\pi, \frac{5}{6}\pi, \frac{7}{6}\pi, \frac{11}{6}\pi\}$.   23. No solution.   25. $\{\frac{1}{6}\pi, \frac{5}{6}\pi\}$.
27. No solution.   29. $\{0, \frac{1}{6}\pi, \frac{5}{6}\pi, \pi, 2\pi\}$.   31. $\{\frac{2}{3}\pi, \pi, \frac{4}{3}\pi\}$.
33. $\{\frac{1}{3}\pi, \frac{5}{3}\pi\}$.   35. $\{\frac{1}{2}\pi, \frac{3}{2}\pi\}$.   37. $\{\frac{1}{2}\pi, \frac{3}{2}\pi\}$.   39. $\{\frac{7}{6}\pi, \frac{11}{6}\pi\}$.

EXERCISE 45.   PAGE 205.

1. $\{\frac{1}{3}\pi, \frac{2}{3}\pi, \frac{4}{3}\pi, \frac{5}{3}\pi\}$.   3. $\{\frac{7}{6}\pi, \frac{3}{2}\pi, \frac{11}{6}\pi\}$.   5. $\{\frac{1}{3}\pi, \frac{2}{3}\pi, \frac{4}{3}\pi, \frac{5}{3}\pi\}$.
7. $\{0, \frac{1}{3}\pi, \pi, \frac{4}{3}\pi, 2\pi\}$.   9. $\{\frac{1}{2}\pi, \frac{2}{3}\pi, \frac{4}{3}\pi, 2\pi\}$.   11. $\{0, \frac{1}{3}\pi, \frac{5}{3}\pi, 2\pi\}$.
13. $\{0, \frac{1}{4}\pi, \pi, \frac{5}{4}\pi, 2\pi\}$.   15. $\{0, \frac{1}{2}\pi, \pi, 2\pi\}$.   17. $\{\frac{3}{2}\pi\}$.
19. $\{\frac{1}{4}\pi, \frac{3}{4}\pi, \frac{5}{4}\pi, \frac{7}{4}\pi\}$.   21. $\{0, \frac{1}{2}\pi, \pi, \frac{3}{2}\pi, 2\pi\}$.
23. $\{\frac{3}{8}\pi, \frac{7}{8}\pi, \frac{11}{8}\pi, \frac{13}{8}\pi\}$.   25. $\{\frac{3}{8}\pi, \frac{5}{8}\pi, \frac{11}{8}\pi, \frac{13}{8}\pi\}$.   27. $\{0, \pi, 2\pi\}$.
29. $\{0, \frac{2}{3}\pi, 2\pi\}$.   31. $\{0, 2\pi\}$.

EXERCISE 46.   PAGE 212.

1. $\frac{1}{3}\pi$.   3. $\frac{1}{4}\pi$.   5. $-\frac{1}{6}\pi$.   7. $\frac{1}{4}\pi$.   9. 0.   11. $\frac{1}{6}\pi$.
13. $-\frac{1}{3}\pi$.   15. $-\frac{1}{4}\pi$.   19. .9.   21. 4.   23. $2x\sqrt{1-x^2}$.
25. $\frac{1}{3}\pi$.   27. $\frac{1}{4}\pi$.   29. $-\frac{1}{4}\pi$.   31. 1.169.   33. 1.257.
35. $-.611$.   37. -1.239.   39. $\frac{1}{3}\sqrt{3}$.   41. $-\frac{1}{2}\sqrt{2}$.
43. $\sin y = -\frac{1}{2}$; $\cos y = \frac{1}{2}\sqrt{3}$; $\tan y = -\frac{1}{3}\sqrt{3}$.
45. $\sin y = \frac{4}{5}$; $\cos y = \frac{3}{5}$; $\tan y = \frac{4}{3}$.
47. $\sin y = -\frac{15}{17}$; $\cos y = \frac{8}{17}$; $\tan y = -\frac{15}{8}$.

EXERCISE 48.   PAGE 219.

11. $f(x) = 4\sin\left(x + \frac{1}{3}\pi\right)$.   13. $(x = 1.3, y = 1.6)$.

EXERCISE 49.   PAGE 225.

1. $\frac{1}{4}\pi$.   3. $\frac{2}{3}\pi$.   5. $65°$.   7. $151.1°$.   9. $36.9°$.   11. $123.7°$.
13. $82.9°$.   15. $82.2°$; $97.8°$.   17. Lines are perpendicular.
19. $26.5°$; $153.5°$.

EXERCISE 50.   PAGE 228.

1. 2.646.   **3.** 3.606.   **5.** $\alpha = 57.1°$; $\beta = 78.5°$; $\gamma = 44.4°$.
7. $\alpha = 25.1°$; $\beta = 96.4°$; $\gamma = 58.4°$.
9. $\alpha = 29.0°$; $\beta = 104.5°$; $\gamma = 46.5°$.

EXERCISE 51.   PAGE 232.

1. $\beta = 80°$; $b = 7.9$; $c = 7.5$.   **3.** $\beta = 138°25'$; $a = 86$; $b = 133$.
5. $\gamma = 21°0'$; $a = 11.2$; $c = 23.9$.   **7.** $\gamma = 2.2°$; $\alpha = 128°$.
9. No solution; inconsistent data.   **11.** $\beta = 142.0°$; $\gamma = 17.5°$.

EXERCISE 53.   PAGE 242.

1. Double in $2\frac{1}{3}$ units of time.   **3.** Double in 3.5 units of time.
5. Half-life is 3.5 t-units.   **11.** .6.

EXERCISE 54.   PAGE 245.

3. $\log_4 H = 3$.   **11.** $\log_{2.5} \frac{25}{4} = 2$.   **13.** $\log_{10} .01 = -2$.   **15.** 25.
17. 1.   **19.** 1000.   **21.** $T = h^2$.   **23.** $N = 1/b$.   **25.** 4.
27. 216.   **29.** 4.   **31.** -1.   **33.** -4.   **35.** -3.   **37.** 6.
39. 2.   **41.** 8.

EXERCISE 55.   PAGE 247.

1. 1.0792.   **3.** $-.1250$.   **5.** -1.4472 because log 1 = 0.
7. 1.6021.   **9.** 1.4771.   **11.** $-.1549$.   **13.** -2.5229.
15. $-.6778$.   **17.** .2007.   **19.** .6111.

EXERCISE 56.   PAGE 252.

1. 3.2686.   **3.** 3.6734.   **5.** 0.9148.   **7.** $9.8773 - 10$.
9. $7.5982 - 10$.   **11.** $5.1667 - 10$.   **13.** 5.2370.   **15.** 6.8154.
17. 290.7.   **19.** .1652.   **21.** 33.48.   **23.** .03403.   **25.** .002891.
27. $1.136(10^4)$.   **29.** $2.107(10^7)$.

EXERCISE 57.   PAGE 255.

1. 24.33.   **3.** 366.3.   **5.** 4.563.   **7.** .08552.   **9.** .02015.
11. $1.073(10^{-4})$.   **13.** 9.665.   **15.** 6.219.   **17.** 4.329.
19. .2943.   **21.** .4840.   **23.** 1.186.

EXERCISE 58.   PAGE 257.

1. $x = 1194$.   **3.** 1.140.   **5.** 7.386.   **7.** 1.430.
9. 4.025; 1.723; −.5799.   **11.** −9.965.

EXERCISE 59.   PAGE 260.

11. $x = 2.2$.   **13.** $x = .78$.   **17.** $k = .23015$.   **19.** $k = -.08665$.

EXERCISE 60.   PAGE 265.

1. $(x = 2 + 6t, y = 4 + 6t)$; midpoint $(5,7)$; trisection points $(4,6)$ and $(6,8)$.
3. $(x = -2 - 6t, y = 3 - 6t)$; midpoint $(-5,0)$; trisection points $(-4,1)$ and $(-6,1)$.
5. $(x = -4 - 6t, y = 5 - 7t)$; midpoint $(-7, 3/2)$; trisection points $(-6, 8/3)$ and $(-8, 1/3)$.
7. $(-1/5, -8/5)$.   **9.** $(27, -22)$.   **13.** $(-8, -1)$.
15. $y = 5$; $5y - 6x = -3$; $3x - 2y = 4$: concurrent at $(14/3, 5)$.
17. $x + 4y = -5$.
19. $x + 2y = 7$; $x - y = -13$; $x - 4y = -33$: concurrent at $(-19/3, 20/3)$.

EXERCISE 61.   PAGE 270.

5. $3/2$.   **7.** 0.

EXERCISE 62.   PAGE 273.

1. $x^2 + y^2 = 4$.   **3.** $16x^2 + 9y^2 = 144$.
5. $9(x - 2)^2 + 4(y - 1)^2 = 36$.   **7.** $y^3 = x^2$.
9. $x^3 = 12x^2 - y^2$.

EXERCISE 63. PAGE 279.

9. Vertices: $(1, 1)$, $(2, 0)$, $(3, 5)$.
11. Vertices: $(1, 0)$, $(0, 2)$, $(-2, 1)$, $(-5, -3)$.
18. A, 20%; B, 68%; C, 12%.

EXERCISE 65. PAGE 285.

1. $(x = 1, y = \sqrt{3})$.   **3.** $(0, 1)$.   **5.** $(-3/2, 3\sqrt{3}/2)$.
7. $(1, \sqrt{3})$.   **9.** $[\sqrt{2}, 45°]$.   **11.** $[2, -\pi/3]$.
13. $[6, 4\pi/3]$.   **15.** $[2, 5\pi/6]$.   **17.** $[5, 143.1°]$.   **19.** $y = 2x$.
21. $y = -3$.   **23.** $x^2 + y^2 = 6y$.   **25.** $x^2 + y^2 + 2y = 0$.
27. $\tan \theta = 3$.   **29.** $r = -4 \sin \theta$.   **31.** $r^2 \cos^2 \theta + 4r^2 \sin^2 \theta = 4$.

EXERCISE 66. PAGE 289.

1. Tangent: $\theta = \pi$.   **3.** Tangent: $\theta = 3\pi/2$.
5. Tangents: $\theta = \pi/6$; $\theta = \pi/2$; $\theta = 5\pi/6$.
7. Tangents: $\theta = 0$; $\theta = \pi/3$; $\theta = 2\pi/3$.
9. Tangents: $\theta = \pi/4$; $\theta = 3\pi/4$.
11. Tangents: $\theta = \pi/4$; $\theta = 3\pi/4$.

EXERCISE 67. PAGE 293.

11. $\sqrt{35}$.   **13.** $\sqrt{17}$.   **15.** $2\sqrt{5}$.

EXERCISE 68. PAGE 297.

27. $\{x = 0, y = 0\}$; etc.

EXERCISE 69. PAGE 301.

1. $\frac{2}{3}, -\frac{2}{3}, 1$.   **3.** $-\frac{3}{7}, -\frac{2}{7}, \frac{6}{7}$.   **5.** $0, 0, 1$.   **7.** $1, 0, 0$.
9. $\frac{14}{15}, -\frac{2}{15}, \frac{1}{3}$.   **11.** $\frac{1}{3}, -\frac{2}{3}, \frac{2}{3}$.   **13.** $-\frac{6}{11}, \frac{2}{11}, \frac{9}{11}$.
15. $\beta = 90°$.   **17.** $\cos \gamma = \pm\frac{2}{3}$.
19. Angles $\{\frac{3}{4}\pi, \frac{1}{4}\pi, \frac{1}{2}\pi\}$.   Cosines $\{-\frac{1}{2}\sqrt{2}, \frac{1}{2}\sqrt{2}, 0\}$.

EXERCISE 70.   PAGE 306.

1. $\{\frac{3}{7}, \frac{2}{7}, -\frac{6}{7}\}$ or $\{-\frac{3}{7}, -\frac{2}{7}, \frac{6}{7}\}$.   **3.** $\{0, 0, -1\}$.

5. $\{-\frac{2}{3}, \frac{11}{15}, -\frac{2}{15}\}$ or $\{\frac{2}{3}, -\frac{11}{15}, \frac{2}{15}\}$.

7. $\{\frac{14}{15}, -\frac{1}{3}, -\frac{2}{15}\}$ or $\{-\frac{14}{15}, \frac{1}{3}, \frac{2}{15}\}$.

9. $\{\frac{12}{13}, 0, -\frac{5}{13}\}$ or $\{-\frac{12}{13}, 0, \frac{5}{13}\}$.   **13.** $-3 : 5 : -6$

15. $-3 : -2 : 4$.   **17.** Parallel. (Why?).   **19.** Perpendicular. (Why?)

27. $110.6°$.   **29.** $65.4°$; $114.6°$.

EXERCISE 71.   PAGE 309.

1. $3x - y + 4z = -3$.   **3.** $4x - y = 5$.   **5.** $2x - 3y + z = -11$.

7. $x - 2y - 3z = 10$.   **9.** $x = -1$.   **11.** $9x - 2y - 6z = 11$.

15. $\{4, \frac{4}{3}, -4\}$.   **17.** $3x - 4y - z = -23$.

19. Perpendicular. (Why?).   **23.** $79.0°$; $101.0°$.

EXERCISE 72.   PAGE 313.

11. $x + z = 2$ and $y - 2z = -2$, and other systems.

EXERCISE 73.   PAGE 316.

1. $4x + 3y = 12$ and $y - 4z = -4$.

3. $\dfrac{x - 2}{-1} = \dfrac{y + 1}{2} = \dfrac{z - \frac{3}{4}}{-3}$.   **5.** $\dfrac{x - 2}{2} = \dfrac{z - 3}{-1}$ and $y = -1$.

7. $\dfrac{x + 3}{1} = \dfrac{z}{-1}$ and $y = 1$.   **9.** $x = 2$ and $y + 3 = z - 2$.

11. $x + 1 = \dfrac{y - 3}{-2}$ and $z = 4$.   **13.** $\dfrac{x - 2}{3} = \dfrac{y + 1}{2} = \dfrac{z - 3}{-1}$.

15. $\dfrac{x - 2}{-3} = \dfrac{y + 1}{-2} = \dfrac{z - 2}{5}$.   **17.** $(0, \frac{5}{2}, 0)$; $(-\frac{2}{3}, 2, 1)$ and $(\frac{2}{3}, 3, -1)$.

19. $(5, -5, 1)$.   **21.** Through $(\frac{1}{2}, 2, -\frac{5}{3})$ with direction $9 : -12 : 2$.

23. One symmetric form is $\dfrac{x - \frac{5}{9}}{2} = \dfrac{y}{3} = \dfrac{z - 1}{-1}$.

EXERCISE 74.   PAGE 319.

1. $(-1, 3, 0)$; $\sqrt{15}$.

EXERCISE 75.   PAGE 323.

1. $25x^2 + 9y^2 + 25z^2 = 225$; etc.

9. A prolate spheroid obtained by revolving the ellipse $4y^2 + z^2 = 4$, in the yz-plane, about the z-axis; or, obtained by revolving the ellipse $4x^2 + z^2 = 4$, in the xz-plane, etc.

11. A cone.

Index